CHEMISTRY PROBLEMS

CHEMISTRY PROBLEMS

SECOND EDITION

M. J. Sienko
CORNELL UNIVERSITY

W. A. BENJAMIN, INC.
Menlo Park, California • Reading, Massachusetts
London • Amsterdam • Don Mills, Ontario • Sydney

CHEMISTRY PROBLEMS, *Second Edition*

Standard Book Number 8053-8808-7 (Paperback Edition)
Library of Congress Catalog Card Number 70-151306
MANUFACTURED IN THE UNITED STATES OF AMERICA
ISBN 0-8053-8808-7
 FGHIJKLMN-AL-79876

PREFACE

THE INCREASING DIVERGENCE of the material coverage in various introductory chemistry courses has placed an added burden on the instructor who wishes to retain a large amount of problem-solving in his course. Newer texts, in attempting to provide coverage of all the diverse topics that may appear in various courses, tend to reduce the amount of space given to problem-solving partly to conserve space and partly because there is growing reliance on supplementary paperbacks to provide extended discussion for the student. The problem-solving book is becoming, therefore, an integral part of the textual course reference as well as a self-study aid for the faltering student. In particular there is growing need for extended lists of problems from which assignments can be given without worry that the solution is obvious from a nearby worked-out example and without worry that the problem is histori-cally old enough to have made its appearance in every fraternity file in the country.

The first edition of this book appeared as a two-volume version entitled respectively *Stoichiometry and Structure* and *Equilibrium.* The two-volume format was made necessary because of the large number of general operation problems included. Subsequently, rigorous pruning of the latter produced a single-volume version entitled *Chemistry Problems.* Class use of the single-volume book has indicated the pruning was too severe in some topics and too mild in others. This revised version is an attempt to redress the balance. Sections on slide rule and significant figures have been restored; some of the more exotic cases of simultaneous equilibria have been deleted. All the problems that are not worked out in the text are new. They are bunched at the ends of chapters to make for ease of assigning. Levels of difficulty are indicated by * easy, ** moderate, and *** difficult. Numerical answers have been furnished for about half of the assigned exercises. Worked-out problems and those intended for assignment are distinguished as "examples" and "exercises," respectively.

Ithaca, New York   M. J. SIENKO

v

CONTENTS

Appendixes

1

REVIEW OF SOME MATHEMATICAL OPERATIONS AND USE OF THE SLIDE RULE

THERE IS A LARGE AMOUNT of information collected in this chapter that is likely to be of general usefulness to the beginning student of chemistry as well as to the student with some experience in the field. Much of the material should already be familiar; some may be new. In any case, a rapid review of all the material is recommended. In this way you will find out what is here and be able to refer back to it when you need intensive study for a later application.

1.1 Significant figures

One of the most abused aspects of chemistry computations is the frequent disregard for the information that is embodied in the numbers obtained from physical measurements. The most common symptom of this carelessness is the carrying out of calculations to too many decimal places. A useful rule to remember, commonly used by scientists, is that, unless indicated to the contrary, digits written down to represent the result of a measurement are considered to be known with certainty except for the last digit, which is assumed to be in doubt. The known digits plus the doubtful one constitute what are called *significant figures*. As a specific example, recording the weight of a sample as 1.369 g implies four significant figures, of which the 1, 3, and 6 are known but the 9 is considered to be in doubt. Sometimes, the doubtful digit is dropped below the line, for example, 1.36_9 instead of 1.369.

Why bother to write down a digit that is doubtful? First, it gives information as to the precision with which the measurement has been carried out and, second, it indicates a best guess for the next digit, which is often quite informative. Thus, writing a weight as 1.369 g implies that the weighing was done on an instrument sensitive to within 0.001 g (in other words, an analytical balance) and not, for example, on one sensitive to within 0.1 g (e.g., a platform balance).

At this point, we might note that there is a great difference between *accuracy* and *precision*. Accuracy concerns itself with the absolute truth of a measurement, whereas precision deals with the detail of the measurement. As an illustration, a chemical weight of 1.369 g implies that the chemist has made the measurement to a precision of ±0.001 g and that he can reproduce the measurement (except possibly for the last digit) by repeating the task on the same instrument. However, the instrument may be inaccurate; that is, its scales and scale markings may have been put on in error, or a part may have become twisted so as to destroy the meaning of the calibrations. In any event, although the weighings can continue to be made with great precision, they may not be very accurate. It is of course the pious hope of all chemists, unfortunately sometimes unjustified, that their instruments are calibrated accurately and that the measurements are made with the precision attributed to the instrument.

As far as we are concerned, the immediate problem with significant figures is to make sure that when we manipulate numbers in a computation we do not distort the information by throwing precision away or apparently adding to it. As a specific illustration of what is involved, we consider the problem of adding or subtracting physical quantities, a case where the role of significant figures is easiest to appreciate.

■ EXAMPLE 1.   Suppose you have an empty beaker weighing 56 g. Into this beaker you put a sample of salt, weighing 0.234 g. What is the total weight of the beaker plus salt?

SOLUTION: The most naive approach to this problem is to add the two numbers 56 plus 0.234 and report the final weight as 56.234 g. However, this is wrong because it implies that you know the final weight to be 56.234 ± 0.001 g, in other words, that you know the total weight to within about 1 part in 56 thousand. Clearly, you know no such thing! Stating that the weight of the empty beaker is 56 g implies you know it to be 56 ± 1 g (1 part in 56). Not only are you in doubt about the digit 6 (it could be 5 or 7, e.g.), but you have no clue to what the decimal places to the right of the 6 have for values. Consequently, any digits you add to these decimal places are completely uncertain and you should not report them, since by so doing you would be pretending to have information you do not possess.

The correct answer can be seen by setting up the problem in the following fashion:

$$
\begin{array}{r}
56.??? \\
+\ \ 0.234 \\
\hline
56.???
\end{array}
$$

Here the question marks in the first line indicate our lack of information about the various decimal places. Even though we may know the 2, 3, and 4 digits in the second line, after the addition our state of information is as shown in the last line. Therefore, we need to report 56 g as the total weight of beaker plus salt. ■

■ EXAMPLE 2.   Suppose you have just opened a bottle containing 453.6 g of chemical. You carefully withdraw 0.234 g of the stuff. What weight of chemical will be left in the bottle?

SOLUTION: Many students find it helpful to underline the doubtful digits as a way of keeping track of where the uncertainties lie. By this device, it is a simple matter to note which column of digits is the first to contain a doubtful number and thereby limits the decimal places that can be reported. If we proceed this way, we would write

$$
\begin{array}{r}
453.\underline{6} \\
-\ \ 0.23\underline{4} \\
\hline
453.\underline{4}
\end{array}
$$

In the number 453.6 in the first line, the 4, 5, and 3 are assumed to be known with certainty, but the 6 is doubtful. This means that any operation involving the column containing the 6—that is, the first column to the right of the decimal—will give rise to a doubtful digit. Furthermore, any column even farther to the right (which gives information in still finer detail) cannot be significant. Realizing this, we can proceed in either to two ways: (a) we can round off the numbers so as to get rid of the columns on which we shall have no final information and then carry through the subtraction, or (b) we can carry through the subtraction as if we knew all the columns and then round off the final answer. Which is better? It makes no difference. They both lead to the same answer (except possibly for the last digit, which, being doubtful anyway, is permitted some variation). Procedure (a) would mean carrying out this operation:

$$
\begin{array}{r}
453.\underline{6} \\
-\ \ 0.\underline{2} \\
\hline
453.\underline{4}
\end{array}
$$

Procedure (b) would mean carrying out this operation:

$$
\begin{array}{r}
453.\underline{600} \\
-\quad 0.23\underline{4} \\
\hline
453.\underline{366}
\end{array}
$$

followed by rounding off to 453.4. ∎

The general rules for rounding off are (1) increase the last retained digit by 1 if the digit to be discarded exceeds 5; (2) leave the last retained digit unchanged if the digit to be discarded is less than 5; (3) if the digit to be discarded is equal to 5, then increase the preceding digit if it is odd, leave it unchanged if it is even. [Rule (3) comes from the fact that odd digits and even digits are equally probable; therefore you will on the average be adding a correction half the time and throwing it away the other half.] As specific examples of rounding off operations, we have the following:

1.1$\underline{84}$ rounds off to 1.1$\underline{8}$ 1.1$\underline{94}$ rounds off to 1.1$\underline{9}$
1.1$\underline{85}$ rounds off to 1.1$\underline{8}$ 1.1$\underline{95}$ rounds off to 1.20
1.1$\underline{86}$ rounds off to 1.1$\underline{9}$ 1.1$\underline{96}$ rounds off to 1.20

Note however that 1.1$\underline{851}$ rounds off to 1.1$\underline{9}$ but 1.1$\underline{850}$ to 1.1$\underline{8}$.

Zeros may add an extra complication because they serve two possible functions: (1) to indicate that a given decimal place has been measured to be zero and (2) to indicate where the decimal point should be put. In case (1) the zeros are counted as significant figures; in case (2) they are not. As examples, in 1009 g there are four significant figures, but in 0.0019 g there are only two. Another way of stating the rule is to say that zeros are counted as significant figures if they are preceded and followed by other digits, but not when they occur at the beginning of a number. At this point, you might object and say that in 0.0019 g, for example, you *know* that the digits preceding the 1 and the 9 are zeros and therefore must be significant. Unfortunately, you cannot reason this way without destroying the whole concept and usefulness of significant figures. The precision to which you know 0.0019 g is ± 0.00001— that is, 1 part in 19, or approximately 5%. If there were four significant figures and you were in doubt only of the last one, you would know your result to roughly 1 part (last digit) in 10,000 (four decimal places), or about 0.01%. Another way of looking at the problem is the following: 0.0019 g can also be expressed as 1.9 mg (milligrams) or as 0.0000019 kg (kilogram) simply by changing the size of the measuring unit in which we express our result. But certainly the way in which we write down the result should have no effect on the precision with which we have made the measurement. In all three cases (0.0019 g, 1.9 mg, and 0.0000019 kg) we know the result to 1 part in 19 and must agree that the number of significant figures is two in each case.

■ EXAMPLE 3.   How many significant figures are there in each of the following: 705 mm; 70.5 cm; 0.705 m; 0.000705 km?

ANSWER: Three in each case. ■

Zeros that occur at the end of a number are a special case because *a priori* we cannot tell whether the zero is put there to indicate a measurement that comes out to be zero in that decimal unit or whether the zero is there merely to fix the position of the decimal point. For example, if I report my weight as 70 kg, you cannot tell whether I mean 70 ± 10 kg (corresponding to one significant figure) or 70 ± 1 kg (corresponding to two significant figures). In the first case, I presumably would have weighed myself on a crude scale to a precision of 10 parts in 70 (or about 15%); in the second case, on a finer scale to a precision of 1 part in 70 (or about 1.5%). How to remove this ambiguity? One way is to use the $\pm$ as was done above. Another way is to use the exponential notation discussed in Section 1.2, whereby numbers are written so that the position of the decimal point is given by the power of 10. Thus, with exponential notation, zeros are not written unless they are significant figures. Specifically, for the example of 70 kg above we would write 7×10^1 if we mean 70 ± 10 kg (the one-significant-figure case) and 7.0×10^1 if we mean 70 ± 1 kg (the two-significant-figure case). The powers of 10 have nothing to do with the number of significant figures. Only the coefficients of the powers of 10 need be considered.

■ EXAMPLE 4.   How many significant figures are there in each of the following: (a) 100 cm^3; (b) 100 ± 1 cm^3; (c) 100 ± 10 cm^3; (d) 1.00×10^2 cm^3; (e) 1.0×10^2 cm^3; (f) 1×10^2 cm^3?

ANSWER: (a) Do not know; (b) 3; (c) 2; (d) 3; (e) 2; (f) 1. ■

Why do we need to concern ourselves so much with being able to tell how many significant figures there are in a number? The answer is: There is a simple rule that states that for multiplication or division the number of significant figures in the answer is equal to the least number of significant figures found in any of the quantities going into the calculation. In other words, if we multiply a two-significant-figure number by a three-significant-figure number, the answer should have but two significant figures.

■ EXAMPLE 5.   How many grams in 0.12 liter of solution containing 0.345 g/liter?

SOLUTION:

$$(0.12 \text{ liter})\left(0.345 \frac{g}{\text{liter}}\right) = 0.041 \text{ g} ■$$

Similarly, if we divide a two-significant-figure number by a three-significant-figure number, or vice versa, the answer again should have two significant figures.

■ EXAMPLE 6.   How large a current is flowing if 0.345 C (coulomb) of electrical charge passes a given point in 0.12 sec?

SOLUTION:

$$\frac{0.345 \text{ C}}{0.12 \text{ sec}} = 2.9 \text{ C/sec, or 2.9 amp} ■$$

In both of these problems, you can easily assure yourself that only two significant figures are permitted in the answer by carrying out the detailed multiplication and division with an underline for each doubtful digit.

<pre>
                                    2.8̲7̲
        0.34̲5̲   0.1̲2̲) 0.34̲5̲
      × 0̇.1̲2̲   2̲4̲
      ──────   ──────
        69̲0̲   10̲5̲
        34̲5̲   9̲6̲
      ──────   ──────
      0.041̲4̲0̲   9̲0̲
                                8̲4̲
                              ──────
</pre>

Rounds off to 0.041   Rounds off to 2.9

A feeling for the physical significance underlying our simple rule can be obtained by considering the following situation. Suppose you have a rectangle, one side of which is x cm long and the other y cm. Evidently, the area of this rectangle xy can only be specified to a precision consistent with that with which you know x or y. If there is uncertainty in either of these dimensions, then no matter how well you know the other dimension, you will be uncertain of the area. Furthermore, the amount of uncertainty in the total area—that is, implicitly, the number of significant figures to which we can specify it—is limited by the least certain of the two sides—that is, the side that we know to fewer significant figures.

■ EXAMPLE 7.   To how many significant figures can you know the areas of the following rectangles: (a) 0.12 cm × 0.345 cm; (b) 0.120 cm × 0.345 cm; (c) 0.120 cm × 0.3 cm; (d) 0.120 cm × 3 mm?

ANSWER: (a) Two; (b) three; (c) one; (d) one. ■

This raises a question: If you are going to carry out a multiplication or division between two numbers, one of which has more significant figures than the other, should you round off before you carry out the operation or after?

The answer is it really makes no difference, although you may have a variation in the last digit doing it one way rather than the other. For example, in
computing the area 0.12 cm × 0.335 cm, we might note that only two significant figures are permitted in the answer, so we first round off our data to two
figures, giving 0.12 cm × 0.34 cm, and then compute the area to be 0.041 cm².
Alternatively, we might first multiply 0.12 cm × 0.335 cm = 0.04020 cm²
and then round off to two figures, getting 0.040 cm² for our area. *Do not
fret that the answers, 0.041 and 0.040, are different. Remember that the last
digit is doubtful!* In general, you improve your guess as to what the last digit is
by carrying through one more significant figure than you really need. This,
in fact, is the basis of the recommended rule: "In multiplying or dividing,
carry through one additional significant figure beyond what is required and
then round off the final answer so it has no more significant figures than the
least number fed into the computation." Applying this rule to our problem
above, for the area 0.12 cm × 0.335 cm, we would prefer to write 0.040 cm²,
rather than 0.041 cm². Note, however, that the latter answer is not "wrong."

In combined operations involving several consecutive multiplications
and/or divisions, the rule on significant figures is the same: "No more in the
answer than the least number fed in." A word of caution, however, if part of
the computation also involves an addition or subtraction. Recall that in adding or subtracting, it is not the *number* of significant figures that is important
but which is the *doubtful decimal place*. Students frequently make the error of
simply looking for which piece of data has the fewest significant figures and
concluding that the answer should have that number of significant figures.
They forget that it might be swallowed up by some other piece of data. The
following problem involves such a trap.

■ EXAMPLE 8.   A given sample contains 2.0 g of hydrogen, 32.1 g of
sulfur, and 64.0 g of oxygen. What is the per cent of sulfur in the sample?

SOLUTION:

Total weight of sample = 2.0 g + 32.1 g + 64.0 g

$$= 98.1 \text{ g}$$

$$\text{Per cent of sulfur} = \frac{\text{weight of sulfur}}{\text{weight of sample}} \times 100$$

$$= \frac{32.1 \text{ g}}{98.1 \text{ g}} \times 100 = 32.7\% \ ■$$

Note that although one of the pieces of data going into the computation,
namely, the weight of hydrogen, is known only to two significant figures,
the final answer is still valid to three significant figures. The point is that the
2.0 g is involved only in an addition step, the result of which comes out to have
three significant figures.

At this stage it would be well to note that our rules for handling significant figures are not completely exact, as any student of differential calculus
can easily show. Strictly speaking, to express the uncertainty in a final
answer, we would need to make a full error analysis considering with the
methods of calculus just how the uncertainties add up. We would need to
consider, for example, that if a quantity comes in squared, its uncertainty has
to be counted double, whereas if it comes in as a square root, its uncertainty
is counted only half. Consideration of such elegancies, unfortunately, are
beyond us and would lead far afield from chemistry. We shall find that the
simple rules given in this section hold in more than 90% of the cases and,
therefore, will suffice. As it is, the handling of significant figures is usually done
in such sloppy fashion that if we do it right in 90% of the cases we will be
considerably improving the present situation.

One final aspect of significant figures might be mentioned here, even
though we shall not make much use of it until we get to computations involving pH. As discussed in Chapter 13, pH involves logarithms. What happens
to significant figures when logs come into the picture? The answer is complex,
but generally speaking the number of significant figures to be written for the
logarithm of a quantity is frequently greater than the number of significant
figures in the quantity itself. More specifically, the number of digits to be
written to the right of the decimal (sometimes called the *mantissa* of the
logarithm) is equal to the number of significant figures in the original quantity. This is true, no matter how many digits appear before the decimal (the
so-called *characteristic* of the logarithm). As an example, the log of 10.7 is
correctly given as 1.029, illustrating how we may go from three significant
figures to four. This whole problem will appear again in Section 1.3 on logarithms and in Chapter 13 on pH.

Not all numbers involved in computations need to be considered in
significant-figure decisions. For example, in the relation $d = 2r$, where d is the
diameter of a circle and r is its radius, the number 2 is exact and is known to
an infinite number of places. This means that the number of significant
figures in d is controlled by the number of significant figures in r.

1.2   *Exponential numbers*

There are two good reasons for using exponential numbers: (1) it avoids
writing lots of zeros for very small and very large numbers, and (2) it gives a
way of expressing significant-figure information with minimal confusion. The
idea of exponential numbers is simply to express quantities as multiples of
powers of 10. Thus, an exponential number, such as 6.02×10^{23}, consists of
two parts: a coefficient (in this case, 6.02) and a power of 10 (in this case,
10^{23}). The coefficient is usually chosen to be somewhere between 1 and 10,
although occasionally specific units corresponding to certain powers of 10

dictate otherwise. The exponent in the power of 10 may be positive or negative. A positive exponent, as in 10^n, means multiplication by 10 n times—that is, moving the decimal n places to the right; a negative exponent, as in 10^{-m}, means division by 10 m times—that is, moving the decimal m places toward the left. The following table lists ways of writing some simple numbers.

$$1 \times 10^0 = 1 \qquad\qquad 1 \times 10^{-1} = 0.1$$
$$1 \times 10^1 = 10 \qquad\qquad 1 \times 10^{-2} = 0.01$$
$$1 \times 10^2 = 100 \qquad\qquad 1 \times 10^{-3} = 0.001$$
$$1 \times 10^3 = 1000 \qquad\qquad 1 \times 10^{-4} = 0.0001$$
$$1 \times 10^4 = 10000 \qquad\qquad 1 \times 10^{-5} = 0.00001$$
$$1 \times 10^5 = 100000 \qquad\qquad 1 \times 10^{-6} = 0.000001$$
$$1 \times 10^6 = 1000000 \qquad 1 \times 10^{-7} = 0.0000001$$
$$1 \times 10^{-8} = 0.00000001$$

(Fractional and decimal powers of 10 are also possible, but they cannot be interpreted in terms of adding or removing zeros or moving a decimal point.) In general, the coefficient is not 1, as in the foregoing table, but some decimal number, which indicates how many significant figures there are in the measured quantity. As indicated in the section on significant figures, the power of 10 has no bearing on the significant figures but simply indicates which way and how many places the decimal would move if the quantity were to be written out.

■ EXAMPLE 9. Write each of the following as an exponential number: (a) 0.0000000181; (b) 22412; (c) 96500 ± 1; (d) $186,000 \pm 1000$; (e) 93 million; (f) four thousand; (g) fifteen thousandths.

ANSWER: (a) 1.81×10^{-8}; (b) 2.2412×10^4; (c) 9.6500×10^4; (d) 1.86×10^5; (e) 9.3×10^7; (f) 4×10^3; (g) 1.5×10^{-2}. ■

Since exponential numbers are products of two numbers, the coefficient and the power of 10, the same rules that hold for arithmetic operations involving products apply.

Addition or *subtraction* of exponential numbers requires that the powers of 10 be the same. If the given numbers do not have the same powers of 10, they must be rewritten, recalling that each time the power of 10 is made more positive by 1 (which corresponds to multiplication by 10), the decimal of the coefficient has to be moved to the left (which corresponds to division by 10). Similarly, if the power of 10 is made more negative, the decimal of the coefficient must be moved to the right.

■ EXAMPLE 10. (a) Add 1.86×10^{-8} cm to 1.86×10^{-7} cm. (b) Subtract 1.86×10^{-8} cm from 1.86×10^{-7} cm.

SOLUTION:

(a) $1.86 \times 10^{-8} = 0.18\underline{6} \times 10^{-7}$
$1.86 \times 10^{-7} = \underline{1.86} \times 10^{-7}$
$\overline{2.0\underline{46} \times 10^{-7}}$

which rounds off to 2.05×10^{-7} cm.

(b) $1.86 \times 10^{-7} = 1.8\underline{6} \times 10^{-7}$
$1.86 \times 10^{-8} = 0.18\underline{6} \times 10^{-7}$
$\overline{1.6\underline{74} \times 10^{-7}}$

which rounds off to 1.67×10^{-7} cm. ■

As far as *multiplication* is concerned, the procedure is to multiply the coefficients together to give a new coefficient and multiply the powers of 10 together to give a new power of 10. Multiplying powers of 10 is achieved by adding their exponents.

■ EXAMPLE 11.   Multiply 2×10^3 by 3×10^{-6}.

SOLUTION:

$$(2 \times 10^3)(3 \times 10^{-6}) = (2 \times 3) \times (10^3 \times 10^{-6})$$
$$= 6 \quad \times \quad 10^{-3} ■$$

In *division* of exponential numbers, the coefficients are divided one into the other and the powers of 10 are combined by subtracting the exponent of the divisor from the exponent of the number divided. Close attention needs to be paid to possible changes of sign.

■ EXAMPLE 12.   Divide 8×10^6 by 2×10^{-3}.

SOLUTION:

$$\frac{8 \times 10^6}{2 \times 10^{-3}} = \frac{8}{2} \times \frac{10^6}{10^{-3}}$$
$$= 4 \times 10^9$$

Note that $10^6 \div 10^{-3}$ requires the exponent (-3) to be subtracted from the exponent (6), or $(6) - (-3) = 9$. ■

■ EXAMPLE 13.   Divide 1.86×10^{-8} by 3.1×10^3.

SOLUTION:

$$\frac{1.86 \times 10^{-8}}{3.1 \times 10^3} = \frac{1.86}{3.1} \times \frac{10^{-8}}{10^3}$$
$$= 0.60 \times 10^{-11}$$
$$= 6.0 \times 10^{-12} ■$$

Note that the answer 0.60×10^{-11} is rewritten as 6.0×10^{-12} in line with the usual recommendation that the coefficient be kept between 1 and 10. To go from 0.60 to 6.0 we need to multiply by 10, so to keep the whole number from changing we must divide the exponential part by 10.

Combined operations involving addition, subtraction, multiplication, and division of exponential numbers follow the same procedure of combining the coefficients separately and the powers of 10 separately. Frequently it is desirable to simplify the coefficient by putting some of its decimal information into the power of 10.

■ EXAMPLE 14.   Calculate the following: Add 9.70×10^{-9} to 1.81×10^{-8}, multiply the result by 6.02×10^{23}, and then divide the whole business by 1.60×10^{-19}.

SOLUTION:

$$\frac{(9.70 \times 10^{-9} + 1.81 \times 10^{-8})(6.02 \times 10^{23})}{1.60 \times 10^{-19}} = ?$$

First add the numbers in the first set of parentheses, remembering to convert to the same power of 10 before doing so. Then combine the coefficients and powers.

$$\frac{(2.78 \times 10^{-8})(6.02 \times 10^{23})}{1.60 \times 10^{-19}} = \frac{(2.78)(6.02)}{1.60} \times \frac{(10^{-8})(10^{23})}{10^{-19}}$$
$$= 10.5 \times 10^{34}$$
$$= 1.05 \times 10^{35}$$

Note that $(10^{-8})(10^{23}) = 10^{-8+23} = 10^{15}$. Also, that

$$\frac{10^{15}}{10^{-19}} = 10^{15-(-19)} = 10^{34} \text{ ■}$$

For calculations requiring that an exponential number be *raised to a higher power*—as, for example, in squaring or cubing a number—the procedure is to raise the coefficient first to the proper power and then raise the power of 10 to the higher power. The latter is done by multiplying the original exponent of 10 by the power to which the whole number is to be raised. Symbolically, the operation can be represented as follows:

$$(a \times b)^n = a^n \times b^n$$

Since b itself is some power of 10, say 10^x, we can also write

$$(a \times 10^x)^n = a^n \times (10^x)^n = a^n \times 10^{nx}$$

■ EXAMPLE 15.   What is the square of 2×10^4?

SOLUTION: "Square" means raising to the second power—that is, multiplying a number by itself.

$$(2 \times 10^4)^2 = 2^2 \times (10^4)^2 = 4 \times 10^8 \blacksquare$$

■ EXAMPLE 16.   What is the cube of 2×10^4?

SOLUTION: "Cube" means raising to the third power—that is, multiplying a number by itself and then the resulting product by the original number again.

$$(2 \times 10^4)^3 = (2 \times 10^4)(2 \times 10^4)(2 \times 10^4)$$
$$= 2^3 \times (10^4)^3 = 8 \times 10^{12} \blacksquare$$

■ EXAMPLE 17.   What is the square of 9.6×10^4?

ANSWER: 9.2×10^9. The square of 9.6 is 92, to two significant figures, and the square of 10^4 is 10^8. The result 92×10^8 is then rewritten as 9.2×10^9. ■

■ EXAMPLE 18.   What is the square of 2×10^{-4}?

SOLUTION:

$$(2 \times 10^{-4})^2 = 2^2 \times (10^{-4})^2$$
$$= 4 \times 10^{-8} \blacksquare$$

■ EXAMPLE 19.   What is 2×10^{-4} taken to the -1 power?

SOLUTION:

$$(2 \times 10^{-4})^{-1} = (2)^{-1} \times (10^{-4})^{-1}$$
$$= 2^{-1} \times 10^4$$
$$= 0.5 \times 10^4 = 5 \times 10^3 \blacksquare$$

Since the meaning of a negative power is to indicate the reciprocal or one-over-the-quantity, 2^{-1} means 1 divided by 2, or 0.5. For the final answer, 0.5×10^4 has been rewritten as 5×10^3.

■ EXAMPLE 20.   What is 2.0×10^{-4} taken to the -2 power?

SOLUTION:

$$(2.0 \times 10^{-4})^{-2} = (2.0)^{-2} \times (10^{-4})^{-2}$$
$$= \frac{1}{(2.0)^2} \times 10^8$$
$$= \frac{1}{4.0} \times 10^8$$
$$= 0.25 \times 10^8 = 2.5 \times 10^7$$

ALTERNATE SOLUTION:

$$(2.0 \times 10^{-4})^{-2} = \frac{1}{(2.0 \times 10^{-4})^2}$$

$$= \frac{1}{4.0 \times 10^{-8}} = \frac{1}{4.0} \times \frac{1}{10^{-8}}$$

$$= 0.25 \times 10^8 = 2.5 \times 10^7 \quad \blacksquare$$

When it becomes necessary to calculate the *square root* or the *cube root* of an exponential number, do the same thing as before, simply noting that taking the square root means raising to the $\frac{1}{2}$ power and taking the cube root means raising to the $\frac{1}{3}$ power.

■ EXAMPLE 21.   What is the square root of 4×10^{-8}?

SOLUTION: Square root of 4×10^{-8} means $(4 \times 10^{-8})^{1/2}$.

$$(4 \times 10^{-8})^{1/2} = (4)^{1/2} \times (10^{-8})^{1/2}$$

$$= 2 \times 10^{-4} \quad \blacksquare$$

In other words, to take the square root of an exponential number $a \times 10^x$, first get the square root of the coefficient a and then get the square root of the power of 10 by dividing the exponent x by 2. What do we do if the exponent is not exactly divisible by 2? Change it so it *is* divisible by 2.

■ EXAMPLE 22.   What is the square root of 4×10^{-9}?

SOLUTION:

$$(4 \times 10^{-9})^{1/2} = (40 \times 10^{-10})^{1/2}$$

$$= (40)^{1/2} \times (10^{-10})^{1/2}$$

$$= 6 \times 10^{-5} \quad \blacksquare$$

Here we have rewritten 4×10^{-9} as 40×10^{-10} so that the power of 10 will be even. Then we proceed to extract the square root. The square root of 40 comes out to be 6.3, but we are allowed only one significant figure, so we write 6. The square root of 10^{-10} is 10^{-5}.

You might have a mental block at this point because it may not be obvious to you that the square root of 40 is 6.3. How do we calculate *that*? Usually, we do not, because (a) most of the handbooks of chemistry and physics contain extensive tables listing square roots and cube roots and (b), as we shall see, it is an easy matter to read square roots and cube roots off a slide rule, once you find out what the gimmick is. Section 1.3 describes one method of getting roots by using log tables.

■ EXAMPLE 23.   What is the cube root of 8×10^{-9}?

SOLUTION: Cube root of 8×10^{-9} means $(8 \times 10^{-9})^{1/3}$.

$$(8 \times 10^{-9})^{1/3} = 8^{1/3} \times (10^{-9})^{1/3}$$
$$= 2 \times 10^{-3} \blacksquare$$

The cube root of the coefficient 8 is 2, and the cube root of 10^{-9} is obtained by dividing the exponent -9 by 3. If the exponent is not exactly divisible by 3, the number should be rewritten as a power of 10 that meets this requirement.

■ EXAMPLE 24.   What is the cube root of 1.86×10^{-8}?

SOLUTION:

$$(1.86 \times 10^{-8})^{1/3} = (18.6 \times 10^{-9})^{1/3}$$
$$= (18.6)^{1/3} \times (10^{-9})^{1/3}$$
$$= 2.65 \times 10^{-3} \blacksquare$$

1.3   Logarithms

In Section 1.2 we considered numbers of the type $a \times 10^x$, where a is the coefficient and x is the power of 10. In all the cases considered, the exponent x was a whole number (e.g., 2, 4, -8, 23) and a was some coefficient that was generally not a whole number (e.g., 6.02, 4.6, 1.86). In a few cases, the coefficient was the whole number 1, and in such cases we sometimes did not even write it down—for example, in Example 17, where we wrote "square of 10^4 is 10^8" instead of "square of 1×10^4 is 1×10^8."

The point about logarithms is that any number can be written as a power of 10 with coefficient unity, provided we allow the power of 10 to take on nonintegral values. This, in fact, is the basis of the definition "a logarithm is an exponent." It is the exponent that would be put on 10 to express the given number. As an example, the number 2 can be expressed as $10^{0.301}$ (read "ten to the power oh-point-three-oh-one"), in which case 0.301 is said to be the logarithm of 2.

Why are logarithms (or logs, as they are frequently called) so useful? They make possible: multiplication of numbers as a simple addition operation; division as a simple subtraction; and computation of powers and roots as a simple multiplying or dividing process. Thus much of the tedious digit writing of arithmetic can be discarded, and operations become faster and less likely to contain errors. How can logarithms do all this? The secret lies in the fact that when two powers of 10 are multiplied, the exponents are added, and when two powers of 10 are divided, the exponents are subtracted. Realizing this, we can appreciate log tables as tabulations of exponents that make the foregoing operations possible.

Some logarithms are immediately obvious without recourse to tables. These are the logarithms of the numbers already expressed as powers of 10. For example, the following list shows the logs of the powers of 10 between $+10$ and -10:

$$\log(10^{10}) = 10 \qquad\qquad \log(10^{-1}) = -1$$
$$\log(10^{9}) = 9 \qquad\qquad\quad \log(10^{-2}) = -2$$
$$\log(10^{8}) = 8 \qquad\qquad\quad \log(10^{-3}) = -3$$
$$\log(10^{7}) = 7 \qquad\qquad\quad \log(10^{-4}) = -4$$
$$\log(10^{6}) = 6 \qquad\qquad\quad \log(10^{-5}) = -5$$
$$\log(10^{5}) = 5 \qquad\qquad\quad \log(10^{-6}) = -6$$
$$\log(10^{4}) = 4 \qquad\qquad\quad \log(10^{-7}) = -7$$
$$\log(10^{3}) = 3 \qquad\qquad\quad \log(10^{-8}) = -8$$
$$\log(10^{2}) = 2 \qquad\qquad\quad \log(10^{-9}) = -9$$
$$\log(10^{1}) = 1 \qquad\qquad\quad \log(10^{-10}) = -10$$
$$\log(10^{0}) = 0$$

simply illustrating that $\log(10^x) = x$, no matter whether x is positive, negative, or zero.

Recalling that multiplication of powers of 10 is carried out by addition of exponents, we see immediately that the log of a product (10^x times 10^y) is simply x plus y.

■ EXAMPLE 25.   What is the log of $(1 \times 10^6)(1 \times 10^5)$?

SOLUTION:

$$(1 \times 10^6)(1 \times 10^5) = 1 \times 10^{11}$$
$$\log(1 \times 10^6)(1 \times 10^5) = \log(1 \times 10^{11}) = 11$$

or, omitting the coefficient one,

$$(10^6)(10^5) = 10^{11}$$
$$\log(10^6)(10^5) = \log(10^{11}) = 11$$

ALTERNATE SOLUTION:

$$\log(10^6)(10^5) = \log(10^6) + \log(10^5)$$
$$= 6 \quad + \quad 5$$
$$= 11 \ ■$$

■ EXAMPLE 26.   What is the log of $(1 \times 10^6)(1 \times 10^{-5})$?

SOLUTION:

$$
\begin{aligned}
\log(10^6)(10^{-5}) &= \log(10^6) + \log(10^{-5}) \\
&= \quad 6 \quad - \quad 5 \\
&= \quad 1
\end{aligned}
$$

Similarly, the log of a quotient $10^x \div 10^y$ is x minus y. ■

■ EXAMPLE 27.   What is the log of $(1 \times 10^6)/(1 \times 10^5)$?

SOLUTION:

$$
\log\left(\frac{1 \times 10^6}{1 \times 10^5}\right) = \log(10^1) = 1
$$

ALTERNATE SOLUTION:

$$
\log\left(\frac{1 \times 10^6}{1 \times 10^5}\right) = \log(1 \times 10^6) - \log(1 \times 10^5)
$$

$$
= 6 - 5 = 1 \ ■
$$

■ EXAMPLE 28.   What is the log of $(1 \times 10^6)/(1 \times 10^{-5})$?

SOLUTION:

$$
\log\left(\frac{1 \times 10^6}{1 \times 10^{-5}}\right) = \log(10^6) - \log(10^{-5})
$$

$$
\begin{aligned}
&= \quad 6 \quad - (-5) \\
&= \quad 6 \quad + 5 = 11 \ ■
\end{aligned}
$$

■ EXAMPLE 29.   What is the log of $\dfrac{(1000)(100)}{0.001}$?

SOLUTION:

$$
\log\frac{(10^3)(10^2)}{10^{-3}} = 3 + 2 - (-3) = 8 \ ■
$$

In the examples above we have restricted ourselves to coefficients that are unity and powers of 10 that are whole numbers. In the general case, digits other than 1 will occur in the coefficient and in such cases we shall need to look up their logs in the log tables. As a sample of what is found in these tables, the following list shows what powers of 10 are needed to express the digits 1 through 9—in other words, what the logarithms are of the digits 1 through 9.

$$1 = 10^0 \qquad\qquad 4 = 10^{0.602} \qquad\qquad 7 = 10^{0.845}$$
$$2 = 10^{0.301} \qquad\quad 5 = 10^{0.699} \qquad\quad 8 = 10^{0.903}$$
$$3 = 10^{0.477} \qquad\quad 6 = 10^{0.778} \qquad\quad 9 = 10^{0.954}$$

Some of these are derivable from the others. For example, since $6 = 2 \times 3 = (10^{0.301}) + (10^{0.477}) = 10^{0.778}$, we can figure out that the logarithm of 6 is the sum of the logs of 2 and 3.

■EXAMPLE 30.   Given that the log of 2 is 0.301, show that the log of 5 is 0.699.

SOLUTION:

$$5 = \frac{10}{2}$$
$$\log 5 = \log \frac{10}{2} = \log (10) - \log (2)$$
$$= 1 - 0.301$$
$$= 0.699 \blacksquare$$

Once we know the logarithms of the various digits and the logarithms of the powers of 10, we are in shape to compute logarithms of exponential numbers and combinations thereof.

■EXAMPLE 31.   What is the log of 2×10^3?

SOLUTION:

$$\log (2 \times 10^3) = \log (2) + \log (10^3)$$
$$= 0.301 \; + \; 3$$
$$= 3.301 \blacksquare$$

■EXAMPLE 32.   What is the log of $(2 \times 10^3)(4 \times 10^8)$?

SOLUTION:

$$\log (2 \times 10^3)(4 \times 10^8) = \log (2) + \log (10^3) + \log (4) + \log (10^8)$$
$$= 0.301 + 3 + 0.602 + 8$$
$$= 11.903 \blacksquare$$

The preceding problem illustrates completely the general rule: "The logarithm of a product $a \times b$ is equal to the sum of log a plus log b." This rule holds even if a and b are products of other numbers. A corollary rule is "The logarithm of a quotient $a \div b$ is equal to the log of a minus the log of b."

■ EXAMPLE 33.   What is the log of $(4 \times 10^8) \div (2 \times 10^3)$?

SOLUTION:

$$\log \frac{4 \times 10^8}{2 \times 10^3} = \log (4) + \log (10^8) - \log (2) - \log (10^3)$$
$$= 0.602 + 8 - 0.301 - 3$$
$$= 5.301 \ ■$$

What do we do if we have negative exponents? Nothing special. The operations are as usual.

■ EXAMPLE 34.   What is the log of 2×10^{-3}?

SOLUTION:

$$\log (2 \times 10^{-3}) = \log (2) + \log (10^{-3})$$
$$= 0.301 + (-3)$$
$$= -2.699 \ ■$$

■ EXAMPLE 35.   What is the log of $(2 \times 10^{-3})(4 \times 10^8)$?

SOLUTION:

$$\log (2 \times 10^{-3})(4 \times 10^8)$$
$$= \log (2) + \log (10^{-3}) + \log (4) + \log (10^8)$$
$$= 0.301 - 3 + 0.602 + 8$$
$$= 5.903 \ ■$$

■ EXAMPLE 36.   What is the log of $(4 \times 10^8) \div (2 \times 10^{-3})$?

SOLUTION:

$$\log \left(\frac{4 \times 10^8}{2 \times 10^{-3}} \right) = \log (4) + \log (10^8) - \log (2) - \log (10^{-3})$$
$$= 0.602 + 8 - 0.301 - (-3)$$
$$= 11.301 \ ■$$

You might be getting the idea that we are putting undue emphasis on calculating logarithms of exponential numbers. Yet, this is precisely what we shall need to do later when we study the pH of solutions in Chapter 13. We shall also need to carry out the reverse process where, given the logarithm, we want to express the number as an integral power of 10. In the following sequence of problems, we look at progressively more complex examples of this operation.

■ EXAMPLE 37.   What is the number whose log is 3? [*Note*: The term "antilogarithm" is sometimes used for "the number whose logarithm is."]

SOLUTION:

$$\log x = 3$$
$$x = 10^3$$

The number is 1×10^3, or 1000. ■

■ EXAMPLE 38.   What is the number whose log is 3.301?

SOLUTION:

$$\log x = 3.301$$
$$x = 10^{3.301}$$
$$= 10^3 \times 10^{0.301}$$
$$= 10^3 \times 2, \text{ or } 2 \times 10^3 ■$$

The method of operating here is to take the given logarithm (3.301), extract the whole number (3), and use that to derive the power of 10 (10^3), and then use the decimal part (0.301) to derive the coefficient (2). The latter operation involves looking in a table of logarithms until you find 0.301 and then noting what number corresponds to it.

■ EXAMPLE 39.   What is the number whose log is -3.301?

SOLUTION:

$$\log x = -3.301$$
$$x = 10^{-3.301}$$
$$= 10^{-3} \times 10^{-0.301}$$
$$= 10^{-3} \times \frac{1}{10^{0.301}}$$
$$= 10^{-3} \times 1/2$$
$$= 10^{-3} \times 0.5 = 5 \times 10^{-4} ■$$

In this solution we note that -3.301 is the sum of two negative numbers, -3 and -0.301. The number whose log is -3 is 10^{-3} and the number whose log is -0.301 is 0.5. The method of solution, which relies on the fact that $10^{-0.301}$ is the same as 1 over $10^{+0.301}$, is not a very good one, because the resulting denominator (in this case, 2) may in other cases not be simple, in which case we would still be left with a messy division. The following method, which is preferred, would avoid any tedious arithmetic even if the numbers were not so simple as they are in this particular case.

■ ALTERNATE SOLUTION:

$$\log x = -3.301$$
$$x = 10^{-3.301}$$

But note that -3.301 can be rewritten as $-4 + 0.699$.

$$x = 10^{-4+0.699}$$
$$= (10^{-4})(10^{0.699})$$

If we look up the antilog of 0.699, we find that it is 5. In other words, $10^{0.699} = 5$.

$$x = (10^{-4})(5)$$
$$= 5 \times 10^{-4} ■$$

Up to this point, all our examples have involved simple one-digit numbers, but in practice the situation is more complicated. For such cases, we need to refer to an expanded table of logarithms or at least have recourse to a slide rule. The procedure for reading logarithms off a slide rule will be taken up in Section 1.4j. For the present, we consider only the tabular problem.

In the back of this book is given a so-called four-place log table. "Four-place" means that the logs are listed in the table to four decimal places and are given directly for all three-digit numbers from 1.00 to 9.99. By interpolation—that is, by estimating values that lie between adjacent ones listed in the table—it is possible to get logs of all the four-digit numbers between 1.000 and 9.999. As an example, once we can read off that the logs of 2.16 and 2.17 are 0.3345 and 0.3365, respectively, we can estimate that the log of 2.164 lies four tenths of the way between these values and is 0.3353. A five-place table would give directly the logs for all the numbers from 1.000 to 9.999 in steps of 0.001. A six-place table would be for steps of 0.0001, and so forth. Put another way, the various log tables list logarithms with varying degrees of fineness. For our purposes, since most of our calculations rarely involve more than four significant figures, the four-place table will be sufficient. The *aficionado* is referred to the ten-place log table compiled by the United States Bureau of Standards.

Given in Figure 1 is a portion of a four-place log table. Unlike most such tables, which omit all the decimal points, this one includes the decimal, both in the numbers and in their logarithms. The two digits in the extreme left column correspond to the first two significant figures of the number whose logarithm is sought; the digits in the column heads across the top give the third significant figure. The intersection of any row and any column corresponds to a three-digit number and the value written there gives its logarithm. As an example, to find the log of 2.16, move down the left column to 2.1, then move across to the column headed 6, where the logarithm is given as .3345. The reverse operation applies for finding antilogarithms. For example, given the logarithm .4048, we note that it is the row marked 2.5 and the column headed by 4, indicating that the corresponding number is 2.54.

	0	1	2	3	4	5	6	7	8	9
2.0	.3010	.3032	.3054	.3075	.3096	.3118	.3139	.3160	.3181	.3201
2.1	.3222	.3243	.3263	.3284	.3304	.3324	.3345	.3365	.3385	.3404
2.2	.3424	.3444	.3464	.3483	.3502	.3522	.3541	.3560	.3579	.3598
2.3	.3617	.3636	.3655	.3674	.3692	.3711	.3729	.3747	.3766	.3784
2.4	.3802	.3820	.3838	.3856	.3874	.3892	.3909	.3927	.3945	.3962
2.5	.3979	.3997	.4014	.4031	.4048	.4065	.4082	.4099	.4116	.4133
2.6	.4150	.4166	.4183	.4200	.4216	.4232	.4249	.4265	.4281	.4298
2.7	.4314	.4330	.4346	.4362	.4378	.4393	.4409	.4425	.4440	.4456
2.8	.4472	.4487	.4502	.4518	.4533	.4548	.4564	.4579	.4594	.4609
2.9	.4624	.4639	.4654	.4669	.4683	.4698	.4713	.4728	.4742	.4757

Fig. 1. Portion of log table showing logarithms for numbers between 2.00 and 2.99

What about four-digit numbers? Suppose you need the log of 2.876, which is not listed directly in the table. You need to interpolate. First look up the log of 2.87 (it comes out to be .4579), then look up the log of 2.88 (it comes out to be .4594). The desired number 2.876 is six tenths of the way between these two; so, we take six tenths of the difference between .4579 and .4594 and add it to .4579. The difference is .0015; six tenths of it is .0009; when added to .4579, it gives .4588. Thus, we find that the log of 2.876 is .4588.

■ EXAMPLE 40.   What is the number whose log is .4000?

SOLUTION: The value .4000 lies between .3997 and .4014 of the table and, in fact, is $(.4000 - .3997)/(.4014 - .3997)$ of the way, or 3/17, or .2.
The antilog of .3997 is 2.51.
The antilog of .4014 is 2.52.
We are .2 of the way between, so our number is 2.512. ■

What if the number does not lie between 1.00 and 9.99? The point to remember is that *any* number can be written as a power of 10, so its coefficient will lie between 1.00 and 9.99.

■ EXAMPLE 41.   What is the log of 2560?

SOLUTION: Rewrite 2560 as 2.56×10^3.

$$\begin{aligned} \log(2560) &= \log(2.56 \times 10^3) \\ &= \log(2.56) + \log(10^3) \\ &= .4082 + 3 \\ &= 3.4082 \;■ \end{aligned}$$

■ EXAMPLE 42.   What is the log of .08206?

SOLUTION:

$$\begin{aligned} \log(.08206) &= \log(8.206 \times 10^{-2}) \\ &= \log(8.206) + \log(10^{-2}) \\ &= .9141 - 2 \\ &= -1.0859 \;■ \end{aligned}$$

(In this case, we need to look at Appendix A for the logs of 8.20 and 8.21 and interpolate between them for 8.206.)

■ EXAMPLE 43.   Given that the log of a number is -2.39. What is the number?

SOLUTION:

$$\begin{aligned} \log x &= -2.39 \\ x &= 10^{-2.39} \\ &= 10^{-3+0.61} \end{aligned}$$

The antilog of .61 is 4.1, so $10^{0.61}$ is 4.1.

$$x = 4.1 \times 10^{-3} \;■$$

In passing, it might be noted, as demonstrated in Example 43, that the number of significant figures changes in a complex way when logs are involved. Specifically, in the logarithm -2.39, the digit before the decimal tells us only about the power of 10 and therefore is not counted in the significant figures. Only the 3 and the 9 count as significant figures, so the final answer should be limited to two significant figures.

At this stage, you may well have the idea that going from numbers to logs or from logs to numbers is all we are interested in. As a matter of fact, this is almost true, because our major log problem will be with pH, where just such operations are the ones involved. However, we should note that logs are also useful in streamlining any computation requiring multiplication and division. In such calculations, the logs are only convenient intermediates in the calculation—that is, numbers are converted to logs and then logs are converted back to numbers.

■ EXAMPLE 44. Ideally, a sample of gas will obey the equation of state $PV = nRT$, where P is the pressure, V is the volume, n is the number of moles, R is a universal constant, and T is the temperature. (Do not worry if you do not know what these quantities mean. This is only an exercise in arithmetic.) If a given sample contains 0.01234 mole at a temperature of 300.0°C and a pressure of 0.5678 atm, what will be the volume, given that R has a value of 82.06 cc-atm?

SOLUTION:

$$V = \frac{nRT}{P}$$

$$\log V = \log\left(\frac{nRT}{P}\right) = \log\frac{(0.01234)(82.06)(300.0)}{0.5678}$$

$$= \log(1.234 \times 10^{-2}) + \log(8.206 \times 10) + \log(3.000 \times 10^{2})$$
$$\quad - \log(5.678 \times 10^{-1})$$

$$= 0.0913 - 2 + 0.9141 + 1 + 0.4771 + 2 - 0.7542 - (-1)$$

$$= 2.7283$$

$$V = 10^{2.7283} = (10^{2})(10^{0.7283}) = (10^{2})(5.349)$$

$$= 5.349 \times 10^{2} \text{ cc} \blacksquare$$

A final point remains to be considered in our discussion of logarithms: the question of roots and powers. As indicated before, a square root corresponds to the $\frac{1}{2}$ power and a cube root the $\frac{1}{3}$ power. In general, the nth root means taking the $1/n$th power. Thus, there is only one general problem to consider: what is the logarithm of a number raised to a power? The rule is simple. It states that the log of a^x is equal to x times the log of a. [We have actually used this several times already. Each time we write $\log 10^x = x$, we imply the following operation: $\log 10^x = x \log 10 = x(1) = x$, based on the fact that the log of 10 is 1.]

■ EXAMPLE 45.   What is the square root of 10?

SOLUTION: Square root of $10 = (10)^{1/2} = x$.

$$\begin{aligned} \log x &= \log (10)^{1/2} \\ &= \tfrac{1}{2} \log (10) \\ &= \tfrac{1}{2} (1) \\ &= 0.500 \\ &= \text{antilog of } 0.500 = 3.16 \blacksquare \end{aligned}$$

■ EXAMPLE 46.   What is the cube root of 10?

SOLUTION: Cube root of $10 = (10)^{1/3} = x$.

$$\begin{aligned} \log x &= \log (10)^{1/3} \\ &= \tfrac{1}{3} \log (10) = \tfrac{1}{3}(1) = 0.333 \\ x &= \text{antilog of } 0.333 = 2.15 \blacksquare \end{aligned}$$

■ EXAMPLE 47.   What is the square root of 2?

SOLUTION: Square root of $2 = 2^{1/2} = x$.

$$\begin{aligned} \log x &= \log (2^{1/2}) \\ &= \tfrac{1}{2} \log 2 = \tfrac{1}{2}(0.3010) = 0.1505 \\ x &= \text{antilog of } 0.1505 = 1.414 \blacksquare \end{aligned}$$

■ EXAMPLE 48.   What is the cube root of 2?

SOLUTION: Cube root of $2 = 2^{1/3} = x$.

$$\begin{aligned} \log x &= \log (2^{1/3}) = \tfrac{1}{3} \log 2 \\ &= \tfrac{1}{3}(0.3010) = 0.1003 \\ x &= \text{antilog of } 0.1003 = 1.260 \blacksquare \end{aligned}$$

In summary, to get the nth root of a number, take the log of the number, divide it by n, and look up the antilog of the result.

■ EXAMPLE 49.   What is the square root of 1.26×10^{-8}?

SOLUTION: You could proceed by getting the logarithm of the entire exponential number, halving the result, then looking up the antilog. However, as noted before (compare Example 21), it is so easy to take the square root of the exponential part that it pays to treat the coefficient and the power of 10 separately.

$$\begin{aligned} (1.26 \times 10^{-8})^{1/2} &= (1.26)^{1/2} \times (10^{-8})^{1/2} \\ &= (1.26)^{1/2} \times 10^{-4} \end{aligned}$$

To get $(1.26)^{1/2}$:

$\frac{1}{2} \log 1.26 = \frac{1}{2}(0.1003) = 0.0502$

Antilog of 0.0502 is 1.12.
So, the square root of 1.26×10^{-8} is 1.12×10^{-4}. ∎

■ **EXAMPLE 50.** Calculate

$$\frac{(2.48 \times 10^6)(1.26 \times 10^{-8})^{1/2}}{3.00 \times 10^{10}}$$

SOLUTION:

$$\log x = \log (2.48 \times 10^6) + \log (1.26 \times 10^{-8})^{1/2} - \log (3.00 \times 10^{10})$$
$$= 6.3945 + \frac{1}{2}(-7.8997) - 10.4771$$
$$= -8.0325$$
$$= -9 + 0.9675$$
$$x = 9.28 \times 10^{-9} \ ∎$$

If, instead of roots, we need to raise the numbers to powers, then we multiply the log of the number by the power and get the antilog of the result.

■ **EXAMPLE 51.** What is the cube of 2.54×10^6?

SOLUTION:

$$x = (2.54 \times 10^6)^3$$
$$\log x = 3 \log (2.54 \times 10^6) = (3)(0.4048 + 6)$$
$$= 19.2144$$
$$x = \text{antilog } 19.2144$$
$$= 10^{19.2144} = (10^{19})(10^{0.2144})$$
$$= (10^{19})(1.64) = 1.64 \times 10^{19} \ ∎$$

■ **EXAMPLE 52.** Calculate the square root of 1.86×10^{-8} times the square of 3.05×10^{-6}.

SOLUTION:

$$x = (1.86 \times 10^{-8})^{1/2}(3.05 \times 10^{-6})^2$$
$$\log x = \frac{1}{2} \log (1.86 \times 10^{-8}) + 2 \log (3.05 \times 10^{-6})$$
$$= \frac{1}{2} (\log 1.86 + \log 10^{-8}) + 2 (\log 3.05 + \log 10^{-6})$$
$$= \frac{1}{2}(0.2695 - 8) + 2(0.4843 - 6)$$
$$= -14.8966$$
$$= -15 + 0.1034$$

$$x = (10^{-15})(10^{0.1034})$$
$$= (10^{-15})(1.27), \text{ or } 1.27 \times 10^{15}$$

(Actually, when there are several consecutive operations of multiplication and/or division, it is usually most efficient to collect all the coefficients together and all the powers of 10 together.)

ALTERNATE SOLUTION:

$$x = (1.86 \times 10^{-8})^{1/2}(3.05 \times 10^{-6})^2$$
$$= (1.86)^{1/2}(3.05)^2 \times (10^{-8})^{1/2}(10^{-6})^2$$
$$= \qquad 12.7 \quad \times \qquad 10^{-16}$$

To calculate the coefficient $(1.86)^{1/2}(3.05)^2$:

$$\tfrac{1}{2}\log(1.86) + 2\log(3.05) = \tfrac{1}{2}(0.2695) + 2(0.4843)$$
$$= 1.1034$$

where the antilog of 1.1034 is $10^{1.1034} = (10^1)(10^{0.1034}) = 12.7$. To calculate the power of 10 portion $(10^{-8})^{1/2}(10^{-6})^2$:

$$(10^{-8})^{1/2}(10^{-6})^2 = (10^{-8 \times 1/2})(10^{-6 \times 2})$$
$$= (10^{-4})(10^{-12}) = 10^{-16}$$

The final combined answer 12.7×10^{-16} is rewritten as 1.27×10^{-15} to conform to the recommendation that the coefficient lie between 1 and 10. ∎

All of the foregoing discussion concerns the use of logarithms referred to the base 10. However, there is nothing magic about using 10 for the base. Any number could be so used, but of course the value of the exponent would need to be consistent. For reasons we cannot pursue here, the most desirable base is the so-called Naperian base or natural base. It has the value 2.718 and is usually designated e. In terms of e, the number 2 can be expressed as $e^{0.693}$ (read "e to the 0.693 power"), so that $2 = 10^{0.301} = e^{0.693}$ represents two equivalent ways of expressing the same number. The exponent on the base e is 2.303 times the exponent on the base 10. Hence, natural logarithms are just equal to 2.303 times the base-10 logarithm of the same number.

Frequently, natural logarithms are designated as ln and base-10 logarithms are designated simply as log. Thus, the statement "the natural logarithm of 2 equals 0.693" can be written ln 2 = 0.693 to distinguish it from the statement "the logarithms of 2 to the base 10 equals 0.301," which can be written log 2 = 0.301. In general, $\ln x = 2.303 \log x$.

∎ EXAMPLE 53.   The entropy of a gas is equal to $R \ln V$ where R is the gas constant [1.99 cal/(deg mole)] and V is the volume of the gas. Calculate the entropy change when a gas expands to double its volume.

SOLUTION:

$$\text{Initial entropy} = R \ln V_{\text{initial}}$$

$$\text{Final entropy} = R \ln V_{\text{final}}$$

$$\text{Increase in entropy} = R \ln V_{\text{final}} - R \ln V_{\text{initial}}$$

$$= R \ln \frac{V_{\text{final}}}{V_{\text{initial}}}$$

$$= R \ln 2$$

$$= R(2.303 \log 2)$$

$$= (1.99)(2.303)(0.301) = 0.912 \ \blacksquare$$

1.4 Use of the Slide Rule

The slide rule is a simple calculating machine based on the principle that the multiplication of two numbers can be achieved by adding two lengths corresponding to the logarithms of the two numbers.

(a) Description of scales

As shown in Figure 2, the essential parts of a slide rule are three in number: (1) a fixed principal part, called the *body*; (2) a sliding part, called the *slide*, which moves in grooves cut into the body; and (3) a bridging, transparent plaque, called the *cursor*, which has a hairline that can be slid along the scales. Different slide rules may differ in the kinds of scales they carry and the relative placing of the scales on the rule. For our purposes, it will be sufficient to consider the following scales:

A and B, for getting squares and square roots
C and D, for multiplication and division
C1, for getting reciprocals
K, for getting cubes and cube roots
L, for getting logarithms
S and T, for getting trigonometric functions

For the slide rule pictured in Figure 2, the S and T scales would be on the back of the slide. The slide can be pulled out completely, turned over, and re-inserted in the grooves so that the S and T scales face forward.

Probably the most useful scales on the slide rule are the C and D scales. These are identical scales, consisting of the digits 1, 2, 3, 4, 5, 6, 7, 8, 9, 1 from left to right as the principal markings, with the length between 1 and 2 on the left end marked off again with the digits 1, 2, 3, 4, 5, 6, 7, 8, 9 as subsidiary markings. Between any two digits, there are additional subdivision lines that could also be marked off with digits except for lack of space.

The first major difficulty a novice encounters in using a slide rule is to translate a given number to a proper position on the rule, and vice versa. The

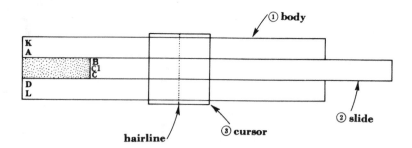

Fig. 2. Slide rule showing component part

point to remember is that, ignoring decimals completely, the various consecutive digits appearing in a number correspond to successively smaller subdivisions on the slide rule. Thus, the number 234 is positioned after the second big division, the third subdivision, and the fourth sub-subdivision. What makes the procedure tricky is that various places on the slide rule are subdivided differently—sometimes into ten pieces, sometimes five, sometimes two. Evidently, care must be taken to interpret the subdivisions correctly, and for this, practice is essential.

The recommended procedure to follow is to note how many minor divisions there are between any two major divisions on the rule, then assign each minor division a numerical value indicating its part of the whole. For example, as shown in Figure 3, which represents a magnified portion of the C or D scale, the space between 2 and 3 has ten large divisions, each of which is subdivided into five smaller subdivisions. If we had room to put the digits on the rule, the large division lines would be numbered 1, 2, 3, 4, 5, 6, 7, 8, 9 and the small subdivision lines 2, 4, 6, 8. In Figure 3 these digits have been supplied, but a glance at the actual slide rule shows that only the big 2 and 3 appear. In other words, you need to furnish the other numbers mentally. Once you know the value of the scale markings, it is relatively easy to translate a given number to a position on the slide rule. The number 234, for example, is located at the dotted line X, following the principle that each successive

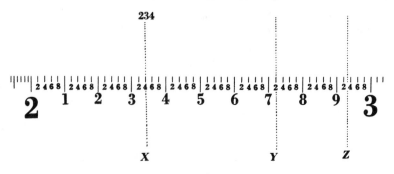

Fig. 3. Portion of slide rule scales C or D between 2 and 3

digit of the given number is positioned according to smaller and smaller subdivisions.

■ EXAMPLE 54.   Read off the numbers corresponding to the dotted lines Y and Z in Figure 3.

ANSWER: Y is 272; Z is 293. ■

At this stage, it should be evident that (on the C and D scales) all numbers starting with 2 are set up somewhere in the portion shown in Figure 3, no matter whether the number is, for example, 2 or 22 or 0.02096 or 2.96×10^{-8}. All numbers starting with 20 would be at the left-hand end of Figure 3 between the big 2 and the small 1; all numbers starting with 200 or 201 would be in the extreme left subdivision between the big 2 and the first tiny 2.

If now you look at the A scale, again at the portion between the digits 2 and 3, you find the scale divisions are different. As shown in Figure 4, there

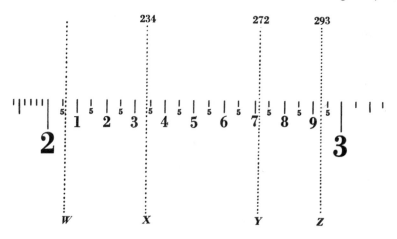

Fig. 4. Portion of slide rule scale A between 2 and 3

are still ten divisions between 2 and 3 but each of these is divided into only two segments instead of five. Therefore, the meanings of the scale markings are as indicated (again, these smaller numbers do not appear on the actual slide rule). The big numbers take care of the first significant figure; the middle-size numbers correspond to the second significant figure; the tiny 5's represent the third significant figure. Obviously, the positioning of the third digit of a number would have to be estimated. The dotted lines marked X, Y, and Z in Figure 4 represent the same numbers as in Figure 3.

■ EXAMPLE 55.   What number corresponds to the dotted line W in Figure 4?

ANSWER: 207 ■

Special features of the various scales will become obvious as we make use of them, but at this point it will suffice to note that each of the scales is progressively more compressed from the left end of the rule to the right, except for the L scale, which is linear throughout. The L scale gives the logarithms and is laid off with equal divisions from one end to the other. It might also be noted that the A scale (and also its twin, the B scale) consists of two D scales compressed into the same length as one normal D scale. The K scale consists of three repeat sequences that are like the D scale. In the following sections, we go through the several slide-rule manipulations useful in chemistry.

(b) Multiplication

The idea here is that one number can be set up on the D scale and the other on the C scale. When the two distances are added up, the product appears on the D scale. The operation works because the distance on the C scale represents the log of the second number. When the two distances are added up, the logs have been added, so in effect a multiplication has been performed.

For ease in describing the manipulations, we shall refer to the digit 1 that appears at the extreme left end or at the extreme right end of a scale as the *index*. When necessary, we shall distinguish these two as the left-hand index or the right-hand index.

■ EXAMPLE 56.   Multiply 234 by 272.

SOLUTION: Move the cursor so the hairline falls on 234 on the D scale. Pull the slide to the right until the left-hand index of the C scale matches the hairline.

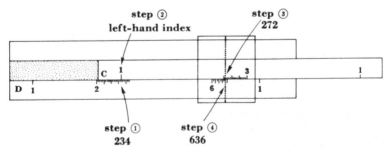

Fig. 5. Slide rule set up for multiplying 234 × 272

Being careful not to disturb the relative position of body and slide, move the cursor to the right until the hairline falls on 272 on the C scale. The answer 636 appears on the D scale.

The final aspect of the rule is shown in Figure 5. The slide rule does not give the decimal, so we need to figure this out separately. This is usually a trivial operation that can be done mentally. Thus, for example, 234 by 272 can be approximated as 2×10^2 by 3×10^2 equal roughly to 6×10^4. We point off our figure 636 to correspond, writing it either as 63,600 or, preferably, 6.36×10^4. ∎

At this point, it might be noted that the number of significant figures obtainable with a slide rule depends on the length of the rule. The usual ten-inch slide rule gives three figures and, in favorable cases involving the left end, four figures. Circular and cylindrical slide rules are available with scales up to sixty inches. These easily give four- and five-figure accuracy.

What do we do in multiplication if the multiplier falls so far to the right on the slide that the cursor cannot be moved to cover it? Such a case would occur if we try to multiply 2.34 by 609. The solutions is to use the right-hand index instead of the left-hand index of scale C. The procedure works out because, once we agree to ignore decimals, logarithms of numbers repeat their values in regular consecutive cycles. In other words, once we go beyond the range of numbers from 1 to 10, say 10 to 100, the logs follow linearly, so it is as if we took scale D and laid it off to extend the slide rule to the right.

∎ EXAMPLE 57.   Multiply 2.34 by 609.

SOLUTION: Set the cursor so the hairline falls on 234 of D scale. Pull the slide to the left so the right-hand index matches the hairline.

Maintaining the relative position of the slide and body, move the cursor to the left until the hairline falls over 609 on the C scale.

The desired result, 143, appears under the hairline on the D scale.

The final view of the slide rule is shown in Figure 6.

With due attention to decimal placing, the answer to the operation 2.34×609 is 1.43×10^3. ∎

[Note: Another way of avoiding the difficulty of running out of scale on the right is to use the A and B scales in the same way we have used the C and D. The disadvantage of this method is that we lose some precision in reading the numbers because the scales are compressed.]

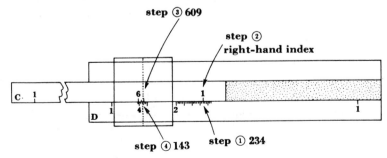

Fig. 6. Slide rule set up for multiplying 234×609

(c) Division

To divide two numbers, set the first number on the D scale. Line it up with the divisor on the C scale. Read the answer on the D scale opposite the index of the C scale.

■ EXAMPLE 58.   Divide 32.1 by 6.02×10^{23}.

SOLUTION: Set the cursor so the hairline falls on 321 on scale D.

Making sure not to move the cursor, pull the slide to the left until the number 602 of the C scale appears under the hairline.

Maintaining the relative position of the slide and body, move the cursor to the right until the hairline matches the index of the C scale.

The answer 533 appears under the hairline on the D scale.

The final arrangement is shown in Figure 7.

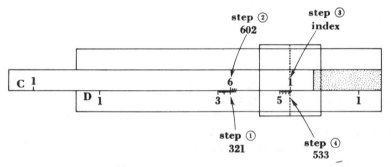

Fig. 7. Slide rule set up for dividing 321 by 602

With the decimal properly placed, the answer is 5.33×10^{-23}. ■

(d) Combined multiplication and division

One of the time-saving features of slide-rule manipulation is that when several consecutive operations need to be performed, the intermediate results need not be read off the slide rule. We simply continue the manipulation of slide and cursor, using each position as the starting point for the next operation.

■ EXAMPLE 59.   Multiply $(3)(96,500) \times (1.89)$.

SOLUTION: Set hairline over 3 of D scale.

Pull slide to left until index of C scale matches hairline.

Move cursor to left until hairline falls over 965 of C scale.

(The answer to 3×965 appears on the D scale, but we do not need to read it so long as we are careful to keep the cursor stationary when we perform the next operation.)

Pull slide to right until left-hand index of C scale matches hairline.

Slide cursor to right until hairline covers 189 of C scale. The answer, 547, appears under the hairline on the D scale.

With proper decimal, the answer is 5.47×10^5. ∎

∎ EXAMPLE 60.   Calculate

$$\frac{22,414}{(82.06)(273)}$$

SOLUTION: Move cursor so hairline falls over 224 of scale D.

Pull slide to left until 8206 of scale C appears under hair line.

Move cursor to right so that hairline falls over right-hand index of scale C.

(The answer to the operation 224 divided by 8206 appears under the hairline on scale D. However, again, we need not read it. Simply divide it by 273.)

Pull slide to right until 273 of scale C appears under the hairline. The answer appears opposite the index (either left-hand or right-hand, in this case) as 100.

To the proper decimal, the answer is 1.00. ∎

(e) Forming the square of a number

This is a simple reading operation where the number is set up on the D (or C scale) and the corresponding square is read off the A (or B) scale. The operation works because the logarithmic scale corresponding to A is twice as compressed as the logarithmic scale corresponding to D. Stated another way, a given number on the D scale corresponds to a certain logarithm, so doubling this logarithm (which is what we do in raising to the second power) should give rise to a logarithm twice as large. Consequently, for the same linear span of logarithms from 0 to 1 corresponding to the D scale we have two full cycles of log from 0 to 2 corresponding to the A scale. This introduces a slight complication in reading because the right half of scale A corresponds to numbers 10 times as great as those in the left half. Some slide rules, in fact, show the right half of scales A (and B) with zeros, as illustrated in Figure 8. If we strike off the zeros from Figure 8, we get the normal aspect of a

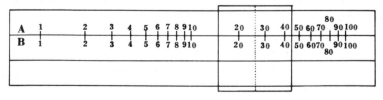

Fig. 8. Slide rule scales A and B showing extra zeros on right half of scale

conventional A or B scale. The zeros do help to fix the decimal point, but since the decimal is normally put in by mental estimation anyway, they really are quite superfluous.

■ EXAMPLE 61.   What is the square of 234?

SOLUTION: Move the cursor so that the hairline falls over 234 of the D scale.
Under the hairline where it crosses the A scale, read off the answer, 548. The operation is illustrated in Figure 9.

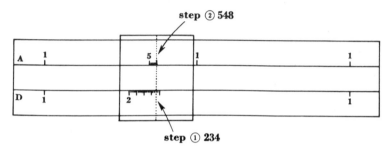

Fig. 9. Slide rule set up for calculating square of 234

To the proper decimal, the answer is 5.48×10^4. ■

■ EXAMPLE 62.   What is the square of 5.86×10^{-8}?

SOLUTION: Set the cursor so the hairline falls over 586 of scale D.
Read off the answer 343 under the hairline where it crosses the A scale.
To adjust the decimal, note that 6×10^{-8} squared is 36×10^{-16}, or 3.6×10^{-15}.
So the square of 5.86×10^{-8} is 3.43×10^{-15}. ■

(f)   Extracting the square root of a number

In the same way that scale A shows the *squares* of numbers from scale D, scale D can be used to show the *square roots* of numbers from scale A. The only special point to remember is that, because powers of 10 should be divisible by 2 when a square root is extracted, the left half of scale A is to be used to set up numbers between 1 and 10 and the right half of scale A, numbers between 10 and 100.

■ EXAMPLE 63.   What is the square root of 2.34?

SOLUTION: Move the cursor to the left half of the rule and set the hairline over 234 of scale A.
Read off the answer 153 under the hairline where it crosses the D scale. The answer is 1.53, as shown in Figure 10. ■

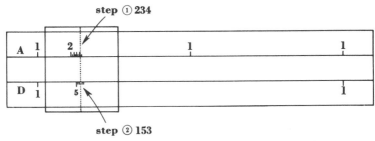

Fig. 10. Slide rule set up for getting square root of 2.34

■ EXAMPLE 64.   What is the square root of 23.4?

SOLUTION: Since this number lies between 10 and 100, we need to move the cursor to the right half of scale A.

Set the cursor so the hairline crosses 234 of scale A.

Read off the answer 484 where the hairline crosses scale D. The setup is shown in Figure 11.

To set the decimal, note that 23.4 is about 25, and the square root of 25 is 5. So, the answer must be in the neighborhood of 5, or 4.84. ■

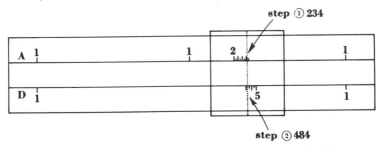

Fig. 11. Slide rule set up for getting square root of 23.4

■ EXAMPLE 65.   What is the square root of 0.100?

SOLUTION: This number is not between 1 and 10 nor between 10 and 100, so it needs to be rewritten. To write it 1.00×10^{-1} is no good, since that would give a power of 10 not divisible by two. The other choice is to write it 10.0×10^{-2}.

Set cursor in middle of slide rule so hairline falls over the central 1 of the A scale.

Read off the answer 316 on the D scale.

Thus, the square root of 10.0×10^{-2} is 3.16×10^{-1}, or 0.316. ■

(g)   Forming the cube of a number

The K scale is a triply condensed version of the D scale, so any number set under the hairline on the D scale shows up raised to the third power under the

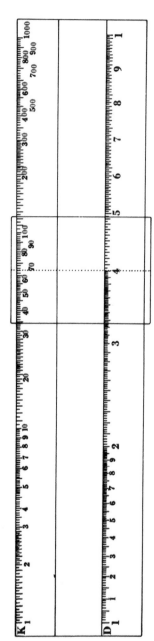

Fig. 12. Slide rule K scale showing extra zeros on middle portion and double zeros on right-hand portion

hairline as it crosses the K scale. The reason this operation succeeds is similar to that for forming squares, namely, a number positioned on the D scale represents the logarithm of that number. If that log is multiplied by 3, the number is effectively raised to the third power. Thus, in the D-scale length that corresponds to logs from 0 to 1 (numbers from 1 to 10) we should have logs from 0 to 3 (numbers from 1 to 1000) on the K scale.

The conventional slide rule shows the K scale with digits 123456789-1234567891234567891 in that order. There are three cycles going from 1 to 1. The first of these corresponds to units; the second, to tens; the third, to hundreds. Some slide rules actually show the zeros in place, as illustrated in Figure 12. If the zeros are not shown, they can be supplied mentally, although the problem is not important until we start to extract cube roots (described in the next section).

■ EXAMPLE 66.   What is the cube of 1.15?

SOLUTION: Set the cursor so the hairline falls on 115 of scale D.

Where the hairline crosses the K scale, read off 152. The operation is illustrated in Figure 13.

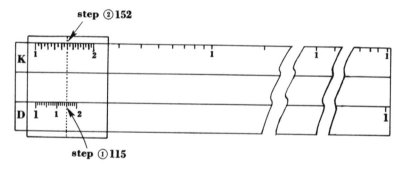

Fig. 13. Slide rule set up for calculating cube of 1.15

To set the decimal, note that approximately 1 raised to the third power is 1.

So, the cube of 1.15 is 1.52. ■

■ EXAMPLE 67.   What is the cube of 2.89×10^{-8}?

SOLUTION: Set the cursor so the hairline crosses scale D at 289.

Look under the hairline at scale K to get 241. The positioning is shown in Figure 14.

To set the decimal we note that 2.89×10^{-8} is about 3×10^{-8} and 3×10^{-8} cubed is 27×10^{-24}, or 2.7×10^{-23}.

Therefore, the cube of 2.89×10^{-8} is 2.41×10^{-23}. ■

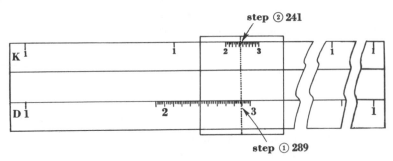

Fig. 14. Slide rule set up for getting cube of 2.89

(h) Extracting the cube root of a number

If the K scale gives the cubes of the D scale, then the D scale gives the cube roots of the K scale. The only special point to remember is that, of the three repeating segments of the K scale, the left one is intended for numbers between 1 and 10, the middle one is used for numbers between 10 and 100, and the right segment is used for numbers between 100 and 1000. Again, the reason for this is that powers of 10 need to be divisible by three if we are to take their cube roots. As a result, coefficients between 1 and 1000 should be expected.

■ EXAMPLE 68.   What is the cube root of 2?

SOLUTION: Set the cursor so the hairline crosses the K scale at 2 (of the left segment).

Where the hairline crosses the D scale, read off the answer 126. The setup is shown in Figure 15.

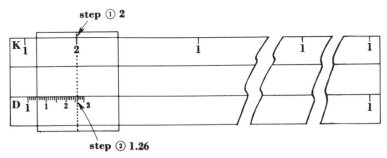

Fig. 15. Slide rule set up for getting cube root of 2

To set the decimal, we note that the cube root of 2 is about 1. Therefore, we conclude that the cube root of 2 is 1.26. ■

■ EXAMPLE 69.   What is the cube root of 20.0?

SOLUTION: Set the cursor so the hairline crosses the K scale at 2 (of the middle segment).

Where the hairline crosses the D scale, read off 271.
To set the decimal, note that 3 cubed is close to 20.
Therefore, the cube root of 20.0 is 2.71. ∎

∎ EXAMPLE 70.   What is the cube root of 0.100?

SOLUTION: This number does not lie between 1 and 1000, so it cannot be set on the K scale without being rewritten. Neither 1.00×10^{-1} nor 10.0×10^{-2} will help, since these powers of 10 are not divisible by three. So, we need to write it as 100×10^{-3}.

To get the cube root of 100, we set the cursor so the hairline crosses the K scale at the 1 that is located at the right end of the middle segment. Where the hairline crosses the D scale we read 464. The setup is shown in Figure 16. To set the decimal we note that 5 cubed is 125, so that 4.64 must be the desired result.

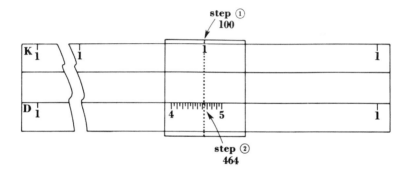

Fig. 16. Slide rule set up for getting cube root of 0.100

Thus, the cube root of 0.100 is 4.64×10^{-1}, or 0.464. ∎

(i)   Calculating the reciprocal of a number

We can calculate the reciprocal of a number—one over a number—by the method described in Section 1.4c using scales D and C. Set the hairline at 1 of the D scale. Pull the slide until the number whose reciprocal is needed shows up on the C scale under the hairline. The answer will then appear on scale D opposite the index of scale C. This operation is equivalent to dividing one by the given number.

Reciprocals, or inverse numbers, are so numerous and useful that most slide rules have a scale on which reciprocals are noted. On our slide rule, scales C and C1 are reciprocal to each other. Be careful, however, to note that the *reciprocal scales are to be read from right to left instead of left to right*. Because the risk of confusion is very great, most slide rules show the reciprocal scales printed with red digits.

To get the reciprocal of a number, set the cursor so the hairline crosses scale C at the desired number. The answer appears on scale C1 where the hairline crosses.

■ EXAMPLE 71.   What is the reciprocal of 1.86?

SOLUTION: Set the cursor so the hairline crosses scale C at 186.
Where the hairline crosses scale C1 read off 538. The setup is shown in Figure 17. Note that the C1 scale digits are marked in descending order from left to right, so the scale divisions must be read from right to left.

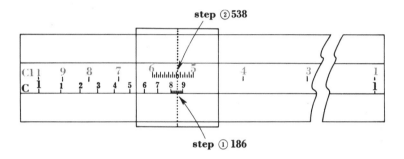

Fig. 17. Slide rule set up for getting reciprocal of 1.86

To set the decimal, we note that 1.86 is about 2 and 1 over 2 is 0.5. So, the desired answer is 0.538. ■

(j)   Finding the logarithm of a number

Opposite each number of the D scale, its logarithm to the base 10 appears on the L scale. The log scale is the whole basis for slide-rule operation, and it is the only linear scale on the slide rule.

To look up a log using the slide rule, set the cursor so the given number on the D scale appears under the hairline. Where the hairline crosses the L scale, read off the logarithm. For a change, the scale divisions of the L scale have uniform value along the whole rule, so there is little chance to go wrong. Also, the decimal point is usually indicated directly on the rule.

■ EXAMPLE 72.   Read off the slide rule the logarithm of 1.86.

SOLUTION: Set the cursor, as shown in Figure 18, so the hairline crosses the D scale at 186.
Where the hairline crosses the L scale, read off the answer 0.270. ■

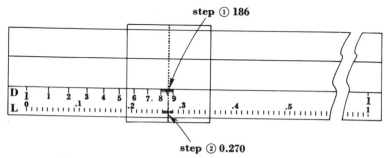

Fig. 18. Slide rule set up for getting logarithm of 1.86

■ EXAMPLE 73. What is the log of 22,400?

SOLUTION: Rewrite the number as 2.24×10^4.

Look up the log of 2.24 by setting the hairline of 224 of the D scale and reading off 0.350 on the L.

Since the log of 10^4 is 4, we can immediately write that the log of 22,400 is 4.350. ■

Of course, we can also do the reverse operation and use scale D to find numbers whose logs are given on scale L.

■ EXAMPLE 74. What is the antilog of 0.450?

SOLUTION: Set the cursor so the hairline crosses the L scale at 0.450.

Look under the hairline at the D scale to get the number 2.82. ■

■ EXAMPLE 75. What is the antilog of -3.862?

SOLUTION: Rewrite -3.862 as $-4 + 0.138$.

Set the cursor so the hairline crosses the L scale at 0.138.

Look under the hairline at the D scale to read off the number 1.37.

Thus, the antilog of 0.138 is 1.37. Since the antilog of -4 is 10^{-4} we can write that the antilog of -3.862 is 1.37×10^{-4}. ■

(k) Finding trigonometric functions

The vaues of the sine, cosine, and tangent can be obtained by use of the S, T, and ST scales in combination with the D scale. The S scale gives the angles for getting the sine; the T scale gives the angles for getting the tangent; the ST scale gives the angles in the small-value range—between $0.6°$ and $5.7°$—where the sine and tangent are almost identical. The black figures on the S scale, when subtracted from $90°$, give the angles for looking up the cosine.

If your slide rule is one in which the slide can be viewed from the back, you need only turn the rule over to use the trigonometric scales. For less

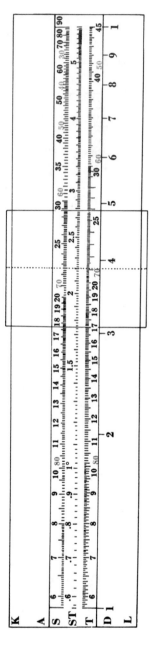

Fig. 19. Angle scales on slide rule

expensive slide rules, where all the markings are on the front of the body, the slide may have to be pulled completely out of the grooves, flipped over, and reinserted in the grooves. When ready for use, the angle scales and the D scale should be visible from the same side, as shown in Figure 19. From there on, the way to operate is to line up the ends of the scales S, ST, and T with the ends of the D scale, set the cursor so the hairline falls on the desired angle on the desired function scale, and read off the value of the function under the hairline where it crosses the D scale. In using the S and T scales, the numbers on the D scale require a decimal (0.) just before the number; in using the ST scale, the numbers on the D scale need a decimal (0.0) before the number.

■ EXAMPLE 76.   What is the sine of 30°?

SOLUTION: Check to see that the index on the slide lines up with the index of the D scale.

Move the cursor so the hairline falls on 30 of the S scale. Be careful to use the black numbers. On many slide rules the S scale has two sets of numbers, one set in black and the other in red. The black numbers are for sines and are read from left to right as normal; the red numbers, denoted by grey tints in Figure 19, are for cosines and are read from right to left.

Where the hairline crosses the D scale, read off the value 500. Put in the decimal and get 0.500 The setup is shown in Figure 20. ■

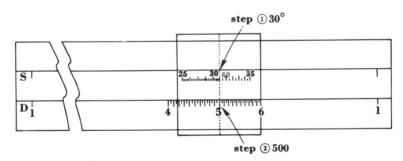

Fig. 20. Slide rule set up for getting sine of 30°

■ EXAMPLE 77.   What is the cosine of 60°?

SOLUTION: Use the red figures of the S scale. You will note that on the S scale, just to the right of the black 30, is a red 60. This means that the same position holds for sin 30° as for cos 60°. This is, of course, in agreement with the relation that cos A = sin(90 − A). The setup is the same as shown in Figure 20. The value from the D scale for cos 90° is again 0.500. ■

■ EXAMPLE 78.   What is the sine of 9.54°?

SOLUTION: Set the cursor so the hairline crosses the S scale at 9.54. Where the hairline crosses the D scale, read off 166. Put in the decimal and get 0.166.

The setup is shown in Figure 21. It applies equally well to the question: What is the cosine of 80.46°? ■

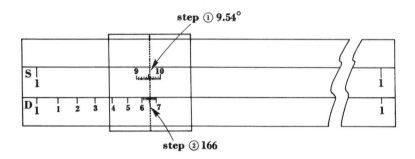

Fig. 21. Slide rule set up for getting sine of 9.54°

■ EXAMPLE 79.   What is the sine of 3.16°?

SOLUTION: Line up the index. Note that the angle is so small that we need to use the ST scale.

Set the cursor so the hairline crosses ST scale at 3.16°.

Under the hairline on the D scale, read off 551.

Recall that using the ST scale requires D values to be preceded by 0.0. If we put in this decimal, we get 0.0551. So, the sine of 3.16° is 0.0551. The setup is shown in Figure 22. ■

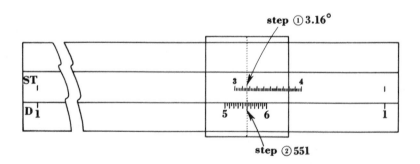

Fig. 22. Slide rule set up for getting sine of 3.16°

■ EXAMPLE 80.   What is the tangent of 30°?

SOLUTION: Use the T scale.
Set the cursor so hairline crosses the T scale at 30°.
Where the hairline crosses the D scale, read off 577.
Put in the decimal and get 0.577. The setup is shown in Figure 23. ■

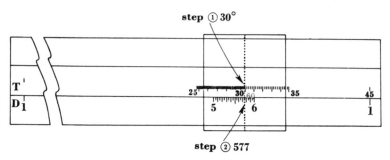

Fig. 23. Slide rule set up for getting tangent of 30°

■ EXAMPLE 81.   What angle has a sine of 0.268?

SOLUTION: Reverse the foregoing procedure.
Set the cursor so the hairline crosses the D scale at 268.
Where the hairline crosses the S scale, read off the angle 15.5°. ■

■ EXAMPLE 82.   What angle has a cosine of 0.0268?

SOLUTION: Set the cursor so hairline crosses D scale at 268.
Using the ST scale (because of the 0.0 preceding the 268), read under the
hairline the angle 1.53°.
This, however, is for sine or tangent. To get the corresponding angle for
the cosine we need to subtract this value 1.53° from 90°, getting 88.47°.
[Note: For angles not lying in the small-value range, the problem is
simpler, since we then only need to use the red numbers of the S scale.] ■

Exercises

*Note: Asterisks indicate level of difficulty—*easy, **moderate, ***most
difficult.*

*1.1. How many significant figures are there in each of the following:
(a) 273°K; (b) 273.15°K; (c) 300.0°K; (d) 300°K; (e) 300 ± 10°K;
(f) 3.0×10^{2}°K?

*1.2. How many significant figures are there in each of the following:
(a) 22414 ml; (b) 22.4 liters; (c) 1.987 cal; (d) 2.303; (e) 82.07 liter-atm;
(f) 96500 ± 10 C?

*1.3. Write each of the following as an exponential number: (a) 0.0254 m;

(b) 0.0030 kg; (c) 5400 ± 10 Å; (d) 0.0000000405 cm;
(e) 2,687,000,000,000,000,000,000,000 molecules.

*1.4. Write out in digit form the following exponential numbers: (a) 4.80×10^{-10}; (b) 1.83×10^5; (c) 10.7×10^{-2}; (d) 0.0020×10^6; (e) 2.020×10^{-8}.

*1.5. Calculate and express in exponential form each of the following:
(a) 2.04×10^{-8} plus 1.12×10^{-9};
(b) 2.04×10^{-8} less 1.12×10^{-9};
(c) 2.04×10^{-8} multiplied by 1.12×10^{-9};
(d) 2.04×10^{-8} divided by 1.12×10^{-9}.

*1.6. To the proper number of significant figures, express in exponential form the result of the following calculation:

$$\frac{(3.06 \times 10^{-3})(8.206 \times 10^{-2})(273.15 + 1.9)}{2.24 \times 10^3}$$

*1.7. Calculate mv^2/r where $m = 9.1 \times 10^{-28}$, v is 3.69×10^8, and r is 0.53×10^{-8}.

**1.8. Calculate the square root of each of the following: (a) 3.0×10^{-8}; (b) 3.0×10^{-9}; (c) 3.29×10^{15}; (d) 6.4×10^{-19}; (e) 2.50×10^{-8} cubed.

**1.9. Calculate the cube root of each of the following: (a) 3.0×10^{-8}; (b) 3.0×10^{-9}; (c) 3.29×10^{15}; (d) 6.4×10^{-19}; (e) 2.50×10^{-8} squared.

**1.10. Calculate the cube root of 8.70×10^{-8} divided by the square root of 2.28×10^3.

**1.11. What is the log of 2.85×10^{-3}?

**1.12. Given that the log of a number is -4.75, what is the number?

**1.13. Take the square root of a number whose log is -2.53 and multiply it by 3.05×10^{-7}.

*1.14. Given that $\log [H^+] = \log K + \log ([HX]/[X^-])$, calculate the value of $[H^+]$ that corresponds to $K = 1.86 \times 10^{-5}$, $[HX] = 1.50$, and $[X^-] = 1.35$.

*1.15. Given that $\varepsilon = \varepsilon° - .059 \log ([H^+]/[H_2]^{1/2})$, calculate the value of ε that corresponds to $\varepsilon° = 0.00$, $[H^+] = 0.0150$, and $[H_2] = 0.66$.

**1.16. Express the number 2.65×10^{-5} as a decimal power of ten.

**1.17. Express the following numbers as exponential numbers with integral powers of ten: (a) $10^{0.301}$; (b) $10^{2.303}$; (c) $10^{0.693}$; (d) $10^{-2.693}$.

**1.18. Given that $\Delta G° = -RT \ln K$, calculate $\Delta G°$ when $R = 1.99$, $T = 273.15$, and $K = 1.86 \times 10^{-5}$.

***1.19. What is the logarithm of 2 expressed to the base 3?

***1.20. With the cursor of a slide rule set so the hairline crosses the T scale at 20, what value (supply the decimal also) appears under the hairline on the K scale and what is its significance?

2

THE GRAM-ATOM

A GRAM-ATOM of an element corresponds to the Avogadro number (6.02×10^{23}) of atoms of the element. How much a gram-atom weighs depends on what element we are talking about. We need to know the atomic weight of the element. If we take a weight in grams equal numerically to the atomic weight, we have 1 gram-atom of that element. Thus, for example, since 47.90 is the atomic weight of titanium, 47.90 g of titanium constitutes 1 gram-atom of titanium and contains 6.02×10^{23} titanium atoms. Similarly, since 1.008 is the atomic weight of hydrogen, 1.008 g of hydrogen constitutes 1 gram-atom of hydrogen and contains 6.02×10^{23} hydrogen atoms.

In this chapter we want to learn by calculation what a gram-atom is and how it fits into chemical problems. To set the stage, we need to review first some of the concepts regarding atomic weight. Atomic weights are simply numbers that tell us something about the relative weights of different atoms. Complications come in because a given element frequently consists of several different kinds of atoms (isotopes) and because the choice of a standard atom for reference is arbitrary.

At present, the standard is the so-called carbon-twelve atom, frequently designated ^{12}C. Natural carbon consists almost exclusively of two kinds of isotopes: carbon-twelve and carbon-thirteen. These differ from each other in the makeup of their nuclei: ^{12}C has 6 protons and 6 neutrons, ^{13}C has 6 protons and 7 neutrons. As might be expected, the ^{13}C atom is heavier than the ^{12}C atom. In fact, it is found experimentally that the ^{13}C atom is 1.08362 times as heavy as the ^{12}C atom. If we say—and this is what is done in setting up the present scale of standards—that the ^{12}C atom will be assigned a weight (or mass) of exactly 12 units (i.e., 12.0000000...), then the ^{13}C atom, being 1.08362 times as heavy, will have to be considered to have a mass of (12) (1.08362), or 13.0034, of these same units. What we are doing, in effect, is defining a unit of mass such that a ^{12}C atom has exactly 12 of these units. Then we proceed to describe all other atoms in terms of these units. The unit is sometimes called the atomic mass unit and may be abbreviated amu. In summary, then, the atomic mass unit is one twelfth of the mass of a carbon-twelve atom, and in terms of this unit the ^{13}C atom is said to have a mass of 13.0034 amu.

Instead of working with individual atoms, the chemist normally works with large numbers of them. In such cases, the chances are very good that his atoms will not all be exactly the same, but will represent a distribution corresponding to the natural abundance. Consequently, the chemist is interested in an *average* atomic weight that reflects the different isotopes and their relative abundances. For example, in the case of carbon, a randomly selected natural sample will contain 98.892% of the ^{12}C variety and 1.108% of the ^{13}C variety. The weighted average mass is what is called the chemical atomic weight of the element.

■ EXAMPLE 83.   Given a sample of carbon in which the relative abundance of isotopes is 98.892% of ^{12}C (mass 12 amu) and 1.108% of ^{13}C (mass 13.0034 amu), what is the average mass of these atoms when weighted according to their abundance?

SOLUTION: Imagine you have 100,000 atoms total.

Of these, 98,892 will have a mass of 12 amu each.

The remaining 1,108 will have a mass of 13.0034 amu each.

The total mass will be $(98,892)(12) + (1,108)(13.0034)$, or 1,201,100 amu. Divided 100,000 ways, this comes out to be 12.011 amu as the average weight of each atom.

[*Note*: In the foregoing calculation, the 12 is an exact number, since it is so defined. This means that the number of significant figures in the answer is decided by the number of significant figures given for the abundance.] ■

On the basis of the calculation given in the example above, we can see why the chemical atomic weight listed in the tables is given as 12.011 for carbon. The atomic weights of the elements are given in most chemistry textbooks; they are listed inside the cover of this book. (The values given in parentheses are *not* weighted averages over all the isotopes, but represent only the mass numbers of the most stable isotopes. In all those cases, the elements are highly radioactive.)

■ EXAMPLE 84.   Natural oxygen consists 99.759% of ^{16}O with mass 15.9949 amu, 0.037% of ^{17}O with mass 16.9991 amu, and 0.204% of ^{18}O with mass 17.9991 amu. With this distribution, calculate the chemical atomic weight of natural oxygen.

SOLUTION: Out of 100,000 atoms

99,759 will have mass 15.9949 amu = 1,595,635
   37 will have mass 16.9991 amu =   630
  204 will have mass 17.9991 amu =   3,670

100,000   1,599,935

The average weight is 1,599,935/100,000, or 15.999.
The doubtful digits have been underlined. ∎

■ EXAMPLE 85.   Natural boron consists only of the isotopes B^{10} (mass 10.0130) and B^{11} (mass 11.0093). If natural boron has an atomic weight of 10.811, what can you conclude about the relative abundance of B^{10} and B^{11} atoms?

SOLUTION: Take 1000 atoms of boron at random.
Let x represent the number that are B^{10} and y the number that are B^{11}:

$x + y = 1000$

Each B^{10} atom weighs 10.0130 amu; x of these B^{10} atoms weigh $(10.0130)(x)$ amu.

Each B^{11} atom weighs 11.0093 amu; y of these B^{11} atoms weigh $(11.0093)(y)$ amu.

The average weight of the boron atoms is 10.811. So, 1000 of them would weigh $(1000)(10.811)$ amu.

Since the weight of the B^{10} atoms plus the weight of the B^{11} atoms must add up to the total weight of the 1000, we can write

$(10.0130)(x) + (11.0093)(y) = (10.811)(1000)$

We know $x = 1000 - y$, so we can substitute $(1000 - y)$ in place of x and get

$(10.0130)(1000 - y) + (11.0093)(y) = (10.811)(1000)$

The solution of this equation is $y = 801$.

Therefore, we conclude that 80.1% of the boron atoms are B^{11} and the remaining 19.9% are B^{10}. ∎

For most chemical considerations, we can ignore the fact that elements frequently consist of several isotopes. For one thing, the chemical behavior of an atom depends very little on small differences in mass such as normally exist between isotopes; for another thing, we generally deal with such astronomically large numbers of atoms that we are justified in working with a weighted average value. In the table of atomic weights, the values for the different elements take into account the various isotope distributions, so when we talk about a gram-atom of an element, we mean a sample of element that weighs as many grams as the number appearing in the atomic weight table. This sample contains a total of 6.02×10^{23} atoms of the element, some of this isotope and some of that, according to the relative abundance. The number in the atomic weight table tells directly what average weight we can assign to each atom so as to be able to describe the whole sample as consisting of 6.02×10^{23} such average atoms.

■ EXAMPLE 86.   Given that the atomic weight of copper is 63.546, what is the weight in grams of one average copper atom?

SOLUTION: One gram-atom of copper weighs 63.546 g.
One gram-atom of copper contains 6.02×10^{23} atoms.
If 6.02×10^{23} atoms weigh 63.546 g, then one atom weighs

$$\frac{63.546 \text{ g}}{6\,02 \times 10^{23} \text{ atoms}} \quad \text{or } 1.06 \times 10^{-22} \text{ g per atom ■}$$

■ EXAMPLE 87.   The isotopic distribution of copper is 69.4% ^{63}Cu and 30.6% ^{65}Cu. How many ^{63}Cu atoms are there in 1 gram-atom?

SOLUTION: One gram-atom contains 6.02×10^{23} atoms.
Of these, 69.4% will be ^{63}Cu.
69.4% of 6.02×10^{23} is 4.18×10^{23}. ■

■ EXAMPLE 88.   How many atoms are there in 1.00 g of copper?

SOLUTION: We know there are always 6.02×10^{23} atoms of any element in 1 gram-atom of that element.
Therefore, if we can calculate what fraction of a gram-atom we have, we know what fraction of 6.02×10^{23} we have.
One gram-atom of copper weighs 63.546 g.
One g of copper is

$$\frac{1.00 \text{ g}}{63.546 \text{ g/gram-atom}} \quad \text{or } 0.0157 \text{ gram-atom}$$

One gram-atom contains 6.02×10^{23} atoms.
0.0157 gram-atom contains

$$(0.0157 \text{ gram-atom}) \left(6.02 \times 10^{23} \frac{\text{atoms}}{\text{gram-atom}} \right)$$

or 9.45×10^{21} atoms. ■

■ EXAMPLE 89.   How many atoms are there in 3.80 g of fluorine?

SOLUTION: The atomic weight of fluorine is 18.9984 amu.
One gram-atom of fluorine weighs 18.9984 g.

$$3.80 \text{ g fluorine is } \frac{3.80 \text{ g}}{18.9984 \text{ g/gram-atom}} \text{ or } 0.200 \text{ gram-atom}$$

One gram-atom of any element contains 6.02×10^{23} atoms.
0.200 gram-atom contains

$$(0.200 \text{ gram-atom}) \left(6.02 \times 10^{23} \frac{\text{atoms}}{\text{gram-atom}}\right)$$

or 1.20×10^{23} atoms. ■

■ EXAMPLE 90.   Given that the atomic weight of bromine is 79.904,
what would be the weight of 4.63×10^{20} bromine atoms?

SOLUTION: One gram-atom contains 6.02×10^{23} atoms.

$$4.63 \times 10^{20} \text{ atoms is } \frac{4.63 \times 10^{20} \text{ atoms}}{6.02 \times 10^{23} \text{ atoms/gram-atom}}$$

or 7.69×10^{-4} gram-atom.
One gram-atom of bromine weighs 79.904 g.
7.69×10^{-4} gram-atom of bromine weighs

$$(7.69 \times 10^{-4} \text{ gram-atom}) \left(79.904 \frac{\text{g}}{\text{gram-atom}}\right) \text{ or } 0.0614 \text{ g} ■$$

■ EXAMPLE 91.   Which of the following samples contains the largest
number of atoms: (a) 6.70 g of iron; (b) 0.11 gram-atom of iron; (c) 7.83×10^{22} atoms of iron?

SOLUTION: We need to convert all these samples to the same units in
order to make comparison. In general, from the standpoints of both instant
recognition and greatest general utility in making calculations, gram-atoms is
preferred over grams and over atoms. The instant recognition comes from
the fact that we always know that 1 gram-atom of anything has 6.02×10^{23}
atoms; the utility comes from the fact that, in specifying the number of
gram-atoms, we are really saying something about the numbers of atoms
without getting involved with large exponents and huge numbers.

(a) Atomic weight of iron is 55.847 amu.

$$6.70 \text{ g} = \frac{6.70 \text{ g}}{55.847 \text{ g/gram-atom}} = 0.120 \text{ gram-atom}$$

(b) 0.11 gram-atom $= 0.11$ gram-atom
(c) One gram-atom contains 6.02×10^{23} atoms.

$$\frac{7.83 \times 10^{22} \text{ atoms}}{6.02 \times 10^{23} \text{ atoms/gram-atom}} = 0.130 \text{ gram-atom}$$

Comparing these by the same measure, (c) represents the largest number of gram-atoms and, therefore, the largest number of atoms. ∎

So far in this chapter the emphasis has been on conversions involving grams, atoms, and gram-atoms. The purpose of the calculations has been to show that, given the atomic weight table and the fact that 1 gram-atom contains 6.02×10^{23} atoms, we can translate information given to us in measurable quantities (e.g., weight) to quantities that are useful for monitoring chemical reaction (e.g., number of atoms). In the following problems we shall see how the latter is done. At this stage, the point to be made is that the concept of gram-atom has an importance that cannot be overemphasized. It is *the* way by which we keep tabs on the number of atoms involved in chemical change. If by now you are still uncertain what a gram-atom is, you should carefully review the material we have just gone through. Otherwise, the going from here on will be slippery indeed.

There are other methods for solving the following problems using chemical equations, as discussed in Chapter 5, but our main purpose at this point is to get a feeling for manipulating gram-atoms. Consequently, you should work out the following problems without recourse to chemical equations or to tricky substitution-in-formula methods. If needed atomic weights are not stated in the problem, look them up inside the cover of the book.

∎ EXAMPLE 92.   You wish to carry out a chemical reaction in which four atoms of cerium (atomic weight 140.12) combine with three atoms of sulfur (atomic weight 32.06). You have 2.50 g of cerium. How many grams of sulfur must you take to satisfy all the cerium atoms?

SOLUTION: Weights tell little about numbers of atoms unless information is put in to take care of the different atomic weights. By converting data given in grams to gram-atoms we are effectively counting the numbers of atoms.

Atomic weight of cerium is 140.12.

One gram-atom of cerium is 140.12 g.

$$2.50 \text{ g cerium} = \frac{2.50 \text{ g}}{140.12 \text{ g/gram-atom}} = 0.0178 \text{ gram-atom}$$

The reaction requires three atoms of sulfur per four atoms of cerium, which is the same as 3 gram-atoms of sulfur per 4 gram-atoms of cerium.

One gram-atom of cerium requires 3/4 gram-atom of sulfur.

0.0178 gram-atom of cerium requires 3/4(0.0178), or 0.0134, gram-atom of sulfur.

Atomic weight of sulfur is 32.06.

One gram-atom of sulfur weighs 32.06 g.

0.0134 gram-atom of sulfur weighs (0.0134 gram-atom)(32.06 g/gram-atom), or 0.430 g. ∎

■ EXAMPLE 93.   When iron is heated in air, it picks up oxygen to the extent of three atoms of oxygen for each two atoms of iron. If 1.50 g of iron is so heated, what will be the total weight of product?

SOLUTION:

$$1.50 \text{ g of iron} = \frac{1.50 \text{ g}}{55.847 \text{ g/gram-atom}} = 0.0269 \text{ gram-atom}$$

Two atoms of iron require three atoms of oxygen.

Two gram-atoms of iron require 3 gram-atoms of oxygen.

0.0269 gram-atom of iron requires (0.0269)(3/2) gram-atom of oxygen.

One gram-atom of oxygen equals 15.9994 g.

(0.0269)(3/2) gram-atom of oxygen equals (0.0269)(3/2)(15.9994), or 0.646 g.

Total weight of product = iron + oxygen

$$= 1.50 + 0.646 = 2.15 \text{ g} ∎$$

■ EXAMPLE 94.   When silver and sulfur are heated together, they combine in such a way that two atoms of silver match each atom of sulfur. If you start with 10.0 g of silver and 1.00 g of sulfur, what is the maximum amount of desired product you can get?

SOLUTION: This is a very common type of problem (sometimes called "excess problem") where two reagents are mixed in more or less arbitrary amounts, but for which the combination requirement is rigidly fixed by the chemistry. In this case, we require two atoms of silver per atom of sulfur, even though this may not be the ratio in which the atoms are made available. In such an event, one of the reagents limits the amount of reaction product, and the other reagent is left over in excess.

The atomic weight of silver is 107.868.
One gram-atom of silver weighs 107.868 g.

$$10.0 \text{ g silver is } \frac{10.0 \text{ g}}{107.868 \text{ g/gram-atom}} \text{ or } 0.0927 \text{ gram-atom}$$

The atomic weight of sulfur is 32.06.
One gram-atom of sulfur weighs 32.06 g.

$$1.00 \text{ g of sulfur is } \frac{1.00 \text{ g}}{32.06 \text{ g/gram-atom}} \text{ or } 0.0312 \text{ gram-atom}$$

If we are going to match two silver atoms for each sulfur atom, we will need 2 gram-atoms of silver for each gram-atom of sulfur. For 0.0312 gram-atom of sulfur, we need (2)(0.0312), or 0.0624, gram-atom of silver. We have more than enough silver, so some of the silver atoms will remain uncombined. The product is thus limited by the sulfur available. To use all the 0.0312 gram-atom of sulfur, we need to use 0.0624 gram-atom of silver.

The product will be made up, then, of 0.0312 gram-atom of sulfur plus 0.0624 gram-atom of silver.

$$0.0312 \text{ gram-atom of sulfur} = (0.0312 \text{ gram-atom}) \left(32.06 \frac{\text{g}}{\text{gram-atom}} \right)$$
$$= 1.00 \text{ g}$$

$$0.0624 \text{ gram-atom of silver} = (0.0624 \text{ gram-atom}) \left(107.868 \frac{\text{g}}{\text{gram-atom}} \right)$$
$$= 6.73 \text{ g}$$

Total product $= 1.00 + 6.73 = 7.73$ g. ∎

■ EXAMPLE 95.   Analysis of chlorophyll shows that it consists 2.68% of magnesium. Given 1.00 g of chlorophyll, how many magnesium atoms does it contain?

SOLUTION: 2.68% of 1.00 g is 0.0268 g.
Thus, the sample contains 0.0268 g of magnesium.
Atomic weight of magnesium is 24.305.
One gram-atom of magnesium weighs 24.305.

$$0.0268 \text{ g magnesium} = \frac{0.0268 \text{ g}}{24.305 \text{ g/gram-atom}} = 0.00110 \text{ gram-atom}$$

One gram-atom contains 6.02×10^{23} atoms.

0.00110 gram-atom contains (0.00110 gram-atom)

$$\times \left(6.02 \times 10^{23} \; \frac{\text{atoms}}{\text{gram-atom}}\right)$$

or 6.62×10^{20} atoms. ∎

∎ EXAMPLE 96.   Morphine is a complex compound that contains, besides other constituents, 67.3% carbon and 4.6% nitrogen. What would be the relative number of carbon and nitrogen atoms in this compound?

SOLUTION: Take 100 g of morphine. Of this, 67.3%, or 67.3 g, will be carbon. 4.6%, or 4.6 g, will be nitrogen.

The atomic weight of carbon is 12.011.

$$67.3 \text{ g of carbon} = \frac{67.3 \text{ g}}{12.011 \text{ g/gram-atom}} = 5.60 \text{ gram-atoms}$$

The atomic weight of nitrogen is 14.0067.

$$4.6 \text{ g of nitrogen} = \frac{4.6 \text{ g}}{14.0067 \text{ g/gram-atom}} = 0.33 \text{ gram-atom}$$

5.60 gram-atoms of carbon per 0.33 gram-atom of nitrogen is the same as 5.60 atoms of carbon per 0.33 atom of nitrogen, or 17. ∎

∎ EXAMPLE 97.   In making transistors, one needs to control the concentration of impurity very carefully. Suppose you wanted to make a germanium transistor containing 1.0×10^{18} boron atoms per cubic centimeter as impurity. If the density of germanium is 5.35 g/cm^3, what relative weights of germanium and boron need to be mixed?

SOLUTION: Let us make 1 cm^3 of product. For this, we need 5.35 g of germanium. (We can safely ignore the tiny effect of the boron impurity on the density.)

We also need 1.0×10^{18} boron atoms.

We know 6.02×10^{23} atoms is 1 gram-atom of anything. 1.0×10^{18} atoms must be

$$\frac{1.0 \times 10^{18} \text{ atoms}}{6.02 \times 10^{23} \text{ atoms/gram-atom}} = 1.7 \times 10^{-6} \text{ gram-atom}$$

One gram-atom of boron is 10.81 g. 1.7×10^{-6} gram-atom of boron is

$$(1.7 \times 10^{-6} \text{ gram-atom}) \times \left(10.81 \; \frac{\text{g}}{\text{gram-atom}}\right) \text{ or } 1.8 \times 10^{-5} \text{ g.}$$

To get the desired product, we need to mix 5.35 g of germanium and 1.8×10^{-5} g of boron. ■

Exercises

*2.1. Natural iron has the following distribution of isotopes: 5.82% ^{54}Fe with mass 53.9396 amu; 91.66% ^{56}Fe with mass 55.9349 amu; 2.19% ^{57}Fe with mass 56.9354 amu; and 0.33% ^{58}Fe with mass 57.9333 amu. What is its chemical atomic weight?

*2.2. Natural bromine is composed only of ^{79}Br and ^{81}Br. The first of these isotopes has a mass of 78.9183 amu and the second, 80.9163 amu. Given that the chemical atomic weight of bromine is 79.904 amu, what is the relative abundance of the two isotopes? (*Answer:* 50.65% ^{79}Br and 49.35% ^{81}Br.)

*2.3. Given that the atomic weight of arsenic is 74.9216, what is the weight in grams of one arsenic atom?

*2.4. How many atoms of nickel are there in 5.63 g of nickel? (*Answer:* 5.77×10^{22}.)

**2.5. How many atoms of copper are there in a piece of copper that weighs the same as a piece of aluminum that contains 4.86×10^{21} aluminum atoms?

**2.6. Arrange the following in order of increasing weight: 10.4 g of sulfur; 0.179 gram-atom of iron; 6.33×10^{25} atoms of hydrogen; 0.771 gram-atom of nitrogen; 3.84×10^{23} atoms of oxygen.

*2.7. An average signature written in pencil weighs about 1 mg. Assuming the black stuff is all carbon, calculate the number of atoms in an average signature.

*2.8. Calculate how many (a) grams, (b) gram-atoms, and (c) atoms of chromium would be needed to deposit a coating that is 0.0075 cm thick on a plate that is 1 m^2 in total area. Density of chromium is 7.20 g/cm^3. (*Answer:* 540 g.)

*2.9. Given an atom X that weighs 2.73×10^{-22} g, what would be the chemical atomic weight of X on the atomic mass unit scale?

*2.10. If atom Y weighs 6.35×10^{-23} g, how many gram-atoms of Y would there be in 1.50 g of Y? (*Answer:* 0.0392 gram-atom.)

**2.11. A bottle contains elements A and B so that there are two atoms of A for every three atoms of B. The total weight of the sample, of which A contributes 13%, is 25.0 g. If the atomic weight of A is 67.0 amu, what would be the atomic weight of B?

*2.12. You wish to make a compound containing one cesium atom for every gold atom. If you have only 1.00 g of gold, how many grams of cesium should you take with it? (*Answer:* 0.675 g.)

**2.13. You wish to make a compound containing copper, zinc, and sulfur atoms in the ratio 1 : 2 : 4. Starting with 63.5 g of each of the elements, what is the maximum amount of compound you can make?

**2.14. LSD is a complex compound whose mass is made up of 74.27% carbon, 7.79% hydrogen, 12.99% nitrogen, and 4.95% oxygen. What per cent of the atoms in LSD are carbon atoms? (*Answer:* 40.8%.)

*2.15. In forming a compound requiring four atoms of P for three atoms of S, how many gram-atoms of S are required by 2.50 g of P?

**2.16. In forming a compound requiring two atoms of Fe and three atoms of S, what is the maximum mass of compound that could be made from 1.00 g of Fe and 1.00 g of S? (*Answer:* 1.87 g.)

**2.17. A given compound contains atoms A, B, and C in the ratio 1 : 3 : 4, respectively. If you had available no more than 0.065 gram-atom of A, 1.35×10^{22} atoms of B, and 2.67 g of C, what is the maximum mass of compound you could make? The atomic weights of A, B, and C are 40.0, 20.0, and 10.0, respectively.

*2.18. Compound X contains P, Q, and R atoms in the ratio 2 : 1 : 3. If you took enough compound to include 2.87×10^{21} atoms of P, you would also include 0.243 g of R. What is the atomic weight of R? (*Answer:* 34.0 amu.)

*2.19. If 0.036 gram-atom of X combines with 0.732 gram of Y to give compound, what would be the atomic weight of Y if the X : Y atomic ratio in the compound is 4 : 3?

**2.20. A given collection of A and B atoms has a total mass of 5.89 g and contains a total of 4.05×10^{22} atoms. The atomic weight of A is just twice that of B, but in the collection B atoms contribute a total mass twice that contributed by A. What is the atomic weight of B? (*Answer:* 73.0.)

3

CHEMICAL FORMULAS

THERE ARE THREE major kinds of chemical formulas: (1) the *simplest* (also called *empirical*); (2) the *molecular*; and (3) the *structural*. The structural formula tells which atoms are connected to which atoms and how these are arranged in space; the molecular formula tells us how many atoms of the various kinds exist in the individual entity we call a molecule, but does not say anything about how they are arranged; the simplest formula tells us only the relative number of atoms of the different sorts in the compound. As a specific example, CH is the simplest formula for benzene, C_6H_6 is its molecular formula, and a hexagonal arrangement of carbon atoms each having an attached hydrogen is its structural formula. All three represent the same material, but they give varying degrees of information about it. The structural formula is the most informative because from it both the molecular and the simplest formulas can be deduced.

In chemistry, it is usually the simplest formula that is discovered first, since it comes directly out of the chemical analysis of the compound. For the molecular and structural formulas, other data are needed—specifically, what the behavior of the material is.

The simplest formula specifies the relative number of atoms of different kinds in the compound. It does this by means of subscripts affixed to the symbols of the various elements. For example, A_xB_y would represent a compound of elements A and B in which there are x atoms of A for every y atoms of B. Because the gram-atom is just another way of counting atoms, the simplest formula also tells the relative number of gram-atoms in the compound. Thus, A_xB_y means there are x gram-atoms of element A per y gram-atoms of B in this compound. In practice, the setting up of simplest formulas usually goes the other way; from the analysis data (which are usually per cent by weight) we first figure out the relative number of gram-atoms and use this same ratio to give the relative number of atoms.

58

■ EXAMPLE 98. Given a compound that contains one atom of magnesium for each two atoms of indium for each four atoms of sulfur. What is its simplest formula?

SOLUTION: The symbols of the elements are Mg, In, and S.
One atom of magnesium would be indicated by subscript 1 on Mg.
Two atoms of indium would be indicated by subscript 2 on In.
Four atoms of sulfur would be indicated by subscript 4 on S.
The result is $Mg_1In_2S_4$, or $MgIn_2S_4$, since the subscript "one" is understood when none is written. ■

■ EXAMPLE 99. What is the simplest formula of a compound that consists of 0.25 gram-atom of cobalt per 0.50 gram-atom of silicon?

SOLUTION: The symbols for cobalt and silicon are Co and Si.
We could designate the compound as $Co_{0.25}Si_{0.50}$, since we are thereby indicating the relative number of gram-atoms of cobalt and silicon in the compound.
Because of the desire to avoid decimals in the subscripts, we usually clear fractions—in this case, by dividing through by 0.25.
The result is $Co_{\frac{0.25}{0.25}}Si_{\frac{0.50}{0.25}}$, or Co_1Si_2, or $CoSi_2$. ■

[Note: This business of clearing fractions is not invariably followed. For example, as we learn more about the solid state, we get more used to formulas such as $Na_{0.31}V_2O_5$ or $WO_{2.90}$. These may have shocked John Dalton in 1805, but they are perfectly acceptable nowadays. The point is we try to have only whole numbers as subscripts, but not if it means getting into ridiculously large values for some of the subscripts.]

■ EXAMPLE 100. What is the simplest formula of a compound that consists of 7.0 g of nitrogen for each gram of hydrogen?

SOLUTION: The data are given as weights. For the formula, we need to know about atoms; so the first thing to do is to translate weights to gram-atoms.

$$7.0 \text{ g nitrogen} = \frac{7.0 \text{ g}}{14.0 \text{ g/gram-atom}} = 0.50 \text{ gram-atom}$$

$$1.0 \text{ g hydrogen} = \frac{1.0 \text{ g}}{1.008 \text{ g/gram-atom}} = 1.0 \text{ gram-atom}$$

Therefore, the simplest formula is $N_{0.50}H_{1.0}$, or NH_2.

(We have cleared fractions by dividing both subscripts by 0.50. Since we are talking relative numbers, the ratio 0.50 to 1.0 is just as informative as the ratio 1 to 2.) ∎

∎ EXAMPLE 101.   What is the simplest formula of a compound that shows the following analysis: 76.86% carbon, 12.90% hydrogen, 10.24% oxygen?

SOLUTION: Assume we have 100 g of compound.
76.86% carbon would mean 76.86 g of carbon.
12.90% hydrogen would mean 12.90 g of hydrogen.
10.24% oxygen would mean 10.24 g of oxygen.
Atomic weight of carbon is 12.011 amu. This means 1 gram-atom carbon weighs 12.011 g.

$$76.86 \text{ g of carbon} = \frac{76.86 \text{ g}}{12.011 \text{ g/gram-atom}} = 6.399 \text{ gram-atoms of C}$$

Atomic weight of hydrogen is 1.0080 amu. This means 1 gram-atom hydrogen weighs 1.0080 g.

$$12.90 \text{ g of hydrogen} = \frac{12.90 \text{ g}}{1.0080 \text{ g/gram-atom}} = 12.80 \text{ gram-atoms of H}$$

Atomic weight of oxygen is 15.9994 amu. This means 1 gram-atom oxygen weighs 15.9994 g.

$$10.24 \text{ g of oxygen} = \frac{10.24 \text{ g}}{15.9994 \text{ g/gram-atom}} = 0.6400 \text{ gram-atom of O}$$

The formula can be written $C_{6.399}H_{12.80}O_{0.6400}$.
If we divide all the subscripts by 0.6400, we get $C_{10}H_{20}O$ as the simplest formula. ∎

∎ EXAMPLE 102.   A given compound consists of 1.20×10^{23} atoms of carbon, 3.61×10^{23} atoms of hydrogen, and 6.02×10^{22} atoms of oxygen. What is its simplest formula?

SOLUTION: For the simplest formula, we need to know the relative number of atoms. As the basis of reference, it is usually most efficient to consider the smallest number, that is, 6.02×10^{22} atoms of oxygen:

$$\frac{1.20 \times 10^{23} \text{ atoms of C}}{6.02 \times 10^{22} \text{ atoms of O}} = 2 \text{ atoms of C per atom of O}$$

$$\frac{3.61 \times 10^{23} \text{ atoms of H}}{6.02 \times 10^{22} \text{ atoms of O}} = 6 \text{ atoms of H per atom of O}$$

$$\frac{6.02 \times 10^{22} \text{ atoms of O}}{6.02 \times 10^{22} \text{ atoms of O}} = 1 \text{ atom of O per atom of O}$$

Thus, we have two carbon atoms and six hydrogen atoms per atom of oxygen. Consequently, the simplest formula is C_2H_6O. ∎

■ EXAMPLE 103.   A given bottle that contains only one pure compound can be described as containing the following ingredients: 0.90 gram-atom of carbon plus 1.445×10^{24} atoms of hydrogen plus 4.8 g of oxygen. What is the compound's simplest formula?

SOLUTION: Translate all the data into gram-atoms.

0.90 gram-atom carbon = 0.90 gram-atom carbon

$$1.445 \times 10^{24} \text{ atoms of hydrogen} = \frac{1.445 \times 10^{24} \text{ atoms}}{6.02 \times 10^{23} \text{ atoms/gram-atom}}$$

$$= 2.40 \text{ gram-atoms of hydrogen}$$

$$4.8 \text{ g of oxygen} = \frac{4.8 \text{ g}}{15.9994 \text{ g/gram-atom}}$$

$$= 0.30 \text{ gram-atom of oxygen}$$

Thus, the gram-atoms of C, H, and O in the compound are in the ratio of 0.90 to 2.40 to 0.30.

We can write $C_{0.90}H_{2.40}O_{0.30}$, or, clearing the decimals by division through with 0.30, C_3H_8O. ∎

In all the problems above, we start with data on the composition of a material and end up writing the simplest formula. This is a typical operation that is routinely performed, for example, when a new compound is first discovered. The reason for making such a fuss over writing the simplest formula is that, once it is obtained, various pieces of information can be drawn from it. Not only is the relative number of atoms immediately obvious, but also by plugging in atomic weight information we get data on weight relations as well. Specifically, we can easily determine what per cent of the total weight is contributed by each of the several elements. This might be useful, for example, in evaluating analytical results.

■ EXAMPLE 104. Given the material copper sulfide, for which the simplest formula is Cu_2S. What is the per cent (by weight) composition of the sample?

SOLUTION: The simplest formula Cu_2S tells us that the compound contains 2 atoms of copper per atom of sulfur, or, stated alternatively, 2 gram-atoms of copper per gram-atom of sulfur.

$$2 \text{ gram-atoms of copper} = (2 \text{ gram-atoms})\left(63.546\frac{g}{\text{gram-atom}}\right)$$
$$= 127.09 \text{ g}$$

$$1 \text{ gram-atom of sulfur} = (1 \text{ gram-atom})\left(32.06\frac{g}{\text{gram-atom}}\right)$$
$$= 32.06 \text{ g}$$

In other words, a representative sample would contain 127.09 g of copper for each 32.06 g of sulfur. The total weight of this sample would be $127.09 + 32.06 = 159.15$ g.

$$\text{Per cent of copper} = \frac{127.09 \text{ g of copper}}{159.15 \text{ g total sample}} \times 100 = 79.86\%$$

$$\text{Per cent of sulfur} = \frac{32.06 \text{ g of sulfur}}{159.15 \text{ g total sample}} \times 100 = 20.14\% ■$$

■ EXAMPLE 105. Given a sample of compound for which the simplest formula is Sc_2O_3. If the sample weighs 5.00 g, how many atoms of scandium does it contain?

SOLUTION: The compound contains 2 gram-atoms of Sc per 3 gram-atoms of O.

$$2 \text{ gram-atoms of Sc} = (2 \text{ gram-atoms})\left(44.956\frac{g}{\text{gram-atom}}\right)$$
$$= 89.912 \text{ g}$$

$$3 \text{ gram-atoms of O} = (3 \text{ gram-atoms})\left(15.9994\frac{g}{\text{gram-atom}}\right)$$
$$= 47.998 \text{ g}$$

Thus, the weight composition of the sample is 89.912 g of scandium per 47.998 g of oxygen.

$$\text{Per cent of scandium} = \frac{89.912 \text{ g of Sc}}{(89.912 + 47.998) \text{ g total}} \times 100 = 65.2\%$$

Given the 5.00-g sample of Sc_2O_3, we know now that 65.2% of it, or (0.652)(5.00) g, is scandium. Once we know how many grams of scandium we have, we can divide by the atomic weight to get the number of gram-atoms, and then multiply by 6.02×10^{23} to get the number of atoms.

$$\frac{(5.00)(0.652) \text{ g}}{44.956 \text{ g/gram-atom}} \times 6.02 \times 10^{23} \frac{\text{atoms}}{\text{gram-atom}} = 4.37 \times 10^{22} \text{ atoms} \blacksquare$$

■ EXAMPLE 106.   Given two minerals of copper for which the simplest formulas are Cu_5FeS_4 and Cu_2S. Which of these, on a per-cent-by-weight-of-copper basis, can be considered a richer mineral for copper?

SOLUTION: In Cu_5FeS_4 we have

$$5 \text{ gram-atoms of Cu} = (5 \text{ gram-atoms})\left(63.546 \frac{\text{g}}{\text{gram-atom}}\right) = 317.73 \text{ g}$$

$$1 \text{ gram-atom of Fe} = (1 \text{ gram-atom})\left(55.85 \frac{\text{g}}{\text{gram-atom}}\right) = 55.85 \text{ g}$$

$$4 \text{ gram-atoms of S} = (4 \text{ gram-atoms})\left(32.06 \frac{\text{g}}{\text{gram-atom}}\right) = 128.24 \text{ g}$$

Out of a total weight of $(317.73 + 55.85 + 128.24)$ g, or 501.82 g, we have 317.73 g of copper.

$$\text{Per cent of Cu} = \frac{317.73 \text{ g of Cu}}{501.82 \text{ g total}} \times 100 = 63.32\%$$

In Cu_2S we have

$$2 \text{ gram-atoms of Cu} = (2 \text{ gram-atoms})\left(63.546 \frac{\text{g}}{\text{gram-atom}}\right) = 127.09 \text{ g}$$

$$1 \text{ gram-atom of S} = (1 \text{ gram-atom})\left(32.06 \frac{\text{g}}{\text{gram-atom}}\right) = 32.06 \text{ g}$$

Out of a total weight of $(127.09 + 32.06)$ g, or 159.15 g, we have 127.09 g of copper.

$$\text{Per cent of Cu} = \frac{127.09 \text{ g of Cu}}{159.15 \text{ g total}} \times 100 = 79.85\%$$

Evidently, Cu_2S is a richer source of Cu than is Cu_5FeS_4. ∎

In this chapter we cannot take up the deduction of *molecular* formulas, since we do not yet have the proper background in properties of materials, specifically, those of gases and solutions. However, once a molecular formula has been determined, per cent composition by weight can be calculated from it in the same way we have done for simplest formulas. This is done in the following problems, where the tip-off that we are dealing with molecular formulas instead of simplest formulas comes from the fact that the subscripts have a common divisor (e.g., C_6H_6 could be written as CH).

∎ EXAMPLE 107.   What is the per cent composition by weight of benzene, C_6H_6?

SOLUTION: The formula says there are 6 gram-atoms carbon for each 6 gram-atoms of hydrogen.

$$6 \text{ gram-atoms of C} = (6 \text{ gram-atoms})\left(12.011\frac{g}{\text{gram-atom}}\right) = 72.066 \text{ g}$$

$$6 \text{ gram-atoms of H} = (6 \text{ gram-atoms})\left(1.0080\frac{g}{\text{gram-atom}}\right) = 6.0480 \text{ g}$$

Total weight is $72.066 + 6.048 = 78.114$ g.

$$\text{Per cent of C} = \frac{72.066 \text{ g of C}}{78.114 \text{ g total}} \times 100 = 92.257\%$$

$$\text{Per cent of H} = \frac{6.0480 \text{ g of H}}{78.114 \text{ g total}} \times 100 = 7.743\% ∎$$

Exercises

*3.1. Write the simplest formula for each of the following compounds: (a) 3 atoms of vanadium per atom of oxygen; (b) 3.18 g of vanadium per gram of oxygen; (c) 0.30 gram-atom of vanadium combined with 9.6 g of oxygen; (d) 1.20×10^{23} atoms of vanadium combined with 1.50 gram-atoms of oxygen.

*3.2. What would be the simplest formula of a compound that analyzes to 88.8% by weight of copper and 11.2% of oxygen? (*Answer*: Cu_2O.)

*3.3. Calculate the per cent by weight of oxygen in each of the following compounds: N_2O; NO; NO_2; N_2O_5.

**3.4. On burning a certain metal M in air, we find that 1.330 g of metal yields 1.970 g of oxide. If the atomic weight of the metal is 50.0 amu, what is the simplest formula of the oxide? (*Answer*: M_2O_3).

**3.5. The analysis of an iron oxide shows 3.354 g of iron and 1.280 g of oxygen. What is the simplest formula of the oxide? When heated in oxygen, the foregoing sample increases in weight to 4.79 g. What is the formula of the new oxide?

*3.6. A given compound contains : 0.782 g of potassium and 0.640 g of oxygen. What is the simplest formula of the compound? (*Answer*: KO_2.)

*3.7. Analysis of a certain impure salt shows 0.0690 g of sodium, 0.093 g of phosphorus, 0.1440 g of oxygen, and 0.102 g of inert material. What is the simplest formula of the salt?

*3.8. In forming the compound Al_2O_3, how many grams of Al_2O_3 can be obtained from 21.6 g of aluminum? (*Answer*: 40.8 g.)

**3.9. If 3.1 g of phosphorus and 76.2 g of iodine are made to react to yield as much PI_3 as possible, what weight of PI_3 will be formed?

**3.10. A sample of iron oxide weighing 1.389 g is heated in a stream of hydrogen until it is reduced completely to metal, which is found to weigh 1.005 g. What is the simplest formula of the oxide? (*Answer*: Fe_3O_4.)

**3.11. When a given oxide of vanadium is heated in vacuum it loses some of its oxygen to form another oxide. The loss in weight so experienced by 1.659 g of the initial oxide is 0.160 g. If the oxygen ratio in the initial and final oxides is 4 : 3, what is the simplest formula of the initial oxide?

*3.12. The simplest formula of morphine is $C_{17}H_{19}NO_3$. What is its per cent by weight of nitrogen? (*Answer*: 4.91%.)

*3.13. The per cent composition by weight of heroin is 68.28% C, 21.66% O, 6.27% H, and 3.79% N. Figure out its simplest formula.

**3.14. Aspirin contains 60.0% carbon and twice as many hydrogen atoms as oxygen atoms. If these are the only elements present in the compound, what is its simplest formula? (*Answer*: $C_9H_8O_4$.)

**3.15. A given mixture consisting only of Na_2SO_4 and K_2SO_4 analyzes to show 50.0% SO_4 by weight. What is the weight ratio of Na_2SO_4 to K_2SO_4 in the mixture?

**3.16. The thermoelectric material Bi_2Te_{3-x} can have its properties drastically altered by making x positive or negative. How much tellurium would you have to add to 5.00 g of Bi_2Te_{3-x} to convert it from an x value of $+0.01$ to -0.01? (*Answer*: 0.0160 g.)

*3.17. The softness and melting point of solder, an alloy of tin and lead, depends on the ratio of tin to lead atoms. What happens to this ratio when a 60% Sn–40% Pb (by weight) solder is converted to 40% Sn–60% Pb (by weight)?

***3.18.** A white powder mixture consists only of $CaTiO_3$, $SrTiO_3$, and $BaTiO_3$. On analysis it shows 23.94% by weight of oxygen and 23.90% by weight of titanium. If the atomic ratio Sr : Ca is 2 : 1, what is Sr : Ba? (*Answer*: 2 : 3.)

***3.19.** Given three elements X, Y, Z. If the per cent by weight of X is 15.4 in XYZ_2, 16.7 in XY_2Z, and 36.4 in X_2YZ, what is the atomic weight of X?

***3.20.** Given three elements A, B, C. If compound AB_2 is 73.7% by weight B and compound BC_2 is 28.0% by weight B, what is the per cent by weight of B in the compound AB_2C_4? (*Answer*: 25.4%.)

4

THE MOLE

IT MAKES a great deal of sense to think of a mole as the Avogadro number of objects. In this respect, we can speak of a mole of H atoms, a mole of H_2 molecules, a mole of electrons, or a mole of ballet dancers, in each case having in mind 6.02×10^{23} of the described articles. However, it is imperative that the nature of the numbered article be clearly specified. Otherwise, the meaning is uncertain. For example, to say a "mole of hydrogen" is ambiguous. It makes a big difference whether we have in mind a mole of hydrogen atoms, a mole of hydrogen molecules, or a mole of hydrogen bombs. In the same way, a "mole of sulfur" could mean different things depending on whether we are counting S_2, S_4, S_6, or S_8 molecules.

In chemistry, the mole takes on a very special importance because numerous properties depend on the number of particles present in a sample (instead of, say, on the total mass of the sample). Furthermore, when one material is dissolved in another, the observed properties of the resulting solution depend on the relative number of particles present. Thus, the mole is an extremely useful concept for counting particles. It counts particles in groups of 6.02×10^{23} in much the same way that the gram-atom counts atoms in groups of 6.02×10^{23} or fractions thereof. The difference is that the mole can be used for counting complex objects (e.g., ballet dancers, nicotine molecules), whereas the gram-atom is reserved for counting atoms.

■ EXAMPLE 108.   Given that the United States' population is 210 million people, how many moles of people are there in the United States?

SOLUTION: One mole is 6.02×10^{23} objects.

$$\frac{210 \text{ million people}}{6.02 \times 10^{23} \text{ people/mole}} = 3.5 \times 10^{-16} \text{ mole of people} \ ■$$

Actually, there is little reason to count people in terms of moles, since the mole is such an enormously large number that we would never have enough people to make up an appreciable fraction of even 1 mole. However, on the atomic and molecular scale we routinely work with quantities of the order of 1 mole of particles.

■ EXAMPLE 109.   Given 1.0×10^{23} sulfur atoms in a sample. (a) How many moles of sulfur atoms do you have? (b) If the sulfur atoms were all grouped in the form of rings containing eight sulfur atoms per ring (molecule), how many moles of S_8 molecules would you have?

SOLUTION: (a) One mole is 6.02×10^{23} particles.

$$\frac{1.0 \times 10^{23} \text{ S atoms}}{6.02 \times 10^{23} \text{ S atoms/mole}} = 0.17 \text{ mole S atoms}$$

(b) There are eight sulfur atoms per S_8 molecule.

$$\frac{1.0 \times 10^{23} \text{ S atoms}}{8 \text{ S atoms/}S_8 \text{ molecule}} = 0.12 \times 10^{23} S_8 \text{ molecules}$$

$$\frac{0.12 \times 10^{23} S_8 \text{ molecules}}{6.02 \times 10^{23} S_8 \text{ molecules/mole}} = 0.020 \text{ mole } S_8 \text{ molecules} ■$$

■ EXAMPLE 110.   Given 0.15 mole of P_4 molecules. (a) How many P_4 molecules is this? (b) How many P atoms is this? (c) How many moles of P atoms do you have?

SOLUTION: (a) One mole is 6.02×10^{23} particles.

$$(0.15 \text{ mole}) \left(6.02 \times 10^{23} \frac{\text{particles}}{\text{mole}}\right) = 9.0 \times 10^{22} P_4 \text{ molecules}$$

(b) One P_4 molecule contains four atoms.

$$(9.0 \times 10^{22} P_4 \text{ molecules}) \times \left(\frac{4 \text{ P atoms}}{P_4 \text{ molecule}}\right) = 3.6 \times 10^{23} \text{ P atoms}$$

(c) One mole is 6.02×10^{23} particles.

$$\left(\frac{3.6 \times 10^{23} \text{ P atoms}}{6.02 \times 10^{23} \text{ P atoms/mole}}\right) = 0.60 \text{ mole P atoms} ■$$

Unfortunately, the business on moles is not so simple as the foregoing discussion implies. However, we start our discussion in the fashion above because it turns out to be helpful to get quickly into your mind the idea that the mole is a concept for counting particles. Complications appear once we realize that in certain cases it may not be so easy to specify what the "particles" are, either because one merges its identity with another by overlapping it or, in many cases, because the "particle" does not have a unique size but continues on and on depending upon how big the sample is. Another complication comes in because, frequently, we may not know what the "particle" consists of. For example, the compound phosphorus pentoxide was assumed to consist of P_2O_5 "particles" until structure studies showed conclusively that the "particle" was P_4O_{10}. Obviously, 1 mole of P_2O_5 is not the same as 1 mole of P_4O_{10}. Because of the foregoing complications (and other more sophisticated ones) we shall have to consider some more-difficult-to-visualize aspects of the mole. Still, you should keep in mind that we are counting particles in one way or another.

The strict definition of mole is that it is equal to the gram-formula-weight. What is a *gram-formula-weight*? Take the atomic weights of the elements as they appear in the formula, multiply each atomic weight by the subscript of the corresponding element in the formula, add the products, and you have the formula-weight. This number is in atomic mass units. If you take it in grams, you have the gram-formula-weight. The procedure is the same no matter whether you have a simplest formula or a molecular formula. Once you can write a formula down in terms of symbols and subscripts and once you have the table of atomic weights available, you can calculate formula-weights.

■ EXAMPLE 111.   What is the formula-weight of H_2O?

SOLUTION: Atomic weight of hydrogen is 1.0080 amu.
You take twice it because of the subscript 2 on H in the formula:

$$(2)(1.0080 \text{ amu}) = 2.0160 \text{ amu}$$

Atomic weight of oxygen is 15.9994 amu.
You take one times it because the subscript on O in H_2O is 1 understood:

$$(1)(15.9994 \text{ amu}) = 15.9994 \text{ amu}.$$

Formula-weight of H_2O
$$= (2)(\text{atomic weight of H}) + (1)(\text{atomic weight of O})$$
$$= (2)(1.0080 \text{ amu}) + 15.9994 \text{ amu}$$
$$= 18.0154 \text{ amu} ■$$

■ EXAMPLE 112.   What is the formula-weight of $Ca(NO_3)_2$?

SOLUTION: The parentheses with the subscript 2 outside means that everything inside the parentheses gets multiplied by two.

Formula-weight of $Ca(NO_3)_2$

= (1)(atomic weight of Ca) + (2)(atomic weight of N)

+ (6)(atomic weight of O)

= 40.08 + (2)(14.007) + (6)(15.999)

= 164.09 amu ■

Once you can have the formula-weight in atomic mass units, you can easily think of taking that many grams of the material, for example, 18.0154 g H_2O or 164.09 g $Ca(NO_3)_2$. In either case, you would be providing yourself with 1 gram-formula-weight of the material. This is what we call 1 *mole*. Thus, 1 mole of H_2O is 18.0154 g; 1 mole of $Ca(NO_3)_2$ is 164.09 g. Note what you need to work with: You must be able to write down a formula; you must put in the atomic weights multiplied by the subscripts; you must choose grams as your weight unit.

(Although we shall not fuss with this, it is often convenient to work with weight units other than grams, as, for example, pounds or tons. In such cases it is efficient to introduce pound-formula-weight or ton-formula-weight, these being the formula-weight expressed in pounds or tons, respectively. Specifically, 1 pound-formula-weight of H_2O is 18.0154 pounds; 1 pound-formula-weight of $Ca(NO_3)_2$ is 164.09 pounds. Furthermore, we could speak of these as 1 pound-mole of H_2O(18.0154 pounds) or 1 pound-mole of $Ca(NO_3)_2$(164.09 pounds). Because of the complication, the purist will insist that *formula-weight* should be taken only as being expressed in atomic mass units. When he means it in *grams*, he will say *gram-formula-weight*. Furthermore, he will insist on saying *gram-mole* instead of *mole* for *gram-formula-weight*, claiming that so far as weight is concerned, the mole can be expressed in any weight unit. We have enough problems as it is and life is too short to worry about getting a vocabulary that most chemists do not use in practice. We shall confine practically all our work to grams, so we shall be dealing almost exclusively with the *gram-mole*. Therefore, it will be safe to assume that when we say *mole* we mean *gram-mole*. If occasionally it becomes necessary to refer to ton-moles or the like, then we shall specify the units.)

■ EXAMPLE 113.   What is the weight in grams of 1 mole of NaCl?

SOLUTION: One mole NaCl is 1 gram-formula-weight.

Formula-weight of NaCl = atomic weight of Na + atomic weight of Cl

$$= 22.9898 \text{ amu} + 35.453 \text{ amu}$$

$$= 58.443 \text{ amu}$$

One mole NaCl = 58.443 g. ■

■ EXAMPLE 114.   What is the weight in pounds of 1 pound-mole of NaCl?

SOLUTION: One pound-mole of NaCl is 1 pound-formula-weight.

Formula-weight of NaCl = 22.9898 amu + 35.453 amu = 58.443 amu

One pound-mole of NaCl = 58.443 pounds. ■

■ EXAMPLE 115.   What is the weight in grams of 1 pound-mole of NaCl?

SOLUTION: Example 114 tells us that 1 pound-mole of NaCl = 58.443 pounds.

1 pound = 454 g

$$(58.443 \text{ pounds}) \left(454 \, \frac{\text{g}}{\text{pound}} \right) = 2.65 \times 10^4 \text{ g} \ ■$$

What now is the connection between a mole, defined as 1 gram-formula-weight, and the Avogadro number of particles? We could say that 1 mole consists of the Avogadro number of "particles," provided we interpret "particle" to mean an entity composed of as many atoms as are shown in the given formula. Such an entity does not necessarily have a real, separate existence. It may have to be invented purely for the sake of counting particles. For example, the material sodium chloride has the simplest formula NaCl. We can easily imagine a unit consisting of one sodium atom and one chlorine atom, even though in solid sodium chloride no particular sodium atom goes with any specific chlorine atom. Instead, each sodium belongs equally to six chlorine neighbors, which in turn belong equally to six sodium neighbors, and so on. The result is a huge, extended structure in which separate pairs— one sodium atom plus one chlorine atom—do not exist. However, for counting purposes, we can imagine that such pairs do exist and we will get away with it, because once a sodium is counted with one chlorine neighbor it will not need to be counted again. To make our reasoning a bit easier to follow we introduce a term, the "formula-unit," to mean specifically this entity (hypothetical or not) which contains as many atoms as are shown in the chemical formula. We thus put a double burden on the formula. In it, a symbol not only stands for an *element* but it also stands for *one atom* of the element; furthermore, the subscript not only stands for the relative number of gram-atoms in the compound but also the number of atoms in the formula-unit.

As specific examples of what the foregoing means, let us look at some typical formulas: H_2O; S_8; $CaCl_2$; P_4O_{10}; $C_{19}H_{22}N_2O$. For H_2O, the formula-unit consists of two hydrogen atoms and one oxygen atom; this entity actually exists and constitutes the water molecule as found in water vapor. For S_8, the formula-unit consists of eight sulfur atoms; this entity also exists and, in fact, is the molecule that makes up solid, rhombic sulfur. For $CaCl_2$, the formula-unit consists of one calcium atom and two chlorine atoms; it does *not* exist as a separate unit in solid calcium chloride. The formula-unit P_4O_{10} consists of four P atoms and ten oxygen atoms; it is the molecule that makes up the white solid called phosphoric anhydride. The formula-unit $C_{19}H_{22}N_2O$ consists of 19 carbon atoms, 22 hydrogen atoms, 2 nitrogen atoms, and 1 oxygen atom; it has a real identity as the molecule of the substance cinchonine (closely related to strychnine).

From all these examples, the point to be learned is that just looking at the chemical formula will not tell us whether there is a real entity consisting only of the atoms shown in the formula. We need to *know* that the chemical formula given is truly a *molecular formula* if we want to attach the label "molecule" to the aggregate represented by the formula. If we have only the *simplest* formula, then we are uncertain. The whole purpose of introducing "formula-unit" is to bypass this uncertainty. In terms of formula-units, no matter whether they are molecules or not, we can always say correctly that "1 mole contains the Avogadro number of formula-units." Only in special cases can we say correctly that "1 mole contains the Avogadro number of molecules."

The foregoing, somewhat lengthy, discussion has been introduced because beginners in chemistry are often enticed into making incorrect use of terms, especially when current practice among chemists is not standardized. This unfortunately is true for the terms "molecule" and "mole," which frequently are imprecisely defined and imprecisely used. The safest procedure (it is the one we shall follow from here on) is to give the formula for which the formula-weight is being calculated and also the formula of the "particle" that is being counted. In some cases, we shall use the term "molecule" instead of "formula-unit," but only in those cases where the identity of the molecule has been established by other evidence. To the beginner, the advice is to use the term "formula-unit" except in those cases where he has accumulated enough chemical know-how to be able to say "molecule." For example, there is no point to saying "the formula-unit H_2" when it is generally known that " the molecule H_2 " is correct. In any case, the other piece of advice is to use the symbols and subscripts to a maximum when ambiguity is likely to occur.

The following summary may be useful in setting the stage for practical calculations involving the mole:

> One mole = 1 gram-formula-weight.
> One mole = 6.02×10^{23} formula-units.
> One formula-unit contains as many atoms as are shown in the formula.

A formula-unit is the same as a molecule in those cases where we are working with the molecular formula.

■ EXAMPLE 116.   How many formula-units are there in 0.10 mole of $Ba(NO_3)_2$? [Note: Like most salts, $Ba(NO_3)_2$ is not a molecular material.]

SOLUTION: One mole contains 6.02×10^{23} formula-units.

$$(0.10 \text{ mole}) \left(6.02 \times 10^{23} \frac{\text{formula-units}}{\text{mole}}\right) = 6.0 \times 10^{22} \text{ formula-units} \ ■$$

■ EXAMPLE 117.   How many atoms of barium are there in 0.10 mole of $Ba(NO_3)_2$?

SOLUTION: One mole contains 6.02×10^{23} formula-units.
One formula-unit of $Ba(NO_3)_2$ contains one barium atom.

$$(0.10 \text{ mole } Ba(NO_3)_2) \left(6.02 \times 10^{23} \frac{\text{formula-units}}{\text{mole}}\right)$$

$$\times \left(\frac{1 \text{ Ba atom}}{\text{formula-unit}}\right) = 6.02 \times 10^{22} \text{ Ba atoms} \ ■$$

■ EXAMPLE 118.   How many atoms of oxygen are there in 0.10 mole of $Ba(NO_3)_2$?

SOLUTION: One formula-unit of $Ba(NO_3)_2$ contains six atoms of oxygen.

$$(0.10 \text{ mole } Ba(NO_3)_2) \left(6.02 \times 10^{23} \frac{\text{formula-units}}{\text{mole}}\right)$$

$$\times \left(6 \frac{\text{atoms oxygen}}{\text{formula-unit}}\right) = 3.6 \times 10^{23} \text{ atoms oxygen.} \ ■$$

■ EXAMPLE 119.   How many gram-atoms of nitrogen are there in 0.10 mole of $Ba(NO_3)_2$?

SOLUTION: One mole contains 6.02×10^{23} formula-units.
0.10 mole contains 6.0×10^{22} formula-units.
One formula-unit of $Ba(NO_3)_2$ contains two nitrogen atoms.
0.10 mole of $Ba(NO_3)_2$ contains

$$(6.0 \times 10^{22} \text{ formula-units}) \times \left(2 \ \frac{\text{N atoms}}{\text{formula-unit}}\right) = 1.2 \times 10^{23} \text{ N atoms}$$

One gram-atom of N $= 6.02 \times 10^{23}$ N atoms.

$$\frac{1.2 \times 10^{23} \text{ N atoms}}{6.02 \times 10^{23} \text{ N atoms/gram-atom}} = 0.20 \text{ gram-atom of N} \ \blacksquare$$

The foregoing problem is an important one because we shall frequently go directly from moles to gram-atoms. As the preceding problem shows, the number of gram-atoms is equal to the number of moles of material times the number of gram-atoms per mole. In the case of $Ba(NO_3)_2$, there are 2 gram-atoms of N per mole of $Ba(NO_3)_2$. In the general case, we simply note that the subscripts not only tell the number of atoms per formula-unit but also the number of gram-atoms per mole. Thus, 1 mole of $Ba(NO_3)_2$ contains 1 gram-atom of barium plus 2 gram-atoms of nitrogen plus 6 gram-atoms of oxygen. Recall that a subscript outside a parenthesis multiplies all subscripts inside the parentheses.

■ EXAMPLE 120.   How many gram-atoms of oxygen are there in 0.15 mole of $Ba(NO_3)_2$?

SOLUTION: One mole of $Ba(NO_3)_2$ contains 6 gram-atoms of oxygen.

$$(0.15 \text{ mole}) \left(\frac{6 \text{ gram-atoms of O}}{\text{mole of } Ba(NO_3)_2}\right) = 0.90 \text{ gram-atom of oxygen} \ \blacksquare$$

■ EXAMPLE 121.   How many atoms of oxygen are there in 5.22 g of $Ba(NO_3)_2$?

SOLUTION: Formula-weight of $Ba(NO_3)_2$ = atomic weight Ba + (2) $\times$ (atomic weight of N) + (6)(atomic weight of O) = 137.34 + (2)(14.0067) + (6)(15.9994) = 261.35 amu.

One mole of $Ba(NO_3)_2$ = 261.35 g.

$$\frac{5.22 \text{ g of } Ba(NO_3)_2}{261.35 \text{ g/mole}} = 0.0200 \text{ mole of } Ba(NO_3)_2$$

One mole $Ba(NO_3)_2$ contains 6 gram-atoms oxygen.

$$(0.0200 \text{ mole}) \left(6 \frac{\text{gram-atoms of O}}{\text{mole}} \right) = 0.120 \text{ gram-atom of O}$$

One gram-atom of oxygen contains 6.02×10^{23} atoms.

$$(0.120 \text{ gram-atom}) \left(6.02 \times 10^{23} \frac{\text{gram-atom}}{\text{atoms}} \right)$$
$$= 7.22 \times 10^{22} \text{ oxygen atoms} \blacksquare$$

■ EXAMPLE 122.   How many grams of $Ba(NO_3)_2$ would you need to take to get 1.00 g of barium?

SOLUTION:

$$\frac{1.00 \text{ g of Ba}}{137.34 \text{ g/gram-atom}} = 0.00728 \text{ gram-atom of Ba}$$

Formula $Ba(NO_3)_2$ shows 1 gram-atom of Ba per mole of $Ba(NO_3)_2$. To get 0.00728 gram-atom Ba, you need to take 0.00728 mole $Ba(NO_3)_2$. One mole $Ba(NO_3)_2$ equals 261.35 g.

$$(0.00728 \text{ mole of } Ba(NO_3)_2) \left(261.35 \frac{\text{g}}{\text{mole}} \right) = 1.90 \text{ g} \blacksquare$$

■ EXAMPLE 123.   Suppose for a given reaction you need 3.0×10^{21} formula-units of NO_3^-. You propose to get this by using $Ba(NO_3)_2$ as the source. What weight of $Ba(NO_3)_2$ is needed?

SOLUTION: One formula-unit of $Ba(NO_3)_2$ provides 2 formula-units of NO_3^-.
To get 3.0×10^{21} formula-units of NO_3^-, we need

$$\left(\frac{3.0 \times 10^{21} \text{ formula-units of } NO_3^-}{2 \text{ formula-units of } NO_3^-/\text{formula-unit of } Ba(NO_3)_2} \right)$$
$$= 1.5 \times 10^{21} \text{ formula-units of } Ba(NO_3)_2$$

One mole $Ba(NO_3)_2$ contains 6.02×10^{23} formula-units of $Ba(NO_3)_2$. To get 1.5×10^{21} formula-units of $Ba(NO_3)_2$, you need

$$\left(\frac{1.5 \times 10^{21} \text{ formula-units of } Ba(NO_3)_2}{6.02 \times 10^{23} \text{ formula-units/mole}} \right) = 2.5 \times 10^{-3} \text{ mole } Ba(NO_3)_2$$

One mole $Ba(NO_3)_2$ equals 261.35 g.

$$(2.5 \times 10^{-3} \text{ mole of } Ba(NO_3)_2) \left(261.35 \frac{g}{mole}\right) = 0.65 \text{ g}$$

[Note: In this case, it would make sense to replace the term "formula-unit NO_3^-" by "ion NO_3^-," or "nitrate ion," since that is what it is. We will say more about this when we discuss dissociation in Chapter 12.] ■

■ EXAMPLE 124.   Given 1.08 g of quinine, $C_{20}H_{24}N_2O_2$, how many molecules is this?

SOLUTION: The fact that the formula is not the simplest that could be written with this ratio of atoms is a good indication that we are dealing with a molecular material for which the makeup of the molecule is known. In other words, in this case, the formula-unit is a bona fide molecule. Formula-weight of quinine = (20)(atomic weight of carbon) + (24)(atomic weight of hydrogen) + (2)(atomic weight of nitrogen) + (2)(atomic weight of oxygen) = (20)(12.011) + (24)(1.008) + (2)(14.007) + (2)(15.999) = 324.424 amu.

One mole of quinine = 324.424 g.

$$\frac{1.08 \text{ g of quinine}}{324.424 \text{ g/mole}} = 0.00333 \text{ mole}$$

One mole contains 6.02×10^{23} molecules.

$$(0.00333 \text{ mole}) \left(6.02 \times 10^{23} \frac{molecules}{mole}\right) = 2.00 \times 10^{21} \text{ molecules} \blacksquare$$

■ EXAMPLE 125.   How many atoms of hydrogen are there in 1.08 g of quinine $C_{20}H_{24}N_2O_2$?

SOLUTION: One mole quinine = 324.424 g.

$$\frac{1.08 \text{ g of quinine}}{324.424 \text{ g/mole}} = 0.00333 \text{ mole}$$

One mole contains 6.02×10^{23} molecules.

$$(0.00333 \text{ mole}) \left(6.02 \times 10^{23} \frac{molecules}{mole}\right) = 2.00 \times 10^{21} \text{ molecules}$$

Each molecule of quinine contains 24 hydrogen atoms.

$$(2.00 \times 10^{21} \text{ molecules}) \left(24 \frac{\text{hydrogen atoms}}{\text{molecule}}\right) = 4.80 \times 10^{22} \text{ H atoms} \blacksquare$$

■ EXAMPLE 126. Suppose you have 0.14 g of radioactive nitrogen. How many molecules of quinine can you make in which all the nitrogen atoms are tagged as radioactive?

SOLUTION: (Given the weight of nitrogen, you can find the number of gram-atoms of nitrogen. Then find the number of moles of quinine this corresponds to, and finally convert to molecules. It is almost always best to work through moles, since the molar relations are usually the easiest to establish.)

Atomic weight of nitrogen is 14.007 amu.

One gram-atom nitrogen weighs 14.007 g.

$$\left(\frac{0.14 \text{ g of nitrogen}}{14.007 \text{ g/gram-atom}}\right) = 0.010 \text{ gram-atom of N}$$

For $C_{20}H_{24}N_2O_2$, there are 2 gram-atoms of N per mole of $C_{20}H_{24}N_2O_2$.

$$\left(\frac{0.010 \text{ gram-atom of N}}{2 \text{ gram-atoms of N/mole of } C_{20}H_{24}N_2O_2}\right)$$
$$= 0.0050 \text{ mole of } C_{20}H_{24}N_2O_2$$

One mole contains 6.02×10^{23} molecules.

$$(0.0050 \text{ mole}) \left(6.02 \times 10^{23} \frac{\text{molecules}}{\text{mole}}\right) = 3.0 \times 10^{21} \text{ molecules} \blacksquare$$

■ EXAMPLE 127. How many gram-atoms of radioactive carbon would you need in order to make 1.0×10^{-6} g of fully tagged quinine? "Fully tagged" means all the carbon atoms would need to be of the radioactive type.

SOLUTION: One mole of $C_{20}H_{24}N_2O_2$ weighs 324 g.

$$\frac{1.0 \times 10^{-6} \text{ g of quinine}}{324 \text{ g/mole}} = 3.1 \times 10^{-9} \text{ mole}$$

One mole $C_{20}H_{24}N_2O_2$ contains 20 gram-atoms of carbon.

$$(3.1 \times 10^{-9} \text{ mole}) \left(20 \frac{\text{gram-atoms of C}}{\text{mole}}\right)$$
$$= 6.2 \times 10^{-8} \text{ gram-atom of carbon} \blacksquare$$

■ EXAMPLE 128.   How many formula-units are there in 0.15 mole of P_2O_5?

SOLUTION: One mole of anything contains 6.02×10^{23} formula-units.

$$(0.15 \text{ mole of } P_2O_5) \left(6.02 \times 10^{23} \frac{\text{formula-units}}{\text{mole}}\right)$$

$$= 9.0 \times 10^{22} \text{ formula-units} \quad ■$$

■ EXAMPLE 129.   How many molecules in 0.15 mole of P_2O_5?

SOLUTION: (This is a trick question. Compare it carefully with Example 128. As a matter of fact, you cannot answer it without putting in the additional information that P_2O_5 is actually composed of the molecules P_4O_{10}. Once you know this, you can see that it takes two of the formula-units P_2O_5 to make one molecule of P_4O_{10}.)
One mole P_2O_5 contains 6.02×10^{23} P_2O_5 units.
0.15 mole P_2O_5 contains 9.0×10^{22} P_2O_5 units.
It takes two P_2O_5 units to make one P_4O_{10} molecule.

$$\frac{9.0 \times 10^{22} \ P_2O_5 \text{ units}}{2P_2O_5 \text{ units}/P_4O_{10} \text{ molecule}} = 4.5 \times 10^{22} \ P_4O_{10} \text{ molecules} \quad ■$$

■ EXAMPLE 130.   How many P_4O_{10} molecules are there in 0.150 g of P_4O_{10}?

SOLUTION: Formula-weight   of $P_4O_{10} = (4)$(atomic   weight   of   P) $+ (10)$(atomic weight of O) $= (4)(30.974) + (10)(15.999) = 283.89$ amu.
One mole of $P_4O_{10} = 283.89$ g.

$$\frac{0.150 \text{ g } P_4O_{10}}{283.89 \text{ g/mole}} = 5.28 \times 10^{-4} \text{ mole of } P_4O_{10}$$

One mole of anything contains 6.02×10^{23} formula-units.

$$(5.28 \times 10^{-4} \text{ mole}) \left(6.02 \times 10^{23} \frac{\text{formula-units}}{\text{mole}}\right)$$

$$= 3.18 \times 10^{20} \text{ formula-units of } P_4O_{10}$$

$$= 3.18 \times 10^{20} \text{ molecules of } P_4O_{10}$$

(because one formula-unit of P_4O_{10} is the same as one molecule of P_4O_{10}). ■

■ EXAMPLE 131.   How many P_4O_{10} molecules are there in 0.150 g of P_2O_5?

SOLUTION: [Note: If you are clever enough, you can figure out without additional pencil-pushing that the answer to this question should be the same as the answer to Example 130. The reasoning goes like this: If the formula-weight is half as great, you have twice as many formula-units; but it takes 2 formula-units to make a molecule, so you end up with the same number of molecules.]

Formula-weight of P_2O_5
      = (2)(atomic weight of P) + (5)(atomic weight of O)
      = (2)(30.974) + (5)(15.999) = 141.94 amu

One mole of P_2O_5 = 141.94 g.

$$\frac{0.150 \text{ g } P_2O_5}{141.94 \text{ g/mole}} = 1.06 \times 10^{-3} \text{ mole of } P_2O_5$$

One mole of anything contains 6.02×10^{23} formula-units.

$$(1.06 \times 10^{-3} \text{ mole})\left(6.02 \times 10^{23} \frac{\text{formula-units of } P_2O_5}{\text{mole}}\right)$$
$$= 6.38 \times 10^{20} \text{ formula-units of } P_2O_5$$

But, it takes two formula-units of P_2O_5 to make one molecule of P_4O_{10}.

$$\frac{6.38 \times 10^{20} \text{ formula-units of } P_2O_5}{2 \text{ formula-units of } P_2O_5/\text{molecule of } P_4O_{10}}$$
$$= 3.19 \times 10^{20} \text{ molecules of } P_4O_{10} \blacksquare$$

■ EXAMPLE 132.   How many atoms of phosphorus are there in 0.150 mole of P_2O_5?

SOLUTION: One mole of P_2O_5 contains 2 gram-atoms of phosphorus.

$$(0.150 \text{ mole of } P_2O_5)\left(2 \frac{\text{gram-atoms of P}}{\text{mole of } P_2O_5}\right) = 0.300 \text{ gram-atom of P}$$

One gram-atom of anything contains 6.02×10^{23} atoms.

$$(0.300 \text{ gram-atom of P})\left(6.02 \times 10^{23} \frac{\text{atoms}}{\text{gram-atom}}\right)$$
$$= 1.81 \times 10^{23} \text{ atoms of P} \blacksquare$$

■ EXAMPLE 133.   How many atoms of phosphorus are there in 0.150 mole of P_4O_{10}?

SOLUTION: One mole of P_4O_{10} contains 4 gram-atoms of phosphorus.

$$(0.150 \text{ mole of } P_4O_{10}) \left(4 \frac{\text{gram-atoms of P}}{\text{mole of } P_4O_{10}}\right) = 0.600 \text{ gram-atom of P}$$

One gram-atom of anything contains 6.02×10^{23} atoms.

$$(0.600 \text{ gram-atom of P}) \left(6.02 \times 10^{23} \frac{\text{atoms}}{\text{gram-atom}}\right)$$

$$= 3.61 \times 10^{23} \text{ atoms of P} ■$$

■ EXAMPLE 134.   How many grams of oxygen are there in 0.150 mole of P_2O_5?

SOLUTION: One mole of P_2O_5 contains 5 gram-atoms of oxygen.

$$(0.150 \text{ mole of } P_2O_5) \left(5 \frac{\text{gram-atoms of O}}{\text{mole of } P_2O_5}\right) = 0.750 \text{ gram-atom of O}$$

Atomic weight of oxygen is 15.999 amu.
One gram-atom of O = 15.999 g.

$$(0.750 \text{ gram-atom of O}) \left(15.999 \frac{\text{g}}{\text{gram-atom}}\right) = 12.0 \text{ g of oxygen} ■$$

■ EXAMPLE 135.   How many grams of oxygen are there in 0.150 mole of P_4O_{10}?

SOLUTION: One mole of P_4O_{10} contains 10 gram-atoms of oxygen.

$$(0.150 \text{ mole of } P_4O_{10}) \left(10 \frac{\text{gram-atoms of O}}{\text{mole of } P_4O_{10}}\right) = 1.50 \text{ gram-atom of O}$$

One gram-atom of O = 15.999 g.

$$(1.50 \text{ gram-atoms of O}) \left(15.999 \frac{\text{g}}{\text{gram-atom}}\right) = 24.0 \text{ g of oxygen} ■$$

■ EXAMPLE 136.   How many moles of P_2O_5 can you make from 2.00 g of phosphorus and 5.00 g of oxygen?

SOLUTION: (Whenever you have weights given, convert to gram-atoms to see quickly which reagent is in excess. Then, it is just another step to figure out how much product will be formed.)

$$\frac{2.00 \text{ g of phosphorus}}{30.974 \text{ g/gram-atom of P}} = 0.0646 \text{ gram-atom of P}$$

$$\frac{5.00 \text{ g of oxygen}}{15.999 \text{ g/gram-atom of O}} = 0.313 \text{ gram-atom of O}$$

To get P_2O_5, the gram-atoms should be in the ratio 2 of P to 5 of O. However, the numbers above indicate that we have only about 1 gram-atom of P for about 5 gram-atoms of O, so the oxygen is evidently in excess. The reaction product is thus limited by the amount of phosphorus. We have to base our calculations on it.

From 2 gram-atoms of P, we can get 1 mole of P_2O_5.

$$(0.0646 \text{ gram-atom of P})\left(\frac{1 \text{ mole of } P_2O_5}{2 \text{ gram-atoms of P}}\right) = 0.0323 \text{ mole of } P_2O_5 \blacksquare$$

■ EXAMPLE 137.   How many moles of P_2O_5 can you make from 5.00 g of phosphorus and 2.00 g of oxygen?

SOLUTION:

$$\frac{5.00 \text{ g of P}}{30.974 \text{ g/gram-atom of P}} = 0.161 \text{ gram-atom of P}$$

$$\frac{2.00 \text{ g of O}}{15.999 \text{ g/gram-atom of O}} = 0.125 \text{ gram-atom of O}$$

To get P_2O_5, we need oxygen to phosphorus in a gram-atom ratio of 5 to 2, or 2.5 times as many gram-atoms of O as of P.

The numbers above indicate that there is nowhere near enough oxygen to satisfy all the given phosphorus, so in this case it will be the oxygen that limits the amount of reaction product.

From 5 gram-atoms of oxygen, we get 1 mole of P_2O_5.

$$(0.125 \text{ gram-atom of O})\left(\frac{1 \text{ mole of } P_2O_5}{5 \text{ gram-atoms of O}}\right) = 0.0250 \text{ mole of } P_2O_5 \blacksquare$$

■ EXAMPLE 138.   How many moles of $MgIn_2S_4$ could you make from 1.00 g of magnesium, 1.00 g of indium, and 1.00 g of sulfur?

SOLUTION:

$$\frac{1.00 \text{ g of Mg}}{24.305 \text{ g/gram-atom of Mg}} = 0.0411 \text{ gram-atom of Mg}$$

$$\frac{1.00 \text{ g of In}}{114.82 \text{ g/gram-atom of In}} = 0.00871 \text{ gram-atom of In}$$

$$\frac{1.00 \text{ g of S}}{32.06 \text{ g/gram-atom of S}} = 0.0312 \text{ gram-atom of S}$$

So far as gram-atoms are concerned, we need Mg:In:S in the ratio 1 : 2 : 4. Based on 0.0411 gram-atom of Mg this would mean 0.0822 gram-atom of In and 0.164 gram-atom of S. We do not have that much In nor that much S, so we cannot base our calculation on Mg.

Based on 0.00871 gram-atom of In, this would mean 0.00436 gram-atom of Mg and 0.0174 gram-atom of S. We have more than enough Mg and S to satisfy these requirements, so this looks OK.

(Just to be sure, look at the needs based on 0.0312 gram-atom of S. We would need 0.00780 gram-atom of Mg and 0.0156 gram-atom of In. We have the Mg but not the In, so we cannot base the calculation on the S.)

Once we decide that the In is the limiting reagent, the rest is easy. To get 1 mole of $MgIn_2S_4$, we need 2 gram-atoms of In.

Therefore, if we have 0.00871 gram-atom of In, we can only get

$$(0.00871 \text{ gram-atom of In})\left(\frac{1 \text{ mole of } MgIn_2S_4}{2 \text{ gram-atoms of In}}\right)$$

$$= 0.00436 \text{ mole of } MgIn_2S_4 \ \blacksquare$$

■ EXAMPLE 139. What is the weight of one molecule of quinine, $C_{20}H_{24}N_2O_2$?

SOLUTION: Formula-weight of $C_{20}H_{24}N_2O_2$ equals 324.424 amu.
One mole of $C_{20}H_{24}N_2O_2$ weighs 324.424 g.
One mole contains 6.02×10^{23} molecules.

$$\frac{324.424 \text{ g/mole}}{6.02 \times 10^{23} \text{ molecules/mole}} = 5.39 \times 10^{-22} \text{ g/molecule} \ \blacksquare$$

■ EXAMPLE 140. Suppose you make a mixture consisting of 1.00 g of N_2O, 1.00 g of CO, and 1.00 g of CO_2. What fraction of the total number of molecules is represented by N_2O molecules? (These are all molecular materials, as is true for most gases.)

SOLUTION: Since the mole gives us a way of counting particles, all we need to do is to determine the fraction of total moles that is N_2O.

$$\frac{1.00 \text{ g of } N_2O}{44.0 \text{ g/mole}} = 0.0227 \text{ mole of } N_2O$$

$$\frac{1.00 \text{ g of CO}}{28.0 \text{ g/mole}} = 0.0357 \text{ mole of CO}$$

$$\frac{1.00 \text{ g of } CO_2}{44.0 \text{ g/mole}} = 0.0227 \text{ mole of } CO_2$$

Total moles in mixture $= 0.0227 + 0.0357 + 0.0227 = 0.0811$

Fraction of total that is $N_2O = \dfrac{0.0227}{0.0811} = 0.280$ ■

■ EXAMPLE 141.   A drop of water is about 0.05 ml. The density of water at room temperature is about 1.0 g/ml. How many H_2O molecules are there in a drop of water?

SOLUTION:

Weight of H_2O in 1 drop $= (0.05 \text{ ml})\left(1.0 \frac{g}{ml}\right) = 0.05$ g.

Formula-weight of H_2O
$= (2)(\text{atomic weight of H}) + \text{atomic weight of O}$
$= (2)(1.0) + 16.0 = 18.0 \text{ amu}$

One mole of $H_2O = 18.0$ g.

$$\frac{0.05 \text{ g of } H_2O}{18.0 \text{ g/mole}} = 0.003 \text{ mole of } H_2O$$

One mole of H_2O contains 6.02×10^{23} H_2O molecules.

$(0.003 \text{ mole of } H_2O)\left(6.02 \times 10^{23} \dfrac{\text{molecules}}{\text{mole}}\right)$

$\qquad = 2 \times 10^{21}$ molecules of H_2O ■

Exercises

*4.1. Phosgene has the molecular formula $COCl_2$. How many molecules of $COCl_2$ are there in a 2.36-g sample?

*4.2. How many moles of $COCl_2$ can you make from 2.36 g of chlorine? (*Answer*: 0.0333 mole.)

*4.3. What is the weight of one molecule of $COCl_2$?

*4.4. How many moles are there in a mixture containing 1.86×10^{22} molecules of $COCl_2$, 2.36×10^{21} molecules of CO, and 4.22×10^{21} molecules of Cl_2? (*Answer*: 0.0418 mole.)

*4.5. How many formula-units are there in 36.9 g of $Al_2(SO_4)_3$?

**4.6. How many molecules of N_2O_2 are there in 1.50 g of NO? (*Answer*: 1.50×10^{22}).

**4.7. How many formula-units of NO are there in 0.072 mole of N_2O_2?

**4.8. What is the maximum number of moles of $Li(NH_3)_4$ that could be made from 1.00 g of Li and 10.00 g of NH_3? (*Answer*: 0.144 mole.)

*4.9. If material X contains 8.62×10^{21} molecules of X in 3.39 g of X, then what is the weight of 1 mole of X?

**4.10. A sample of amphetamine contains 0.225 gram-atom of carbon, 0.328 g of hydrogen, and 1.50×10^{22} atoms of nitrogen per 1.50×10^{22} molecules of amphetamine. Deduce the chemical formula. (*Answer*: $C_9H_{13}N$.)

**4.11. A sample composed of a mixture of CO molecules and CO_2 molecules contains a total of 3.66×10^{-3} mole and weighs 0.122 g. What per cent of the molecules are CO?

**4.12. A sample composed of a mixture of CO molecules and CO_2 molecules analyzes to show 66.0% by weight of oxygen. What is the ratio of $CO : CO_2$ molecules in the mixture? (*Answer*: 1.19.)

*4.13. Krypton fluoride, KrF_x, is a white molecular solid containing 0.749 g of fluorine per 5.93×10^{21} molecules of KrF_x. What is the value of x?

*4.14. A given mixture composed of 0.028 mole of X and 0.043 mole of Y weighs 2.843 g. Another mixture composed of 0.043 mole of X and 0.028 mole of Y weighs 3.482 g. What is the weight of 1 mole of X? (*Answer*: 65.8 g.)

**4.15. The dye phenolphthalein weighs 5.29×10^{-22} g/molecule. If the per cent by weight composition is 75.46% carbon, 4.43% hydrogen, and 20.11% oxygen, what is the molecular formula?

**4.16. The drug marihuana owes its activity to tetrahydrocannabinol, which contains 70% as many carbon atoms as hydrogen atoms and 15 times as many hydrogen atoms as oxygen atoms. The number of moles in a gram of tetrahydrocannabinol is 0.00318. What is the molecular formula? (*Answer*: $C_{21}H_{30}O_2$.)

*4.17. How many formula-units are there in 1.50 g of $Na_{0.83}WO_3$?

*4.18. How many gram-atoms of oxygen are there in 3.68 g of $Al_2(SO_4)_3$? (*Answer*: 0.129.)

4.19 A 4.88-g sample of CH_4, C_2H_6, and C_6H_6 mixture contains a total of 0.100 mole and analyzes to 88.5% carbon by weight. How many moles are there of each compound in the mixture?

4.20. Given three unknown elements P, Q, and R. If the compound PQ_2R_3 is 10.5% by weight P; the compound P_3Q_2R is 35.3% by weight P; and 0.15 mole of PQ_4 weighs 31.5 g, what are the atomic weights of P, Q, and R? (*Answer*: P = 23.4 amu; Q = 46.7 amu; R = 35.3 amu.)

5

CHEMICAL REACTIONS AND CHEMICAL EQUATIONS

IN ADDITION to the symbols and formulas we have just considered in detail, the other major shorthand device used by the chemist is the chemical equation.

5.1 Writing equations

A chemical equation is a way of describing a chemical reaction. It uses formulas and symbols not only to represent the nature of the materials used up and formed but also specific molar amounts of these materials. Coefficients, or numbers preceding the formulas and symbols, tell how many moles of each material are involved. Thus, for example, in the chemical equation

$$C_3H_8 + 5O_2 \longrightarrow 3CO_2 + 4H_2O$$

the formulas C_3H_8, O_2, CO_2, and H_2O stand, respectively, for the substances propane, oxygen, carbon dioxide, and water; the coefficients 1 (understood to precede C_3H_8), 5 (preceding O_2), 3 (before CO_2), and 4 (before H_2O) indicate how many moles of these substances are involved in the reaction. The arrow, which is read "reacts to give," is sometimes replaced by an equal sign ($=$) or double arrows ($\rightleftarrows$). The single arrow ($\rightarrow$) is generally used when we wish to emphasize the change going from left to right, that is, the disappearance of the stuff on the left (the "reactants") and the appearance of the stuff on the right (the "products"). The equal sign ($=$) is preferred by some when the quantitative aspects of chemical reaction are being emphasized, that is, that the mass of starting material used up equals the mass of product material formed. The double arrows ($\rightleftarrows$) call attention to the reversibility of the reaction and are particularly useful when we are considering the equilibrium state, that is, the state where the forward chemical reaction is just offset by the reverse chemical reaction (see Chapter 11). In this chapter we use the single arrow, since we shall be concerned with 100% conversion from the state of affairs represented by the left of the equation to the state of affairs represented by the right of the equation.

For a chemical equation to represent correctly a chemical reaction, the following conditions must be met:

1. The formulas shown on the left and on the right of the equation must correspond to the observed experimental facts as to what is used up in the reaction and what is produced in the reaction.

2. The coefficients shown on the left and right must correspond to a *balanced equation*, that is, one that is consistent with conservation of mass and with conservation of charge. In special applications, which we shall take up in Chapter 8, it may be of interest to include in the balancing pertinent information on the energy changes in the reaction.

With these two conditions satisfied, the chemical equation gives us efficiently information as to what happens to atoms, molecules, and formula-units in a reaction, but, more important, it enables us to deduce weight relations applying to the same reaction. First we take up the various methods for balancing equations and then the several kinds of calculations that can be done using balanced equations.

How do we write a balanced equation? First, get the observed facts: "so-and-so reacts with so-and-so to give," and so forth. Then write down the correct formulas of the materials, used-up substances on the left and produced substances on the right. Be particularly careful that the subscripts of the formulas are correctly written, also that any superscript charges on ionic species are not omitted. Supply any missing species needed to give a complete reaction and insert properly chosen coefficients to give a balanced equation.

How do we select the proper coefficients? There are three general methods: (a) hit-or-miss, sometimes more elegantly referred to as "balancing by inspection;" (2) balancing via oxidation-number change (what this is, we shall consider shortly); and (3) balancing via half-reaction, sometimes referred to as the "ion-electron method." Which of these methods is best to use depends on what kind of reaction it is, and how complicated it is likely to be. Some chemical reactions are quite simple in that they involve only pairing or unpairing of species or perhaps shifting of partners. For example, the combining of magnesium ion (Mg^{2+}) and fluoride ion (F^-) to form the insoluble salt magnesium fluoride (MgF_2) can be represented by the equation

$$Mg^{2+} + 2F^- \longrightarrow MgF_2(s).$$

[The notation (s) means solid state; (g) would stand for gaseous state; (ℓ), liquid state. No designation on an ion generally means "in aqueous solution."] Given the raw information that $Mg^{2+} + F^- \rightarrow MgF_2(s)$, the coefficient 2 for F^- could be deduced by noting that two fluorine atoms appear on the right, so two fluorine atoms ought to appear on the left. In other words, mass (or atoms) must be conserved. Alternatively, we could note that MgF_2 on the right side of the equation is electrically neutral. On the left, since Mg^{2+} is doubly positive, we would need twice as many singly negative particles (F^-) to preserve electrical neutrality.

Another example: Sodium sulfate (Na_2SO_4) dissolves in water to give sodium ions (Na^+) and sulfate ions (SO_4^{2-}). Because the starting material Na_2SO_4 contains two Na^+ per one SO_4^{2-}, as indicated by the subscript 2 in the formula Na_2SO_4, we can conclude that the dissolving reaction liberates two Na^+ per SO_4^{2-}. Therefore, as a balanced equation, we can write

$$Na_2SO_4(s) \longrightarrow 2Na^+ + SO_4^{2-}$$

In both of these examples, the chemical change can be visualized, respectively, as the coming together or the going apart of ions. So-called double-substitution, or metathesis reactions, can be similarly represented. For example, mixing silver sulfate (Ag_2SO_4) with barium chloride ($BaCl_2$) in water produces solid barium sulfate ($BaSO_4$) and solid silver chloride ($AgCl$). The balanced equation

$$Ag_2SO_4 + BaCl_2 \longrightarrow BaSO_4(s) + 2AgCl(s)$$

shows a coefficient 2 preceding the formula of the product $AgCl$. Since there are two silver atoms on the left of the equation, there must be two on the right; similarly, since there are two chlorine atoms on the left of the equation, there must be two on the right. The reaction can be visualized as one in which the two Ag^+ ions, originally paired with the SO_4^{2-} ion in Ag_2SO_4, have exchanged places with the Ba^{2+} that is originally paired with the two chloride ions. It would not be cricket to write the product as Ag_2Cl_2, since we are given the information that it is to be written $AgCl$. Therefore, we need the coefficient 2 to precede $AgCl$ to conserve the atomic balance.

■ EXAMPLE 142.   Write a balanced equation for the combining of aluminum ion (Al^{3+}) and sulfate ion (SO_4^{2-}) to form solid aluminum sulfate, $Al_2(SO_4)_3$.

SOLUTION: First we write the given information in symbols:

$$Al^{3+} + SO_4^{2-} \longrightarrow Al_2(SO_4)_3(s)$$

Two Al atoms on the right means two Al atoms on the left, so put a 2 before Al^{3+}.

Three SO_4^{2-} groups on the right means three SO_4^{2-} groups on the left, so put a 3 before SO_4^{2-}.

The balanced equation then reads

$$2Al^{3+} + 3SO_4^{2-} \longrightarrow Al_2(SO_4)_3(s)$$

To check, we can note that not only are the numbers of each kind of atom equal on both sides of the equation but also the net charge is the same on both sides. For the latter, two Al^{3+} ions contribute (two times three) positive charges and the three SO_4^{2-} ions contribute (three times two) negative charges to the left-hand side of the equation. This makes $(2)(+3)$ plus $(3)(-2)$, or $+6 - 6 = 0$, or neutral, which agrees with the right-hand side of the equation. ■

■ EXAMPLE 143. Write a balanced equation for the formation of ferrous ion (Fe^{2+}), ammonium ion (NH_4^+), and sulfate ion (SO_4^{2-}) from the dissolving of the solid, ferrous diammonium disulfate, $Fe(NH_4)_2(SO_4)_2$.

SOLUTION: The given information is expressed symbolically:

$$Fe(NH_4)_2(SO_4)_2(s) \longrightarrow Fe^{2+} + NH_4^+ + SO_4^{2-}$$

Two NH_4 groups on the left mean two on the right, so put a 2 before NH_4^+.

Two SO_4 groups on the left mean two on the right, so put a 2 before SO_4^{2-}.

The balanced equation reads

$$Fe(NH_4)_2(SO_4)_2(s) \longrightarrow Fe^{2+} + 2NH_4^+ + 2SO_4^{2-}$$

It checks out properly, since the left side is neutral and the right side is $(+2) + (2)(+1) + (2)(-2) = 0$, corresponding to one dipositive ion plus two monopositive ions plus two dinegative ions, or zero, also. ■

■ EXAMPLE 144. When $Al(OH)_3$ reacts with H_2SO_4 under the proper conditions, the products are $Al_2(SO_4)_3$ and H_2O. Write a balanced equation for the reaction.

SOLUTION: Write the raw information, noting that H_2O can be written HOH:

$$Al(OH)_3 + H_2SO_4 \longrightarrow Al_2(SO_4)_3 + HOH ■$$

The subscript 2 on Al on the right indicates two Al atoms in the product. There must be two Al atoms in the starting material, so put a 2 before $Al(OH)_3$. The subscript 3 on SO_4 on the right indicates three SO_4 groups in the product. There must be three SO_4 groups in the starting material, so put a 3 before H_2SO_4.

Note now that the $2Al(OH)_3$ on the left of the partially balanced equation says there are two times three OH groups (two, from the coefficient, times the subscript 3), or six OH groups, in the starting material. There must be six OH groups in the product, so on the right side write a 6 before HOH.

The final equation reads

$$2Al(OH)_3 + 3H_2SO_4 \longrightarrow Al_2(SO_4)_3 + 6HOH$$

These are almost trivial cases of equation balancing because they simply require proper counting of atoms and groups of atoms. They are typical of the equations that describe reactions in which no electron shifting occurs. More likely to cause trouble are those reactions in which electrons are shifted from one atom to another, so-called oxidation–reduction, or redox, reactions. For these oxidation–reduction reactions, balancing by inspection is not only likely to be a hit-or-miss project but also may lead to a wrong answer. Therefore, it is advisable that a systematic approach be employed.

5.2   Oxidation numbers

First we look at the oxidation-number method of balancing equations. For this, we need to define briefly what an oxidation number is and how it is used. (Sometimes oxidation numbers are referred to as "valence numbers" or "oxidation states.") An oxidation number is simply a number assigned to an atom in an element or in a compound, following quite specific rules, so as to keep tabs on electron shifts in chemical reactions. The oxidation number in a given case may turn out to be positive ($+$), negative ($-$), or zero; it may be a whole number, or a fraction. The rules are basically derived from an agreement to count shared electrons with the more electronegative atom unless the two sharing atoms are identical, in which case the shared electrons are split equally in the counting process. (See your textbook for a more complete discussion.) The rules are

1. In the elemental, or uncombined, state the atoms are assigned an oxidation number of zero.
2. In compounds, the oxidation number of fluorine is always assigned to be -1.

3. In compounds, the group I elements (Li, Na, K, Rb, Cs, and Fr) have an oxidation number of $+1$.

4. In compounds, the group II elements (Be, Mg, Ca, Sr, Ba, and Ra) have an oxidation number of $+2$.

5. In compounds, hydrogen is generally assigned the oxidation number of -1.

6. In compounds, oxygen is generally assigned the oxidation number -2. The exceptions are the fluorine–oxygen compounds, where the fluorine assignment -1 takes precedence, and the peroxides, where oxygen is given an oxidation number of -1.

7. For neutral species, the sum of the oxidation numbers times the number of each kind of atom adds up to zero.

8. For charged species (ions), the sum of the oxidation numbers times the number of each kind of atom adds up to the net charge on the ion.

■ EXAMPLE 145.   What is the oxidation number of S in H_2SO_4?

SOLUTION: Oxidation number of hydrogen is $+1$.
Oxidation number of oxygen is -2.
Two hydrogens contribute $(2)(+1)$, or $+2$.
Four oxygens contribute $(4)(-2)$, or -8.
Net contribution by hydrogens and oxygens is $+2 -8$, or -6.
Compound H_2SO_4 is neutral, so we need a $+6$ to counteract the -6.
This $+6$ must be contributed by the sulfur. Since there is but one S atom, its oxidation number must be $+6$. ■

■ EXAMPLE 146.   What is the oxidation number of S in $H_2S_2O_7$?

SOLUTION: Two H's contribute $(2)(+1)$, or $+2$.
Seven O's contribute $(7)(-2)$, or -14.
Net contribution by H's and O's is $+2 -14$, or -12.
We need a $+12$ contribution from the S's.
There are two S atoms per formula-unit, so each S atom contributes $1/2(+12)$, or $+6$.
Thus, the oxidation number of S in $H_2S_2O_7$ is $+6$. ■

■ EXAMPLE 147.   What is the oxidation number of S in $Na_2S_2O_3$?

SOLUTION: Oxidation number of Na is $+1$.
Two Na's contribute $(2)(+1)$, or $+2$.
Oxidation number of O is -2.
Three O's contribute $(3)(-2)$, or -6.
Net contribution by Na's and O's is $+2 -6$, or -4.
We need a contribution of $+4$ from the S's to come out neutral.

There are two S's per formula-unit, so each S atom contributes $1/2(+4)$, or $+2$.

Thus, the oxidation number of S in $Na_2S_2O_3$ is $+2$. ■

■ EXAMPLE 148. What is the oxidation number of S in $S_4O_6^{2-}$?

SOLUTION: Oxidation number of O is -2.
Six O's contribute $(6)(-2)$, or -12.
The net charge on the ion is -2.
What must we have with -12 to get a net of -2? We need $+10$.
This $+10$ must be contributed by the four S's.
Each S atom contributes $+10/4$, or $+2.5$.
Thus, the oxidation number of S in $S_4O_6^{2-}$ is $+2.5$. ■

■ EXAMPLE 149. What is the oxidation number of S in $Ca(HSO_3)_2$?

SOLUTION: The oxidation number of Ca is $+2$.
Since the compound is neutral, each HSO_3 must contribute a charge of -1. In HSO_3^-, one H and the three O's contribute $(1)(+1) + (3)(-2)$, or $1-6$, or -5. To come out with a net charge of -1 on the whole complex ion HSO_3^-, the S must contribute $+4$. There is but one S atom per formula-unit, so the oxidation number of S in $Ca(HSO_3)_2$ is $+4$. ■

From the foregoing examples, the procedure to follow is to assign first those oxidation numbers you know, using either the rules given above or knowledge of the group behavior in the periodic table (e.g., Ca is a group II element and forms Ca^{2+} ions), and then figure out what the other atom must be to be consistent with rules and electrical charge balance.

For shorthand convenience, it is frequently desirable to write the oxidation number just below the atom for which it applies. (Remember that oxidation number is *per atom*.) Then, if necessary, under that you can write the total contribution to the charge by all the atoms of that type—in other words, oxidation number times the number of atoms. This notation, which is illustrated in the next problems, will be handy when we start balancing equations using oxidation numbers.

■ EXAMPLE 150. What is the oxidation number of C in $H_2C_2O_4$?

SOLUTION:

$H_2C_2O_4$
$+1?\ -2$ ⟵⎯⎯ oxidation numbers (i.e., per atom)
$+2?\ -8$ ⟵⎯⎯ total charge = atoms × oxidation number

From the $+2$ and -8, we conclude that the total charge contribution by the carbon must be $+6$. There are two carbons, so each is $+3$. Therefore, the oxidation number of C in $H_2C_2O_4$ is $+3$. ∎

At this point it should be evident that the oxidation number is only a formal way of keeping track of the electrical charge in compounds. In most cases, proper application of the rules leads to a unique assignment of oxidation number for each element in a compound. However, this is not always the case. For example, in the peroxide compound Na_2MoO_8, a $+6$ assignment for Mo is obtainable only provided we know that this is a peroxide and, therefore, that the oxygen is to be regarded as -1. Suppose we did not have this information. Then we would assume -2 for oxygen, eventually leading to an oxidation number of $+14$ for Mo. Now this is a ridiculous oxidation number for molybdenum, seeing where it is in the periodic table and knowing something about how oxidation numbers run. But how could *you* be expected to know this? The answer is you would not and, in fact, you could make a good case for choosing to assign Mo in Na_2MoO_8 an oxidation number of $+14$ instead of $+6$. The point is that in some cases, there is not a unique assignment possible and a choice may rest on other information. The fortunate part is that it *does not matter which choice you make*, provided you are consistent in assigning the other atoms. Specifically, in Na_2MoO_8 if you say O is -1 then Mo has to be $+6$, but if you say O is -2, then Mo has to be $+14$. So far as balancing equations is concerned, either choice would work out.

∎ EXAMPLE 151.   What is the oxidation number of C in NaSCN?

SOLUTION: The oxidation number of Na is $+1$.

Therefore, the sum of the oxidation numbers of S, C, and N must add up to -1.

But the number of possibilities for C appear unlimited, since our rules do not fix either the S or the N uniquely.

You might not know it, but oxidation numbers of S generally run from -2 to $+6$ and those of N from -3 to $+5$. Correspondingly, there are 81 possible combinations ranging from -2 for S and -3 for N (in which case, C would come out to be $+4$, which is quite reasonable) to the other extreme, where S is taken to be $+6$ and N is taken to be $+5$ (in which case C would come out to be -12, which most chemists would consider unreasonable). However, for balancing equations it does not matter which value of oxidation number is taken, $+4$ or -12 or anything in between, so long as we are consistent with the other atoms and stick with them in the course of the reaction. This ambiguity in the choice of some oxidation numbers is one principal reason for preferring the half-reaction method of balancing equations, which, as we shall see later, dispenses with oxidation numbers completely. ∎

5.3   Balancing equations by use of oxidation numbers

The underlying principle here is that electrical charge must be conserved in the course of a chemical reaction, so any increase in oxidation number must be compensated by a decrease. To balance an equation, all we need do is assign oxidation numbers to those atoms which undergo change in oxidation state and then match the increase against the decrease. Then, the coefficients of all the other reactants and products are adjusted consistent with conservation of mass and charge. In more sophisticated problems, only the oxidizing agent (substance in which oxidation number goes down) and the reducing agent (substance in which oxidation number goes up) are given along with their products, so the equation may have to be completed (i.e., missing reactants and products supplied) as well as balanced. In the following problems, the oxidation numbers involved are written just below the symbol of the corresponding atom. Up or down arrows are used to show how many units of electronic charge must be shifted in or out to account for the observed change in oxidation number.

■ EXAMPLE 152.   Balance the equation $?CH_4 + ?O_2 \rightarrow ?CO_2 + ?H_2O$ using oxidation numbers.

SOLUTION: The oxidation number of C changes from -4 in CH_4 to $+4$ in CO_2.

The oxidation number of O changes from 0 in O_2 to -2 in CO_2 and H_2O. This can be summarized schematically as follows:

$$?CH_4 + ?O_2 \longrightarrow ?CO + ?H_2O$$
$$-4 0 +4 \ -2 \ -2$$

The C atom, in going from -4 to $+4$, has to get rid of 8 units of negative charge. We can indicate this by showing an arrow going downward from the -4 carbon and labeling it $8e^-$ (which stands for eight electronic charges). The O atom, in going from 0 to -2, has to gain 2 units of negative charge. We can indicate this by an arrow going upward to the zero oxygen. The tableau looks like this:

$$?CH_4 + ?O_2 \longrightarrow ?CO_2 + ?H_2O$$
$$-4 0 +4 \ -2 -2$$
$$\downarrow 8e^- \ \uparrow 2e^-$$

Next we note that, whereas only one carbon atom need be considered (the subscript of C being unity in CH_4), two oxygen atoms need to be considered (because the subscript is 2 in O_2). This means that, although each O atom goes from 0 to -2, two such changes need to be considered. We can indicate this by putting a " $\times 2$ " next to the " $2e^-$."

$$?CH_4 + ?O_2 \longrightarrow ?CO_2 + ?H_2O$$

$$\begin{array}{cccc} -4 & 0 & +4\,-2 & -2 \\ \downarrow 8e^- & \uparrow 2e^- \times 2 & & \end{array}$$

So now we have $8e^-$ moving out and $2e^- \times 2$ moving in, which is unsatisfactory if we want to keep electrical balance. We can correct the situation by taking twice as many of the $2e^- \times 2$ changes as of the $8e^-$ changes. This means multiplying the whole O_2 business by two:

$$CH_4 + 2O_2 \longrightarrow ?CO_2 + ?H_2O$$

$$\begin{array}{cccc} -4 & 0 & +4\,-2 & -2 \\ \downarrow 8e^- & \uparrow 2 \times 2e^- \times 2 & & \end{array}$$

At this stage, the problem is essentially solved because we have determined that CH_4 and O_2 molecules react in the ratio $1:2$. (One CH_4 gives up $8e^-$'s; two O_2's pick up $2 \times 2 \times 2 = 8e^-$.) All we need do now is fix up the rest of the equation, being careful however not to disturb this $1:2$ ratio.

For the conclusion of the problem, all the tableau involving oxidation numbers is to be ignored and we work only from

$$CH_4 + 2O_2 \longrightarrow ?CO_2 + ?H_2O$$

One C on the left requires one C on the right, so we put a 1 (understood) before CO_2.

Four O atoms on the left require four O atoms on the right. Two of these have just been provided in the $1CO_2$; the other two can come from the H_2O if we put a 2 before H_2O.

The final balanced equation is

$$CH_4 + 2O_2 \longrightarrow CO_2 + 2H_2O$$

Actually, in this case, you probably could have balanced it by inspection more quickly. ■

■ EXAMPLE 153. Balance the following by the oxidation number method:

$$?C_3H_8 + ?O_2 \longrightarrow ?CO + ?H_2O$$

$$\begin{array}{cccc} (3)(-8/3) & 0 & (3)(+2) & -2 \\ \downarrow 14e^- & \uparrow 2e^- \times 2 & & \end{array}$$

Each carbon atom goes from $-8/3$ to $+2$. Because $8/3$ is such an awkward number to work with, we would be better off to consider three atoms as a group; in other words, multiply this particular oxidation-number change by 3. On the left of the equation, this will be done automatically for us when we take into account the subscript 3 on C in C_3H_8. These three carbons together are worth $(3)(-8/3)$, or -8. On the right, we need to be careful to insert the same factor of three. We do this by putting a 3 before the CO, noting that three such carbon atoms are worth $(3)(+2)$, or $+6$. Thus, for C_3H_8 going to 3CO, the charge change would be from -8 to $+6$, corresponding to a loss of $14e^-$.

So, the problem is really

$$C_3H_8 + ?O_2 \longrightarrow 3CO + ?H_2O$$
$$\downarrow 14e^- \quad \uparrow 4e^-$$

We can square away the electron loss and electron gain if we take two of the $14e^-$ changes $(=28e^-)$ for seven of the $4e^-$ changes $(=28e^-)$—in other words, take $2C_3H_8$ for every $7O_2$. However, if we take $2C_3H_8$ on the left, we need to have 6CO on the right.

$$2C_3H_8 + 7O_2 \longrightarrow 6CO + ?H_2O$$

To get the coefficient for H_2O, we can either count H atoms (16H on the left means 16H on the right) or count O atoms (14 oxygen on the left means 14 oxygen on the right, of which 6 are already furnished by 6CO.) In either case, $8H_2O$ comes out.

The final equation is

$$2C_3H_8 + 7O_2 \longrightarrow 6CO + 8H_2O \; \blacksquare$$

■ EXAMPLE 154.   Balance the following by the oxidation-number method:

$$?KMnO_4 + ?H_2C_2O_4 \longrightarrow ?K_2CO_3 + ?MnO_2 + ?H_2O + ?CO_2$$

SOLUTION: The Mn goes from $+7$ in $KMnO_4$ to $+4$ in MnO_2. This is a $3e^-$ change.

The C goes from $+3$ in $H_2C_2O_4$ to $+4$ in K_2CO_3 and in CO_2. This looks like a $1e^-$ change, but it needs to be taken twice, because of the subscript 2 on C in $H_2C_2O_4$.

To balance a $3e^-$ gain against a $2e^-$ loss, we need to take two of the former with three of the latter. This gives us

$$2KMnO_4 + 3H_2C_2O_4 \longrightarrow ?K_2CO_3 + ?MnO_2 + ?H_2O + ?CO_2$$

Since there are two K atoms on the left, we need to put a 1 before K_2CO_3 on the right.

Since there are two Mn atoms on the left, we need to put a 2 before MnO_2 on the right.

Since there are six H atoms on the left (in $3H_2C_2O_4$), we need to put a 3 before H_2O on the right.

Finally, the six C atoms on the left (in $3H_2C_2O_4$) require six C atoms on the right, one of which is already accounted for in the $1K_2CO_3$; so we need to put a 5 before the CO_2 on the right.

The final equation is

$$2KMnO_4 + 3H_2C_2O_4 \longrightarrow K_2CO_3 + 2MnO_2 + 3H_2O + 5CO_2$$

It can be checked out by counting oxygen atoms (2×4 plus $3 \times 4 = 20$ on the left; 3 plus 2×2 plus 3 plus $5 \times 2 = 20$ on the right), which we have not used in establishing the balanced equation. ■

The great majority of chemical equations describe reactions in aqueous solutions, and frequently the given information is not complete as to all reactants and all products. What is usually specified is the oxidizing and the reducing agent, what they go to in the course of the reaction, and whether the solution is acidic or basic. The rest of the information (whether the equation involves H_3O^+ or OH^- or H_2O and on which side of the equation these need to be put) has to be figured out. The following systematic procedure is recommended:

1. Assign oxidation numbers to the atoms that change.

2. Choose the proper ratio of oxidizing to reducing agent so the oxidation-number change is balanced.

3. Adjust the coefficients of the products to correspond to the coefficients selected in step 2.

4. Count up the oxygen atoms on both sides and add H_2O to the side that is deficient in oxygen.

5. Count up the hydrogen atoms on both sides and add H^+ to the side that is deficient in hydrogen. It is important that step 4 precede step 5.

6. If in *acidic* solution, convert each H^+ into H_3O^+ and add an equal number of H_2O molecules to the opposite side of the equation.

7. If in *basic* solution, convert each H^+ into H_2O and add an equal number of OH^- ions to the opposite side of the equation.

8. Cancel any duplications that appear on both sides of the equation.

■ EXAMPLE 155.   Given the change $Cr_2O_7^{2-} + H_2SO_3 \to Cr^{3+} + HSO_4^-$ occurring in acidic solution. Complete and balance the equation for the reaction, using the oxidation-number method.

SOLUTION: The Cr goes from $+6$ in $Cr_2O_7^{2-}$ to $+3$ in Cr^{3+}. This is a $3e^-$ gain. There are two of them because of the subscript 2 on Cr in $Cr_2O_7^{2-}$. In effect, this is a $6e^-$ change.

The S goes from $+4$ in H_2SO_3 to $+6$ in HSO_4^-. This is a $2e^-$ loss. To balance a $6e^-$ gain against a $2e^-$ loss, we need to take one of the former for three of the latter. Therefore, the ratio of $Cr_2O_7^{2-}$ to H_2SO_3 must be taken 1 to 3. We thus have

$$Cr_2O_7^{2-} + 3H_2SO_3 \longrightarrow \ ?Cr^{3+} + ?HSO_4^-$$

Two chromium atoms on the left require a coefficient of 2 to be placed before Cr^{3+} on the right. Likewise, the three sulfur atoms on the left can be balanced by putting a 3 before the HSO_4^- on the right. We now have

$$Cr_2O_7^{2-} + 3H_2SO_3 \longrightarrow 2Cr^{3+} + 3HSO_4^-$$

Counting up the oxygen atoms gives $7 + (3)(3) = 16$ on the left and $(3)(4) = 12$ on the right. The fact that the right side has four less oxygen atoms than the left can be fixed up by adding $4H_2O$ to the right side of the equation. It now reads

$$Cr_2O_7^{2-} + 3H_2SO_3 \longrightarrow 2Cr^{3+} + 3HSO_4^- + 4H_2O$$

Counting up the hydrogen atoms, we have $(3)(2)$, or 6, on the left and $3 + (4)(2)$, or 11, on the right. The left side has five less hydrogen atoms, and we can fix it by adding $5H^+$ to the left. The result is

$$5H^+ + Cr_2O_7^{2-} + 3H_2SO_3 \longrightarrow 2Cr^{3+} + 3HSO_4^- + 4H_2O$$

Now convert the $5H^+$ on the left to $5H_3O^+$ and add 5 more molecules of H_2O to the right side. The final balanced equation is

$$5H_3O^+ + Cr_2O_7^{2-} + 3H_2SO_3 \longrightarrow 2Cr^{3+} + 3HSO_4^- + 9H_2O$$

It can be checked by comparing the net charge on the left and the right. On the left we have 5 monopositive ions plus 1 dinegative plus 3 neutral species, which adds up to a net charge of $+3$; on the right side we have 2 tripositive plus 3 mononegative plus 9 neutral, which adds up to a net charge of $(2)(+3) + (3)(-1)$, or $+3$, also. ∎

■ EXAMPLE 156.   The   reaction $CrO_4^{2-} + SO_3^{2-} \rightarrow Cr(OH)_4^- + SO_4^{2-}$ occurs in basic solution. Write a complete balanced equation for the reaction, using the oxidation-number method.

SOLUTION: The Cr goes from $+6$ in CrO_4^{2-} to $+3$ in $Cr(OH)_4^-$. This is a $3e^-$ gain.

The S goes from $+4$ in SO_3^{2-} to $+6$ in SO_4^{2-}. This is a $2e^-$ loss.

To balance electron gain against electron loss, we take two of the former against three of the latter.

This gives us

$$2CrO_4^{2-} + 3SO_3^{2-} \longrightarrow \quad ?Cr(OH)_4^- + ?SO_4^{2-}$$

Two Cr atoms on the left require a coefficient 2 before the $Cr(OH)_4^-$ on the right; three S atoms on the left require $3SO_4^{2-}$ on the right. We now have

$$2CrO_4^{2-} + 3SO_3^{2-} \longrightarrow 2Cr(OH)_4^- + 3SO_4^{2-}$$

Counting up oxygens gives $(2)(4) + (3)(3) = 17$ on the left and $(2)(4) + (3)(4) = 20$ on the right. The left side is deficient by three oxygen atoms, so we stick $3H_2O$ on the left side of the equation. It now reads

$$3H_2O + 2CrO_4^{2-} + 3SO_3^{2-} \longrightarrow 2Cr(OH)_4^- + 3SO_4^{2-}$$

Counting up hydrogens gives $(3)(2) = 6$ on the left and $(2)(4) = 8$ on the right. The left side is deficient by two hydrogen atoms, so we stick $2H^+$ on the left side of the equation. The result is

$$2H^+ + 3H_2O + 2CrO_4^{2-} + 3SO_3^{2-} \longrightarrow 2Cr(OH)_4^- + 3SO_4^{2-}$$

This equation is completely balanced but is *wrong* because it is not valid for basic solution. To make it right, we can add $2OH^-$ to the left to chew up the $2H^+$, but to leave the balancing undisturbed we must simultaneously add $2OH^-$ to the right. On the left, the $2H^+$ and $2OH^-$ give $2H_2O$, which can be combined with the $3H_2O$ to give $5H_2O$. On the right, two additional OH^-'s appear. The final equation, valid for basic solution, is

$$5H_2O + 2CrO_4^{2-} + 3SO_3^{2-} \longrightarrow 2Cr(OH)_4^- + 3SO_4^{2-} + 2OH^-$$

The net charge checks out: On the left, $(5)(0) + (2)(-2) + (3)(-2) = -10$; on the right, $(2)(-1) + (3)(-2) + (2)(-1) = -10$. ∎

5.4 Balancing equations by use of half-reactions

As indicated at the end of Section 5.2, there is frequently ambiguity in the choice of oxidation number to be assigned. It is possible to get around the whole concept, which is an artificial one, anyhow. This is what the half-reaction method does. It separates the oxidizing and reducing agents from each other, shows each one going to its respective product, and writes a balanced half-reaction for each one separately. The only unorthodox feature is that electrons appear as chemical species in each of the half-reactions. Normally, electrons do not exist separately as a species in solutions, so in this sense we are still using an artificiality. However, it does allow us to monitor charge changes in oxidation–reduction reactions, and it does so without making any use of oxidation numbers.

The procedure of half-reactions is particularly useful for solutions but it may be used also for reactions in the gas phase and in the solid state. In such cases, the interpretation of the half-reaction as a real change may not be tenable, but still we can use it to balance equations and that, after all, is our purpose at this point.

The steps to follow are these:

1. Separate the oxidizing and the reducing agent.
2. Show the oxidizing agent going to its reduced form.
3. Show the reducing agent going to its oxidized form.
4. Make sure the atoms other than H and O are balanced on the two sides of each half-reaction. If necessary, adjust coefficients.
5. Count up the number of oxygen atoms on the left and right of each half-reaction, and add H_2O to the side deficient in oxygen.
6. Count up the number of hydrogen atoms on the left and right and add H^+ to the side deficient in hydrogen. (It is important that step 5 precede step 6.)
7. Count up the net charge on left and right and add electrons (e^-) to the side deficient in negative charge.
8. If the reaction is specified to be in *acidic solution*, replace each H^+ by H_3O^+ and add an equal number of H_2O molecules to the other side of the half-reaction. Cancel any H_2O duplications that appear on both sides of the half-reactions.
9. If the reaction is specified to be in *basic solution*, replace each H^+ by H_2O and add an equal number of OH^- ions to the other side of the half-reaction. Cancel H_2O duplication.

At this stage you should have two half-reactions, each of which is completely balanced so far as atoms and charges are concerned, the only odd point being that e^- appears on the right of one half-reaction and on the left of the other.

10. To get a balanced equation, multiply each half-reaction all the way through by an appropriate number so that when the two half-reactions are added up, the electrons can be canceled out.
11. Add up the half-reactions and cancel any duplication of species on the left and right sides.

■ EXAMPLE 157.   Balance the equation $?C_3H_8 + ?O_2 \rightarrow ?CO + ?H_2O$ by the method of half-reactions.

SOLUTION: One half-reaction involves the change $C_3H_8 \rightarrow CO$. To write a balanced half-reaction for this change we note that three carbon atoms on the left require three carbon atoms on the right, so we put a 3 before the CO. This gives us

$$C_3H_8 \longrightarrow 3CO$$

Counting up oxygen atoms, we find none on the left and three on the right, so the left side is deficient by three oxygen atoms. We can fix it up by putting $3H_2O$ on the left side. This gives us

$$C_3H_8 + 3H_2O \longrightarrow 3CO$$

Counting up hydrogen atoms, we find $8 + (3)(2) = 14$ on the left and none on the right. We, therefore, add $14H^+$ to the right side, giving

$$C_3H_8 + 3H_2O \longrightarrow 3CO + 14H^+$$

All the atoms are now balanced (so, mass is conserved) and all we have to worry about is the charge. If we add up the charge, we find one neutral C_3H_8 plus three neutral H_2O on the left, which adds up to zero charge, and three neutral CO plus 14 monopositive H^+ on the right, which adds up to $+14$. To make sure that electrical charge is preserved when the reaction goes from left to right, we need to have the same net charge on both sides of the half-reaction. We achieve this by adding $14e^-$ to the right side. The final half-reaction looks like this:

$$C_3H_8 + 3H_2O \longrightarrow 3CO + 14H^+ + 14e^-$$

The other half-reaction involves the change $O_2 \rightarrow H_2O$. To get a balanced half-reaction for this, we note first that there are two oxygen atoms on the left and one on the right. Therefore, we need to multiply the H_2O by two, giving us

$$O_2 \longrightarrow 2H_2O$$

Counting up oxygen atoms shows balance in this respect. Counting up hydrogen atoms shows a deficiency of four on the left, so we add $4H^+$ to the left side of the half-reaction. This gives us

$$O_2 + 4H^+ \longrightarrow 2H_2O$$

To balance charge, we note that the left side has $0 + (4)(+1) = +4$, and the right side, zero. So, we add $4e^-$ to the left, getting

$$O_2 + 4H^+ + 4e^- \longrightarrow 2H_2O$$

We now have two half-reactions, both balanced as to atoms and charge conservation. One is for the reducing agent (C_3H_8) and the other is for the oxidizing agent (O_2):

reducing agent: $C_3H_8 + 3H_2O \longrightarrow 3CO + 14H^+ + 14e^-$

oxidizing agent: $O_2 + 4H^+ + 4e^- \longrightarrow 2H_2O$

If these two half-reactions are to be part of the same overall reaction, the electron production of the first one has to be matched by the electron consumption of the second one. This will be the case if seven of the second occur for every two of the first. So, we multiply the first all the way through by two and the second by seven. The result is

reducing agent: $2C_3H_8 + 6H_2O \longrightarrow 6CO + 28H^+ + 28e^-$

oxidizing agent: $7O_2 + 28H^+ + 28e^- \longrightarrow 14H_2O$

Now we can add these, keeping all terms to the left or right of the arrow as they occur. The sum is

$$2C_3H_8 + 6H_2O + 7O_2 + 28H^+ + 28e$$
$$\longrightarrow 6CO + 28H^+ + 28e^- + 14H_2O$$

Canceling out duplications right and left, we strike out the 28 e's and also the 28 H^+'s from each side. Furthermore, there are $6H_2O$ on the left and $14H_2O$ on the right. Six of these are duplicated and can be deducted from each side, leaving only $8H_2O$ on the right. The final result is

$$2C_3H_8 + 7O_2 \longrightarrow 6CO + 8H_2O$$

which is a completely balanced equation, identical to that obtained on page 96. Note that it has been obtained here without introducing oxidation numbers. ■

■ EXAMPLE 158.   Given the change $Cr_2O_7^{2-} + H_2SO_3 \rightarrow Cr^{3+} + HSO_4^-$ occurring in acidic solution, complete and balance the equation for the reaction, using the half-reaction method.

SOLUTION: In one half-reaction, the change is $Cr_2O_7^{2-} \rightarrow Cr^{3+}$. Two chromium atoms on the left require a 2 before Cr^{3+}.

$$Cr_2O_7^{2-} \longrightarrow 2Cr^{3+}$$

There are seven oxygen atoms on the left, none on the right, so we add $7H_2O$ to the right, getting

$$Cr_2O_7^{2-} \longrightarrow 2Cr^{3+} + 7H_2O$$

There are 14 H atoms on the right, none on the left, so we add $14H^+$ to the left, leading to

$$Cr_2O_7^{2-} + 14H^+ \longrightarrow 2Cr^{3+} + 7H_2O$$

As to charge, the left side has one dinegative species plus 14 monopositive ones, corresponding to $-2 + 14 = +12$; the right side, two tripositive plus seven neutral $= (2)(+3) = +6$. The left side needs $6e^-$. We have

$$Cr_2O_7^{2-} + 14H^+ + 6e^- \longrightarrow 2Cr^{3+} + 7H_2O$$

Finally, because the reaction is specified to be in acidic solution, we change the $14H^+$ on the left to $14H_3O^+$ and add 14 more H_2O to the right. The final result is

$$Cr_2O_7^{2-} + 14H_3O^+ + 6e^- \longrightarrow 2Cr^{3+} + 21H_2O$$

In the other half-reaction, the change is $H_2SO_3 \rightarrow HSO_4^-$. The sulfur atoms are already balanced. As to oxygen atoms, there are three on the left and four on the right, so we make up the deficit by adding $1H_2O$ to the left. This gives us

$$H_2SO_3 + H_2O \longrightarrow HSO_4^-$$

As to hydrogen atoms, there are now four on the left and one on the right. To redress the balance, we add $3H^+$ to the right, getting

$$H_2SO_3 + H_2O \longrightarrow HSO_4^- + 3H^+$$

For charge, the left side is neutral but the right side has a net charge of $(-1) + (+3) = +2$. We fix it up by adding $2e^-$ to the right side.

$$H_2SO_3 + H_2O \longrightarrow HSO_4^- + 3H^+ + 2e^-$$

Finally, because the reaction is specified to be in acidic solution, we change the $3H^+$ on the right to $3H_3O^+$ and add three more H_2O to the left. The final result is

$$H_2SO_3 + 4H_2O \longrightarrow HSO_4^- + 3H_3O^+ + 2e^-$$

The $Cr_2O_7^{2-}$ half-reaction uses up $6e^-$, the H_2SO_3 half-reaction produces $2e^-$. Therefore, we must take three of the latter for every one of the former. Multiplying the $Cr_2O_7^{2-}$ half-reaction by one and the H_2SO_3 half-reaction by three gives us

$$Cr_2O_7^{2-} + 14H_3O^+ + 6e^- \longrightarrow 2Cr^{3+} + 21H_2O$$
$$3H_2SO_3 + 12H_2O \longrightarrow 3HSO_4^- + 9H_3O^+ + 6e^-$$

$$Cr_2O_7^{2-} + 14H_3O^+ + 6e^- + 3H_2SO_3 + 12H_2O$$
$$\longrightarrow 2Cr^{3+} + 21H_2O + 3HSO_4^- + 9H_3O^+ + 6e^-$$

Duplications are $6e^-$, $12H_2O$, and $9H_3O^+$. If we strike these out from both sides, we are left with

$$Cr_2O_7^{2-} + 5H_3O^+ + 3H_2SO_3 \longrightarrow 2Cr^{3+} + 9H_2O + 3HSO_4^-$$

which agrees with what we got by the oxidation-number method on page 98. ■

■ EXAMPLE 159.   Using half-reactions, deduce the balance equation for the change $CrO_4^{2-} + SO_3^{2-} \rightarrow Cr(OH)_4^- + SO_4^{2-}$ in basic solution.

SOLUTION: In one half-reaction, the change is $CrO_4^{2-} \rightarrow Cr(OH)_4^-$.
The chromium atoms are balanced.
The oxygen atoms are balanced.
As to hydrogen, there are none on the left but four on the right, so we stick $4H^+$ on the left, getting

$$4H^+ + CrO_4^{2-} \longrightarrow Cr(OH)_4^-$$

As to charge, the left side is $(4)(+1) + (-2) = +2$ but the right side is -1. So we need to add $3e^-$ to the left. We get

$$3e^- + 4H^+ + CrO_4^{2-} \longrightarrow Cr(OH)_4^-$$

This is a balanced half-reaction but is no good for basic solution. We need to get rid of the $4H^+$. We can do this by replacing the $4H^+$ on the left by $4H_2O$ and adding $4OH^-$ to the right. The result is

$$3e^- + 4H_2O + CrO_4^{2-} \longrightarrow Cr(OH)_4^- + 4OH^-$$

which is not only balanced but also is valid for basic solution.

In the other half-reaction, the change is $SO_3^{2-} \to SO_4^{2-}$.
The sulfur atoms are balanced.

As to oxygen atoms, the left side is deficient by one oxygen, so we need to add $1H_2O$ to the left. The result is

$$H_2O + SO_3^{2-} \longrightarrow SO_4^{2-}$$

As to hydrogen atoms, the right side is now deficient by two, so we stick $2H^+$ on the right, getting

$$H_2O + SO_3^{2-} \longrightarrow SO_4^{2-} + 2H^+$$

For balancing the charge, we note that the left side has one neutral plus one dinegative, or -2, charge. The right side has one dinegative plus two monopositive, or 0, charge. To square away -2 with 0, we need to add $2e^-$ to the right. The result

$$H_2O + SO_3^{2-} \longrightarrow SO_4^{2-} + 2H^+ + 2e^-$$

is, however, not valid for basic solution. We get rid of the $2H^+$ by replacing them with $2H_2O$ and adding $2OH^-$ to the opposite side of the half-reaction. The final balanced half-reaction for basic solution is $2OH^- + SO_3^{2-} \to SO_4^{2-} + H_2O + 2e^-$.

To combine the two half-reactions

oxidizing agent: $3e^- + 4H_2O + CrO_4^{2-} \longrightarrow Cr(OH)_4^- + 4OH^-$

reducing agent: $2OH^- + SO_3^{2-} \longrightarrow SO_4^{2-} + H_2O + 2e^-$

we need to multiply the first by two and the second by three. This gives

$$6e^- + 8H_2O + 2CrO_4^{2-} \longrightarrow 2Cr(OH)_4^- + 8OH^-$$

$$6OH^- + 3SO_3^{2-} \longrightarrow 3SO_4^{2-} + 3H_2O + 6e^-$$

which adds up to

$$6e^- + 8H_2O + 2CrO_4^{2-} + 6OH^- + 3SO_3^{2-}$$
$$\longrightarrow 2Cr(OH)_4^- + 8OH^- + 3SO_4^{2-} + 3H_2O + 6e^-$$

Cancellation of $6e^-$, $3H_2O$, and $6OH^-$ from both sides leaves

$$5H_2O + 2CrO_4^{2-} + 3SO_3^{2-} \longrightarrow 2Cr(OH)_4^- + 2OH^- + 3SO_4^{2-}$$

the same as obtained on page 99 by the oxidation-number method. ∎

The great advantage of the half-reaction method really shows up in those cases where oxidation numbers are difficult to assign (e.g., SCN^-) or where the changes in oxidation number are apt to be miscounted (e.g., HO_2^- going to H_2O). If there are no false moves, the oxidation-number method of balancing equations is likely to be faster. However, the half-reaction method is less likely to lead to errors. In the old days, people used to object to the half-reaction method on the grounds that a half-reaction by itself cannot occur and therefore has no physical reality. This kind of objection is fast dying out under the impact of new electrochemical devices such as the fuel cell, in which an overall reaction of the type $C_3H_8 + 5O_2 \rightarrow 3CO_2 + 4H_2O$ is made to occur so one half-reaction occurs at one electrode and the other half-reaction at the other electrode.

5.5   Calculations involving equations

Once a balanced chemical equation has been written, it can be used as a summary not only of the net atomic change in the course of reaction but also (after atomic weight information has been put in) of the weight changes that occur when reactants are converted to products. This comes about because the coefficients in the balanced equation tell not only how many formula-units (atoms, molecules, or formula-units) are involved but also how many moles of each come into play. Thus, for the balanced equation

$$2C_3H_8 + 7O_2 \longrightarrow 6CO + 8H_2O$$

the coefficients 2, 7, 6, and 8 tell us (a) that *two molecules* of C_3H_8 plus *seven molecules* of O_2 react to give *six molecules* of CO plus *eight molecules* of H_2O or (b) that *two moles* of C_3H_8 plus *seven moles* of O_2 react to give *six moles* of CO plus *eight moles* of H_2O. The second part of the statement comes from the fact that once the coefficients in the equation are established, they fix once and for all the relative number of particles involved. If we multiply through by any number, the coefficients still have relative validity. Specifically, if we multiply through by the Avogadro number, we convert the information of the chemical equation from individual particles to large groups such as are of the order of magnitude encountered in typical samples in a laboratory. Because our main emphasis is on weight calculations involving typical samples, we shall concentrate on the molar aspects of chemical equations.

The two points that will be of use to us are these: (1) the coefficients in the balanced equation tell us the number of moles of each reactant and each product when the reaction takes place as written, and (2) we need to consider only those reactants and products asked about in the calculation, assuming that the other reactants are available in needed amount and other products may also form. Thus, given

$$2C_3H_8 + 7O_2 \longrightarrow 6CO + 8H_2O$$

we could calculate how much O_2 is needed for a given amount of C_3H_8, using the $7 : 2$ molar ratio only, ignoring the rest of the equation. Similarly, we could calculate how much CO comes from a given amount of C_3H_8 using only the $6 : 2$ molar ratio. Stated another way, you do not have to use *all* the information implicit in the equation to solve a particular problem.

Before we undertake typical problems, we might summarize the way information is stored in a specific equation.

Given

$$2C_3H_8 + 7O_2 \longrightarrow 6CO + 8H_2O$$

This means

$$\begin{pmatrix} 2 \text{ molecules} \\ C_3H_8 \end{pmatrix} + \begin{pmatrix} 7 \text{ molecules} \\ O_2 \end{pmatrix} \begin{matrix} \text{react} \\ \text{to give} \end{matrix} \begin{pmatrix} 6 \text{ molecules} \\ CO \end{pmatrix} + \begin{pmatrix} 8 \text{ molecules} \\ H_2O \end{pmatrix}$$

It also means

$$\begin{pmatrix} 2 \text{ moles} \\ C_3H_8 \end{pmatrix} + \begin{pmatrix} 7 \text{ moles} \\ O_2 \end{pmatrix} \begin{matrix} \text{react} \\ \text{to give} \end{matrix} \begin{pmatrix} 6 \text{ moles} \\ CO \end{pmatrix} + \begin{pmatrix} 8 \text{ moles} \\ H_2O \end{pmatrix}$$

Furthermore

$$2 \text{ moles of } C_3H_8 = 2 \times \text{gram-formula-weight of } C_3H_8$$
$$= 2 \times (44.097 \text{ g}) = 88.194 \text{ g}$$

$$7 \text{ moles of } O_2 = 7 \times \text{gram-formula-weight of } O_2$$
$$= 7 \times (31.999 \text{ g}) = 223.99 \text{ g}$$

$$6 \text{ moles of CO} = 6 \times \text{gram-formula-weight of CO}$$
$$= 6 \times (28.010 \text{ g}) = 168.06 \text{ g}$$

$$8 \text{ moles of } H_2O = 8 \times (18.015 \text{ g}) = 144.12 \text{ g}$$

Putting this information into the foregoing statement in moles, we can say

$$\begin{pmatrix} 88.194 \text{ g} \\ C_3H_8 \end{pmatrix} + \begin{pmatrix} 223.99 \text{ g} \\ O_2 \end{pmatrix} \begin{matrix} \text{react} \\ \text{to give} \end{matrix} \begin{pmatrix} 168.06 \text{ g} \\ CO \end{pmatrix} + \begin{pmatrix} 144.12 \text{ g} \\ H_2O \end{pmatrix}$$

Obviously, the latter numbers are harder to keep in your head and to manipulate. Because the molar numbers are easier to work with, most chemists carry out their calculations on a molar basis, even if it means converting data given in grams to number of moles first. For the novice, it is extremely important to get into the habit of thinking in terms of moles and setting up calculations in these terms.

■ EXAMPLE 160.   How many grams of O_2 are required to oxidize 1.00 g of C_3H_8 to CO and H_2O?

SOLUTION: (Figure out how many moles of C_3H_8 you are given. Then from the equation decide how many models of O_2 are required. Finally, compute how many grams of O_2 this amounts to.)
One mole of C_3H_8 weighs 44.097 g.

$$1.00 \text{ g of } C_3H_8 = \frac{1.00 \text{ g}}{44.097 \text{ g/mole}} = 0.0227 \text{ mole of } C_3H_8$$

From the equation

$$2C_3H_8 + 7O_2 \longrightarrow 6CO + 8H_2O$$

we note that 7 moles of O_2 are required per 2 moles of C_3H_8. Therefore, for 0.0227 mole of C_3H_8 we need

$$(0.0227 \text{ mole of } C_3H_8) \left(\frac{7 \text{ moles of } O_2}{2 \text{ moles of } C_3H_8} \right) = 0.0794 \text{ mole of } O_2 .$$

One mole of $O_2 = 32.00$ g.

$$0.0794 \text{ mole of } O_2 = (0.0794 \text{ mole}) \left(\frac{32.00 \text{ g}}{\text{mole}} \right) = 2.54 \text{ g of } O_2 \text{ ■}$$

■ EXAMPLE 161.   How many grams of CO are produced when 3.42 g of C_3H_8 is oxidized by O_2 to CO and H_2O?

SOLUTION: (Figure out how many moles of C_3H_8 you have. Then from the equation decide how many moles of CO are produced. Finally, calculate how many grams of CO this is.)
One mole of $C_3H_8 = 44.097$ g.

$$3.42 \text{ g of } C_3H_8 = \frac{3.42 \text{ g}}{44.097 \text{ g/mole}} = 0.0776 \text{ mole of } C_3H_8$$

From the equation

$$2C_3H_8 + 7O_2 \longrightarrow 6CO + 8H_2O$$

we note that for each 2 moles of C_3H_8 used up, 6 moles of CO are produced. Therefore, if 0.0776 mole of C_3H_8 is used up, there will be produced

$$(0.0776 \text{ mole of } C_3H_8) \left(\frac{6 \text{ moles of CO}}{2 \text{ moles of } C_3H_8} \right) = 0.233 \text{ mole of CO}$$

One mole of CO weighs 28.010 g.

$$0.233 \text{ mole of CO} = (0.233 \text{ mole}) \left(28.010 \frac{g}{\text{mole}} \right) = 6.53 \text{ g of CO} \blacksquare$$

■ EXAMPLE 162.   When C_3H_8 is oxidized by O_2 to CO and H_2O, how many grams of H_2O will be produced at the same time as 3.43 g of CO?

SOLUTION: (Figure out how many moles of CO are produced. From the equation decide how many moles of H_2O form at the same time. Convert to grams.)
One mole of CO = 28.010 g.

$$3.43 \text{ g of CO} = \frac{3.43 \text{ g}}{28.010 \text{ g/mole}} = 0.122 \text{ mole of CO}$$

From the equation

$$2C_3H_8 + 7O_2 \longrightarrow 6CO + 8H_2O$$

we note that 8 moles of H_2O are produced every time 6 moles of CO form. Therefore, when 0.122 mole of CO is produced, we will get simultaneously

$$(0.122 \text{ mole of CO}) \left(\frac{8 \text{ moles of } H_2O}{6 \text{ moles of CO}} \right) = 0.163 \text{ mole of } H_2O$$

One mole of H_2O = 18.015 g.

$$0.163 \text{ mole of } H_2O = (0.163 \text{ mole}) \left(18.015 \frac{g}{\text{mole}} \right) = 2.94 \text{ g of } H_2O \blacksquare$$

■ EXAMPLE 163.   Suppose that 2.00 g of C_3H_8 and 7.00 g of O_2 are allowed to react to the maximum possible extent to form CO and H_2O. How many grams of CO will be formed?

SOLUTION: (This is an excess problem where one of the reagents limits the extent of the reaction, the other one being present in excess. To solve it, convert the data given into moles. Then compare the given number of moles with that required in the equation. Decide which reagent is present in excess. Then work with the limiting one to determine how many moles of CO can be formed. Convert to grams.)

$$2.00 \text{ g of } C_3H_8 = \frac{2.00 \text{ g}}{44.097 \text{ g/mole}} = 0.0454 \text{ mole of } C_3H_8$$

$$7.00 \text{ g of } O_2 = \frac{7.00 \text{ g}}{32.00 \text{ g/mole}} = 0.219 \text{ mole of } O_2$$

The equation

$$2C_3H_8 + 7O_2 \longrightarrow 6CO + 8H_2O$$

says that 7 moles of O_2 are needed for 2 moles of C_3H_8. For 0.0454 mole of C_3H_8 we would then need

$$(0.0454 \text{ mole of } C_3H_8) \left(\frac{7 \text{ moles of } O_2}{2 \text{ moles of } C_3H_8} \right) = 0.159 \text{ mole of } O_2$$

This is the minimum amount of O_2 needed. We actually are given 0.219 mole of O_2, which is considerably in excess. Therefore, we conclude O_2 is more than sufficient. The amount of product is limited by the 0.0454 mole of C_3H_8.

The equation

$$2C_3H_8 + 7O_2 \longrightarrow 6CO + 8H_2O$$

tells us we get 6 moles of CO for each 2 moles of C_3H_8 used. Therefore, if we use up only 0.0454 mole of C_3H_8, we would get

$$(0.0454 \text{ mole of } C_3H_8) \left(\frac{6 \text{ moles of CO}}{2 \text{ moles of } C_3H_8} \right) = 0.136 \text{ mole of CO}$$

One mole of CO = 28.010 g.

$$0.136 \text{ mole of CO} = (0.136 \text{ mole}) \left(28.010 \frac{\text{g}}{\text{mole}} \right) = 3.81 \text{ g of CO} \blacksquare$$

■ EXAMPLE 164. Suppose that 2.00 g of C_3H_8 and 7.00 g of O_2 are allowed to react to the maximum possible extent to form CO_2 and H_2O. How many grams of CO_2 will be formed?

SOLUTION: (The method is similar to that used in Example 163 except we must not overlook the important fact that one of the products is different. The pertinent equation in this problem is different from the one used in Example 163.)

$$2.00 \text{ g of } C_3H_8 = \frac{2.00 \text{ g}}{44.097 \text{ g/mole}} = 0.0454 \text{ mole of } C_3H_8$$

$$7.00 \text{ g of } O_2 = \frac{7.00 \text{ g}}{32.00 \text{ g/mole}} = 0.219 \text{ mole of } O_2$$

The equation for the reaction is

$$C_3H_8 + 5O_2 \longrightarrow 3CO_2 + 4H_2O$$

It tells us we need 5 moles of O_2 per mole of C_3H_8. For 0.0454 mole of C_3H_8 we would have to use up

$$(0.0454 \text{ mole of } C_3H_8)\left(\frac{5 \text{ moles of } O_2}{1 \text{ mole of } C_3H_8}\right) = 0.227 \text{ mole of } O_2$$

But we do not have this much O_2 available; we are given only 0.219 mole o O_2. In other words, we do not have enough O_2 to chew up all the C_3H_8, so it must be the C_3H_8 that is now in excess. Any computations based on C_3H_8 would be incorrect, since the 0.219 mole of O_2 is the limiting reagent.

Once we decide which reagent is limiting, the rest is easy. The equation

$$C_3H_8 + 5O_2 \longrightarrow 3CO_2 + 4H_2O$$

says that for every 5 moles of O_2 used up, 3 moles of CO_2 form. If we use up only 0.219 mole of O_2, we get

$$(0.219 \text{ mole of } O_2)\left(\frac{3 \text{ moles of } CO_2}{5 \text{ moles of } O_2}\right) = 0.131 \text{ mole of } CO_2$$

One mole of $CO_2 = 44.010$ g.

$$0.131 \text{ mole of } CO_2 = (0.131 \text{ mole})\left(\frac{44.010 \text{ g}}{\text{mole}}\right) = 5.77 \text{ g of } CO_2 \ ■$$

▪ EXAMPLE 165. Suppose we mix 1.00×10^{-3} mole of Ag^+ and 1.00×10^{-3} mole of CrO_4^{2-}. Suppose, further, they react as much as possible to precipitate Ag_2CrO_4. How many grams of Ag_2CrO_4 could be formed?

SOLUTION: (Write the equation for the reaction. Figure out which reagent is limiting the product. Calculate moles of product consistent with limiting reagent. Convert to grams.)
The equation

$$2Ag^+ + CrO_4^{2-} \longrightarrow Ag_2CrO_4$$

says that we need to use twice as many moles of Ag^+ as of CrO_4^{2-}. If we took 1.00×10^{-3} mole of CrO_4^{2-} to be limiting, we would need 2.00×10^{-3} mole of Ag^+. But we do not have that much, so that approach is wrong. It must be that the 1.00×10^{-3} mole of Ag^+ is limiting. It would require only 0.500×10^{-3} mole of CrO_4^{2-}, which is easily there in excess. The equation says we get 1 mole of Ag_2CrO_4 for each 2 moles of Ag^+ used up. If we use up all the 1.00×10^{-3} mole of Ag^+ provided, we would get

$$(1.00 \times 10^{-3} \text{ mole of } Ag^+)\left(\frac{1 \text{ mole of } Ag_2CrO_4}{2 \text{ moles of } Ag^+}\right)$$

$$= 0.500 \times 10^{-3} \text{ mole of } Ag_2CrO_4.$$

One mole of $Ag_2CrO_4 = 331.73$ g.

$$0.500 \times 10^{-3} \text{ mole of } Ag_2CrO_4 = (0.500 \times 10^{-3} \text{ mole})\left(331.73 \frac{g}{mole}\right)$$

$$= 0.166 \text{ g of } Ag_2CrO_4 \text{ ▪}$$

▪ EXAMPLE 166. Suppose that 1.00 g of $Cr_2O_7^{2-}$ is oxidized in acidic solution by excess SO_2 to form HSO_4^- and Cr^{3+}. What is the minimum number of moles of H_3O^+ that must be provided?

SOLUTION: (Write the balanced equation for the reaction. Calculate how many moles of $Cr_2O_7^{2-}$ you will use up. From the equation, decide how many moles of H_3O^+ must be provided.)
For the change

$$SO_2 + Cr_2O_7^{2-} \longrightarrow HSO_4^- + Cr^{3+}$$

the balanced equation can be figured out by the methods discussed earlier
in this chapter to be

$$5H_3O^+ + 3SO_2 + Cr_2O_7^{2-} \longrightarrow 3HSO_4^- + 2Cr^{+3} + 6H_2O$$

We are given 1.00 g of $Cr_2O_7^{2-}$.
One mole of $Cr_2O_7^{2-}$ weighs 215.988 g.

$$1.00 \text{ g of } Cr_2O_7^{2-} = \frac{1.00 \text{ g}}{215.988 \text{ g/mole}} = 0.00463 \text{ mole of } Cr_2O_7^{2-}$$

The equation tells us we use 5 moles of H_3O^+ for each mole of $Cr_2O_7^{2-}$.
Therefore, if we use up 4.63×10^{-3} mole of $Cr_2O_7^{2-}$, we will need to use up
simultaneously

$$(0.00463 \text{ mole of } Cr_2O_7^{2-})\left(\frac{5 \text{ moles of } H_3O^+}{1 \text{ mole of } Cr_2O_7^-}\right) = 0.0232 \text{ mole of } H_3O^+ \quad \blacksquare$$

■ EXAMPLE 167.   Suppose that H_2O_2 is to be oxidized by MnO_4^- in
acidic solution to form O_2 and Mn^{2+}. How many grams of O_2 could you get
from 1.50×10^{-3} mole of MnO_4^- and 1.50×10^{-3} g of H_2O_2?

SOLUTION: (Convert the data "grams of H_2O_2" into moles. Compare
with the balanced equation to see which reagent is in excess. Compute moles
of O_2 determined by the limiting reagent. Convert to grams.)

One mole of $H_2O_2 = 34.014$ g.

$$1.50 \times 10^{-3} \text{ g of } H_2O_2 = \frac{1.50 \times 10^{-3} \text{ g}}{34.014 \text{ g/mole}} = 4.41 \times 10^{-5} \text{ mole of } H_2O_2$$

For the change $H_2O_2 + MnO_4^- \rightarrow Mn^{2+} + O_2$ in acidic solution, the
balanced equation comes out to be

$$6H_3O^+ + 5H_2O_2 + 2MnO_4^- \longrightarrow 2Mn^{2+} + 14H_2O + 5O_2$$

It tells us that 5 moles of H_2O_2 require 2 moles of MnO_4^- in the reaction.
For 4.41×10^{-5} mole of H_2O_2, we would need

$$(4.41 \times 10^{-5} \text{ mole of } H_2O_2)\left(\frac{2 \text{ moles of } MnO_4^-}{5 \text{ moles of } H_2O_2}\right)$$

$$= 1.76 \times 10^{-5} \text{ mole of } MnO_4^-$$

We are given 1.50×10^{-3} mole of MnO_4^-, so we have plenty of MnO_4^- in excess of the 1.76×10^{-5} mole required. The product O_2 is determined by the limiting reagent, H_2O_2.

The equation tells us 5 moles of H_2O_2 lead to 5 moles of O_2. If we use up the 4.41×10^{-5} mole of H_2O_2 that is available, then we should get

$$(4.41 \times 10^{-5} \text{ mole of } H_2O_2)\left(\frac{5 \text{ moles of } O_2}{5 \text{ moles of } H_2O_2}\right)$$

$$= 4.41 \times 10^{-5} \text{ mole of } O_2$$

One mole of $O_2 = 31.999$ g.

$$4.41 \times 10^{-5} \text{ mole of } O_2 = (4.41 \times 10^{-5} \text{ mole})\left(31.999 \frac{g}{\text{mole}}\right)$$

$$= 1.41 \times 10^{-3} \text{ g of } O_2 \ \blacksquare$$

■ EXAMPLE 168. When a basic solution of I^- is exposed to air, the I^- is gradually converted to IO_3^- by the oxidizing action of O_2 going to OH^-. Given a solution that initially contains 1.50×10^{-3} mole of I^-, what fraction of the I^- could be converted to IO_3^- by using 1.50×10^{-3} mole of O_2?

SOLUTION: (Write the balanced equation for the reaction. From it, decide how many moles of I^- could be chewed up by 1.50×10^{-3} mole of O_2. Figure out what fraction this is of the I^- made available initially.)

For the change $I^- + O_2 \rightarrow IO_3^- + OH^-$ in basic solution, the net equation comes out to be

$$2I^- + 3O_2 \longrightarrow 2IO_3^-$$

This tells us we need 3 moles of O_2 per 2 moles of I^-. If we use 1.50×10^{-3} mole of O_2, we would need to use up

$$(1.50 \times 10^{-3} \text{ mole of } O_2)\left(\frac{2 \text{ moles of } I^-}{3 \text{ moles of } O_2}\right) = 1.00 \times 10^{-3} \text{ mole of } I^-$$

We are given 1.50×10^{-3} mole of I^-. Therefore, the fraction of I^- used up is

$$\frac{1.00 \times 10^{-3} \text{ mole of } I^- \text{ used}}{1.50 \times 10^{-3} \text{ mole of } I^- \text{ available}} = 0.667 \ \blacksquare$$

Exercises

*5.1. What is the oxidation number of Fe in each of the following compounds: FeO; Fe_2O_3; $BaFeO_4$; Fe_3O_4?

*5.2. What is the oxidation number of X in each of the following: X_2; X_2O; XO_2; HXO_2; HXO_3; $NaXO_4$? (*Answer:* 0; +1; +4; +3; +5; +7.)

*5.3. What is the oxidation number of the underlined atom in each of the following: $H_2\underline{S}$; $\underline{Al}H_3$ (a hydride); $\underline{K}_2O_2$ (a peroxide); $\underline{Ba}O_2$ (a peroxide); $\underline{Cu}_2O$; $Na_3\underline{P}O_4$; $Li\underline{Al}H_4$ (hydride); $H_2\underline{P}O_2$?

*5.4. What is the oxidation number of nitrogen in each of the following: NH_4^+; NH_3; NO_2^-; $N_2H_5^+$; NO_3^-; $N_2O_2^{2-}$; NH_3OH^+? (*Answer:* −3; −3; +3; −2; +5; +1; −1.)

*5.5 Write the formula for an oxide of X in which X has the following oxidation numbers: +1; +2; +3; +4; +5; +3/4.

*5.6. Balance the following, using the oxidation-number method: $?KNO_3 + ?C \rightarrow ?CO_2 + ?KNO_2$. (*Answer:* $2KNO_3 + C \rightarrow CO_2 + 2KNO_2$.)

*5.7. Balance the following, using the oxidation-number method: $?K_2CrO_4 + ?S \rightarrow ?Cr_2O_3 + ?K_2SO_4 + ?K_2O$.

*5.8. What is the value of n in the following balanced equation: $4nClO_3 + 5Br_2 \rightarrow 10BrO_n + 4Cl_nO$?

*5.9. Balance the following using the oxidation-number method: $?KMnO_4 + ?C_{12}H_{22}O_{11} \rightarrow ?CO_2 + ?MnO + ?H_2O + ?K_2CO_3$.

**5.10. Balance the following using the oxidation-number method: $?HCNS + ?KClO_3 \rightarrow ?CO_2 + ?NO_2 + ?SO_2 + ?KCl + ?H_2O$.   (*Answer:* $6HCNS + 13KClO_3 \rightarrow 6CO_2 + 6NO_2 + 6SO_2 + 13KCl + 3H_2O$.)

*5.11. Given the change $Cl^- + MnO_2 \rightarrow Cl_2 + Mn^{2+}$ in acidic solution. Complete and balance the equation for the reaction using the oxidation-number method.

*5.12. Given the change $Cl_2 + Mn(OH)_2 \rightarrow Cl^- + MnO_2$ in basic solution. Complete and balance the equation for the reaction using the oxidation-number method.

*5.13. Given the change $N_2O_4 + N_2O_4 \rightarrow NO_3^- + NO$ in acidic solution. Complete and balance the equation for the reaction using the oxidation-number method.

*5.14. Given the change $N_2O_4 + N_2O_4 \rightarrow NO_3^- + NO$ in basic solution. Complete and balance the equation for the reaction using the oxidation-number method. (*Answer:* $4OH^- + 3N_2O_4 \rightarrow 4NO_3^- + 2NO + 2H_2O$.)

*5.15. Given the change H_2O_2 (peroxide) + $BrO_3^- \rightarrow Br_2 + O_2$ in acidic solution. Complete and balance the equation, using the oxidation-number method.

*5.16. Given the change HO_2^- (peroxide) $+ Br^- \rightarrow BrO_3^- + H_2O$ in basic solution. Complete and balance the equation, using the oxidation-number method.

*5.17. Write the complete balanced half-reaction for the conversion $ClO_3^- \rightarrow Cl^-$ in acidic solution.

*5.18. Write the complete balanced half-reaction for the conversion $ClO_3^- \rightarrow Cl^-$ in basic solution. (*Answer*: $6e^- + 3H_2O + ClO_3^- \rightarrow Cl^- + 6OH^-$.)

*5.19. Write the complete balanced half-reaction for the conversion $O_3 \rightarrow H_2O$ in acidic solution.

*5.20. Using the half-reaction method, complete and balance the equation for the change $BrO_3^{2-} + SO_2 \rightarrow HSO_4^- + Br^-$ in acidic solution.

*5.21. Using the half-reaction method, complete and balance the equation for the change $BrO_3^- + SO_3^{2-} \rightarrow SO_4^{2-} + Br^-$ in basic solution.

*5.22. Using the half-reaction method, complete and balance the equation for the change $Fe_3O_4 + ClO^- \rightarrow FeO_4^{2-} + Cl^-$ in basic solution. (*Answer*: $5ClO^- + 6OH^- + Fe_3O_4 \rightarrow 3FeO_4^{2-} + 3H_2O + 5Cl^-$.)

*5.23. Using the half-reaction method, complete and balance the equation for the change HO_2^- (peroxide) $+ Ni(OH)_2 \rightarrow Ni_2O_3 + Cl^-$ in basic solution.

**5.24. Using the half-reaction method, complete and balance the equation for the change $XO_2 + Y(OH)_4^- \rightarrow X_2O_3 + YO_4^{2-}$ in basic solution.

*5.25. How many grams of Al are required to reduce 2.65 g of Fe_2O_3 to form Fe and Al_2O_3?

*5.26. How many grams of NO_2 would be produced by the oxidation of 0.065 mole of NO with excess O_2? (*Answer*: 3.0 g.)

**5.27. A mixture consisting of CO and CO_2 weighs 3.69 g. On being oxidized with O_2 to get complete conversion to CO_2, it gives a final product that weighs 4.23 g. What is the molar ratio of CO to CO_2 in the original mixture?

*5.28. When hydrogen peroxide decomposes it releases oxygen gas according to the reaction $2H_2O_2 \rightarrow 2H_2O + O_2(g)$. What will be the loss in weight shown by 1 kg of a 30% by weight H_2O_2 sample that decomposes by 50%? (*Answer*: 71 g.)

*5.29. Balance the equation for the reaction of Mg_3N_2 with H_2O to form $Mg(OH)_2$ and NH_3. What is the maximum number of grams of NH_3 that can be produced from 1.00 g of Mg_3N_2 and 1.00 g of H_2O?

*5.30. The fermentation of sugar, $C_6H_{12}O_6$, produces ethyl alcohol, C_2H_5OH, and CO_2. How many tons of CO_2 would be produced as by-product accompanying the formation of one million tons of ethyl alcohol? (*Answer*: 0.96 million tons.)

**5.31. When ZnS is treated with acid, H_2S and Zn^{2+} are formed; FeS undergoes a similar reaction. Given a 1.00-g sample that consists of an unknown mixture of ZnS and FeS, what is the per cent by weight composition of the sample if treatment with acid generates 0.366 g of H_2S?

*5.32. Given the change $SO_2 + MnO_4^- \rightarrow Mn^{2+} + HSO_4^-$ in acidic solution. How many moles of SO_2 would you need to take to reduce the MnO_4^- in 2.60 g of $KMnO_4$ by the foregoing reaction? (*Answer*: 0.0411 mole.)

**5.33. Given the change $S_2O_3^{2-} + ClO^- \rightarrow Cl^- + SO_4^{2-}$ in basic solution. How many electrons would be transferred if 6.00 g of ClO^- and 7.00 g of $S_2O_3^{2-}$ were allowed to react to the fullest by the foregoing reaction?

**5.34. Addition of Ba^{2+} to SO_3^{2-} produces a precipitate of $BaSO_3$, soluble in excess nitric acid. Addition of Ba^{2+} to SO_4^{2-} produces a precipitate $BaSO_4$ that is insoluble in excess nitric acid. Suppose you are given a mixed sample of Na_2SO_3 and Na_2SO_4. Dissolving in water and treating with Ba^{2+} produces a precipitate weighing 4.687 g. Addition of nitric acid decreases the weight of precipitate to 3.026 g. How big is the original sample and what is its composition? (*Answer*: 2.80 g, 34.3% by weight of Na_2SO_3, 65.7% Na_2SO_4.)

**5.35. When $M_2S_3(s)$ is heated in air, it converts to $MO_2(s)$. If in such a reaction a 4.00-g sample of M_2S_3 shows an increase in weight of 0.13 g, what is the atomic weight of M?

***5.36. When the polysulfide ion S_n^{2-} is oxidized by BrO_3^- in basic solution, the products are SO_4^{2-} and Br^-. For a particular value of n, it is observed that the ratio of BrO_3^- to OH^- consumed is 0.548. What is that value of n?

6

THE EQUIVALENT

THE EQUIVALENT comes up in two important senses. In acid–base theory, it denotes the amount of material that furnishes or uses up 1 mole of hydrogen ions (i.e., $6.02 \times 10^{23} H^+$); in oxidation–reduction theory, it denotes the amount of material that furnishes or uses up 1 mole of electrons (i.e., $6.02 \times 10^{23} e^-$). Although most stoichiometry calculations can be performed without introducing equivalents, your work will be made so much easier and faster that you would be foolish not to take the time to master the concept. The chief advantage of working with equivalents is that you will not have to write the balanced equation for the problem, since the same information is embodied in the very definition of an equivalent. The other point that argues for equivalents is that much solution work is built around the concept, in fact, it is hard to walk into a chemistry laboratory without finding some bottles labeled in terms of equivalents.

Strictly speaking, we should use the term "gram-equivalent-weight" instead of "equivalent," but in common parlance the terms are equally acceptable, so we pick the simpler one.

6.1   Acids and bases

An equivalent of an acid is defined as the amount of material that can furnish 1 mole of H^+; an equivalent of a base is defined as the amount of material that can pick up 1 mole of H^+. Since OH^- is the usual vehicle for picking up H^+, an alternate definition of an equivalent of base would be the amount of material that can furnish 1 mole of OH^-. By the nature of the definitions, 1 equivalent of any acid can neutralize 1 equivalent of any base.

For the neutralization reaction

$$HCl + NaOH \longrightarrow NaCl + H_2O$$

118

1 mole of HCl reacts with 1 mole of NaOH. We can visualize this as coming about because the mole of H^+ (from HCl) combines with the mole of OH^- (from NaOH). So, for the neutralization part, we are really interested in the fact that 1 mole of HCl neutralizes 1 mole of NaOH by virtue of matching 1 mole of H^+ against 1 mole of OH^-. Such matching of H^+ with OH^- is assured when we require equal number of equivalents of acid and base.

In the reaction

$$H_2SO_4 + 2NaOH \longrightarrow Na_2SO_4 + 2H_2O$$

2 moles of NaOH are required per mole of H_2SO_4, because each mole of H_2SO_4 furnishes 2 moles of H^+. Two moles of NaOH is the same as 2 equivalents of base, so we need 2 equivalents of acid to neutralize. These 2 equivalents of acid are furnished in the 1 mole of H_2SO_4.

In the reaction

$$H_3PO_4 + 3NaOH \longrightarrow Na_3PO_4 + 3H_2O$$

1 mole of H_3PO_4 is neutralized by 3 moles of NaOH. Three moles of NaOH is the same as 3 equivalents of base, so we need 3 equivalents of acid to neutralize it. These 3 equivalents of acid are furnished in the 1 mole of H_3PO_4.

The only real trick in figuring out equivalents of acids and bases is to know how many H^+ or OH^- can be furnished by a given molecule. In the vast majority of cases, this can be deduced from the chemical formula: HCl gives one H^+, H_2SO_4 gives $2H^+$, H_3PO_4 gives $3H^+$. In these cases, we assume that all the H^+ available is actually used. However, this need not always be the case. For example, in the reaction

$$H_2SO_4 + NaOH \longrightarrow NaHSO_4 + H_2O$$

only one of the H's of the H_2SO_4 is used up. In terms of equivalents *for this particular reaction*, the H_2SO_4 is worth only half as much as if both H's were used. If only one H of H_2SO_4 is neutralized, we have to say that 1 mole of H_2SO_4 furnishes 1 mole of H^+—that is, 1 mole of H_2SO_4 is 1 equivalent. If both H's of H_2SO_4 are neutralized, we have to say that 1 mole of H_2SO_4 furnishes 2 moles of H^+—that is, 1 mole of H_2SO_4 is two equivalents. In most cases, we shall use up all the H^+ available, and we shall call this "full reaction" or "full neutralization." Otherwise, we shall specify how many H^+ are used. In case of doubt, it is safest to write the equation for the reaction in which the reagent is being used.

The other pitfall is that sometimes, even for full neutralization, the chemical formula may be misleading. As an example, 1 mole of H_3PO_2

looks as if it could furnish 3 moles of H^+. Yet, if you try the experiment, you will find that only 1 mole of H^+ can be pulled out. Two of the three hydrogens are bound differently and cannot be neutralized by conventional acid–base reaction. This can be indicated, as is sometimes done, by writing the formula as $H(H_2PO_2)$ instead of H_3PO_2. Fortunately, most of the common acids and bases do not have this complication and we can deduce their equivalent relations, at least for full reaction, directly from the chemical formula.

■ EXAMPLE 169.   How many grams of HCl would there be in 1 equivalent of HCl?

SOLUTION: By definition, 1 equivalent furnishes 1 mole of H^+. To get 1 mole of H^+ from HCl, we need to take 1 mole of HCl.
One mole of HCl weighs 36.461 g.
Therefore, to get 1 mole of H^+ from HCl, we need to take 36.461 g of HCl.
Thus, 1 equivalent of HCl = 36.461 g of HCl. ■

■ EXAMPLE 170.   You are given 0.15 mole of H_2SO_4. How many equivalents of H_2SO_4 do you have (for full reaction)?

SOLUTION: One mole of H_2SO_4 gives 2 moles of H^+.
By definition, this is 2 equivalents.
So, 1 mole of H_2SO_4 = 2 equivalents.

0.15 mole of H_2SO_4 = 0.30 equivalent ■

■ EXAMPLE 171.   You are given 0.15 equivalent of H_3PO_4 (full reaction). How many grams of H_3PO_4 do you have?

SOLUTION: One mole of H_3PO_4 gives 3 moles of H^+.
Therefore, 1 mole of H_3PO_4 is 3 equivalents.

$$(0.15 \text{ equivalents}) \left(\frac{1 \text{ mole}}{3 \text{ equivalents}} \right) = 0.050 \text{ mole}$$

One mole of H_3PO_4 = 98.00 g.

$$(0.050 \text{ mole of } H_3PO_4) \left(98.00 \ \frac{g}{\text{mole}} \right) = 4.9 \text{ g} ■$$

■ EXAMPLE 172.   You are given 1.00 g of HCl. How many equivalents is this?

SOLUTION: One mole of HCl gives 1 mole of H^+.
Therefore, 1 mole of HCl = 1 equivalent.
One mole of HCl = 36.461 g = 1 equivalent.

$$\frac{1.00 \text{ g of HCl}}{36.461 \text{ g/equivalent}} = 0.0274 \text{ equivalent} \quad \blacksquare$$

■ EXAMPLE 173.   You are given 3.65 g of H_2SO_4. For full reaction, how many equivalents of H_2SO_4 is this?

SOLUTION: One mole of H_2SO_4 gives 2 moles of H^+
Therefore, 1 mole of H_2SO_4 = 2 equivalents.
One mole of H_2SO_4 weighs 98.08 g.
Two equivalents of H_2SO_4 weigh 98.08 g.
One equivalent of H_2SO_4 weighs (1/2)(98.08 g) = 49.04 g.

$$\frac{3.65 \text{ g of } H_2SO_4}{49.04 \text{ g/equivalent}} = 0.0744 \text{ equivalent} \quad \blacksquare$$

■ EXAMPLE 174.   How many equivalents are there in 3.66×10^{22} molecules of HCl?

SOLUTION: One mole of HCl contains 6.02×10^{23} molecules.

$$\frac{3.66 \times 10^{22} \text{ molecules}}{6.02 \times 10^{23} \text{ molecules/mole}} = 0.0608 \text{ mole}$$

One mole of HCl furnishes 1 mole of H^+.
By definition, this is 1 equivalent.
So, 0.0608 mole of HCl is 0.0608 equivalent. ■

■ EXAMPLE 175.   You are given 20.5 g of an unknown acid with the statement that this amount of acid will furnish $2.05 \times 10^{22} H^+$. What is the weight of 1 equivalent of your unknown acid?

SOLUTION: (Find out how many moles of H^+ you have. This defines how many equivalents you have. Divide the weight given by this number of equivalents.)

$$\frac{2.05 \times 10^{22} \text{ } H^+}{6.02 \times 10^{23} \text{ } H^+/\text{mole}} = 0.0341 \text{ mole of } H^+$$

To get 0.0341 mole of H^+, you must have 0.0341 equivalents of acid in your sample.

0.0341 equivalent of acid = 20.5 g.

$$\frac{20.5 \text{ g}}{0.0341 \text{ equivalent}} = 601 \text{ g/equivalent} \blacksquare$$

■ EXAMPLE 176.   You are given 2.00 g of NaOH. How many equivalents of base is this?

SOLUTION: One mole of NaOH gives 1 mole of OH^-.
By definition, this is 1 equivalent of base.
One mole of NaOH weights 39.997 g.
One mole of NaOH = 39.997 g = 1 equivalent.

$$\frac{2.00 \text{ g of NaOH}}{39.997 \text{ g/equivalent}} = 0.0500 \text{ gram-equivalent} \blacksquare$$

■ EXAMPLE 177.   You are given 2.00 g of $Ca(OH)_2$. How many equivalents of base is this (full reaction)?

SOLUTION: One mole of $Ca(OH)_2$ gives 2 moles of OH^-.
By definition, this is 2 equivalents of base.
One mole of $Ca(OH)_2$ weighs 74.09 g.

74.09 g $Ca(OH)_2$ = 2 equivalents

37.04 g $Ca(OH)_2$ = 1 equivalent

$$\frac{2.00 \text{ g of } Ca(OH)_2}{37.04 \text{ g/equivalent}} = 0.0540 \text{ equivalent.} \blacksquare$$

■ EXAMPLE 178.   An unknown base gives 0.030 mole of OH^- per 0.78 g of base. What is the weight of 1 equivalent of the base?

SOLUTION: One equivalent of any base gives 1 mole of OH^-. To get 0.030 mole of OH^-, you must have 0.030 equivalent of base.
Therefore, 0.78 g of base = 0.030 equivalent.

$$\frac{0.78 \text{ g}}{0.030 \text{ equivalent}} = 26 \text{ g/equivalent} \blacksquare$$

■ EXAMPLE 179.   You are given 0.032 equivalent of HCl. How many equivalents of base X do you need to neutralize this?

SOLUTION: It takes 1 equivalent of any acid to neutralize 1 equivalent of any base. Therefore, if you have 0.032 equivalent of acid, you must get 0.032 equivalent of base.
In this case, you need 0.032 equivalent of base X. ■

■ EXAMPLE 180.   You are given 2.00 g of HNO_3. How many grams of NaOH do you need to neutralize this?

SOLUTION: One mole HNO_3 = 1 equivalent = 63.01 g.

$$\frac{2.00 \text{ g of } HNO_3}{63.01 \text{ g/equivalent}} \ 0.0317 \text{ equivalent of acid}$$

It takes 1 equivalent of base to neutralize 1 equivalent of acid.

If you have 0.0317 equivalent of acid, you need 0.0317 equivalent of NaOH.

One mole of NaOH = 1 equivalent of NaOH = 39.997 g.

$$(0.0317 \text{ equivalent NaOH})\left(39.997 \ \frac{g}{\text{equivalent}}\right) = 1.27 \text{ g} ■$$

■ EXAMPLE 181.   A given sample contains 0.206 equivalent of $Ca(OH)_2$. Assuming complete reaction, how many grams of H_3PO_4 would be required to neutralize the sample?

SOLUTION: You have 0.206 equivalent of base; therefore, you need 0.206 equivalent of acid.

For H_3PO_4, 1 mole gives 3 moles H^+. This means 1 mole H_3PO_4 is the same as 3 equivalents.

To get 0.206 equivalent H_3PO_4, we need

$$(0.206 \text{ equivalent})\left(\frac{1 \text{ mole}}{3 \text{ equivalents}}\right) = 0.0687 \text{ mole of } H_3PO_4$$

One mole of H_3PO_4 weighs 97.995 g.

$$(0.0687 \text{ mole of } H_3PO_4)\left(97.995 \ \frac{g}{\text{mole}}\right) = 6.73 \text{ g} ■$$

■ EXAMPLE 182.   What is the weight of 1 equivalent of H_3PO_4 in each of the following reactions:

(I) $H_3PO_4 + NaOH \longrightarrow NaH_2PO_4 + H_2O$

(II) $H_3PO_4 + 2NaOH \longrightarrow Na_2HPO_4 + 2H_2O$

(III) $H_3PO_4 + 3NaOH \longrightarrow Na_3PO_4 + 3H_2O$

SOLUTION: In reaction (I) only one H out of three of H_3PO_4 is neutralized. So, for reaction (I), 1 mole of H_3PO_4 furnishes but 1 mole of H^+ (to react with 1 mole of NaOH). This means, so far as reaction (I) is concerned, that 1 mole of H_3PO_4 is the same as 1 equivalent.

One mole of $H_3PO_4 = 97.995$ g.

One equivalent of H_3PO_4 for reaction (I) = 97.995 g.

In reaction (II), where 2 moles of NaOH are used, each mole of H_3PO_4 must furnish 2 moles of H^+. So, for reaction (II) 1 mole of H_3PO_4 is the same as 2 equivalents.

One mole $H_3PO_4 = 97.995$ g = 2 equivalents.

One equivalent = (1/2)(97.995) = 48.998 g.

In reaction (III), 3 moles of NaOH react per mole of H_3PO_4, so each mole of H_3PO_4 must furnish 3 moles of H^+. For reaction (III), 1 mole of H_3PO_4 is the same as 3 equivalents.

One mole $H_3PO_4 = 97.995$ g = 3 equivalents.

One equivalent = (1/3)(97.995) = 32.665 g. ∎

6.2   Equivalents in oxidation-reduction

An oxidizing agent picks up electrons; a reducing agent supplies them. One equivalent of oxidizing agent is the weight of material required to pick up 1 mole of electrons; 1 equivalent of reducing agent is the weight of material required to furnish 1 mole of electrons. One equivalent of any oxidizing agent just matches 1 equivalent of any reducing agent so far as electron transfer is concerned.

Consider the reaction

$$Zn + Cl_2 \longrightarrow ZnCl_2$$

where 1 mole of Zn reacts with 1 mole of Cl_2. In this reaction, the Zn can be considered to go from the 0 oxidation state to the +2 oxidation state. In doing so, each Zn atom gives up two electrons. One mole of Zn atoms (6.02×10^{23} Zn atoms) must consequently give up 2 moles of electrons ($2 \times 6.02 \times 10^{23}$ electrons). We have defined 1 equivalent as the amount that gives up 1 mole of electrons, so if we give up 2 moles of electrons, we must have present 2 equivalents of reducing agent. In other words, 1 mole of Zn equals 2 equivalents of reducing agent.

Simultaneously, in the reaction, the chlorine can be considered to go from the 0 state to the −1 state. This means a gain of one electron. Since there are two chlorine atoms per Cl_2 molecule, each Cl_2 molecule picks up two electrons. One mole of Cl_2 molecules (6.02×10^{23} Cl_2 molecules) must pick up 2 moles of electrons ($2 \times 6.02 \times 10^{23}$ electrons). By definition, this is

2 equivalents of oxidizing agent. So, we have the result that when 1 mole of Zn reacts with 1 mole of Cl_2, 2 equivalents of reducing agent (Zn) are reacting with 2 equivalents of oxidizing agent (Cl_2).

Take a more complicated case, such as Zn reacting with MnO_4^- to give Zn^{2+} plus Mn^{2+}. Again the Zn goes from 0 to +2, so 1 mole of Zn equals 2 equivalents. In MnO_4^- the Mn goes from +7 to +2 in Mn^{2+}. This is a gain of five electrons. So far as $MnO_4^- \rightarrow Mn^{2+}$ is concerned, 1 mole of MnO_4^- accepts 5 moles of electrons; consequently, 1 mole of MnO_4^- is 5 equivalents of oxidizing agent. To make sure that electron gain and electron loss are balanced, the number of equivalents of oxidizing and reducing agent must be identical. If we take 1 mole of Zn (=2 equivalents), then we need to take 2 equivalents of MnO_4^-. The latter is 2/5 of a mole of MnO_4^-. So the molar ratio of Zn to MnO_4^- is 1 : (2/5), as can easily be checked by writing the balanced equation

$$5Zn + 2MnO_4^- + 16H_3O^+ \longrightarrow 5Zn^{2+} + 24H_2O + 2Mn^{2+}$$

From the examples above, it should be evident that in any oxidation–reduction reaction:

1. Total electron gain = total electron loss.
2. Number of equivalents used of oxidizing agent = number of equivalents used of reducing agent.
3. Weight of 1 equivalent of reducing agent = weight of 1 mole divided by electron loss.
4. Weight of 1 equivalent of oxidizing agent = weight of 1 mole divided by electron gain.

How these points work out in specific cases is illustrated in the following problems.

■ EXAMPLE 183.   When elemental Fe is oxidized to FeO, what is the weight for 1 equivalent of Fe?

SOLUTION: Fe goes from 0 in Fe to +2 in FeO.
Each Fe atom gives up two electrons.
Each mole of Fe gives up 2 moles of electrons.
This, by definition, is 2 equivalents of reducing agent.
Therefore, 1 mole of Fe = 2 equivalents.
One mole of Fe weighs 55.847 g.
Two equivalents = 55.847 g.
One equivalent = (1/2)(55.847) = 27.923 g. ■

■ EXAMPLE 184.   When elemental Fe is oxidized to Fe_2O_3, what is the weight for 1 equivalent of Fe?

SOLUTION: Fe goes from 0 in Fe to + 3 in Fe_2O_3.
Each Fe atom gives up three electrons.
Each mole of Fe gives up 3 moles of electrons.
This, by definition, is 3 equivalents of reducing agent.
Therefore, 1 mole of Fe = 3 equivalents.
One mole of Fe weighs 55.847 g.
Three equivalents = 55.847 g.
One equivalent = (1/3)(55.847) = 18.616 g. ■

■ EXAMPLE 185.   When elemental Fe is oxidized to Fe_3O_4, what is the weight for 1 equivalent of Fe?

SOLUTION: Fe goes from 0 in Fe to +8/3 in Fe_3O_4.
Each Fe atom can be considered to give up 8/3 of an electron.
Each mole of Fe gives up 8/3 of a mole of electrons.
By definition, this is 8/3 of an equivalent.
One mole of Fe = 8/3 equivalent.
One mole of Fe weighs 55.847 g.

8/3 equivalent = 1 mole = 55.847 g

1 equivalent = (3/8)(55.847) = 20.943 g

[*Note*: In these three preceding problems elemental iron has been oxidized to three different products: FeO, Fe_2O_3, and Fe_3O_4. The equivalent weight of iron is, respectively, 27.9 g, 18.6 g, and 20.9 g. These represent the weights that would yield 6.02×10^{23} electrons. If you now ask "How much oxygen would be required to produce the foregoing three products from 27.9, 18.6, and 20.9 g of Fe?" the answer would be "the same," since it is just the amount of oxygen required to pick up 6.02×10^{23} electrons.] ■

■ EXAMPLE 186.   When elemental oxygen picks up electrons to form oxides, how many grams of O_2 are required to pick up 6.02×10^{23} electrons?

SOLUTION: Each O atom picks up two electrons in going from the 0 state of the element to the −2 state in oxides.
Each O_2 molecule picks up 4 electrons.
Each mole of O_2 picks up 4 moles of electrons.

6.02×10^{23} electrons = 1 mole of electrons

To pick up 1 mole of electrons, we need 1/4 mole of O_2.
One mole of $O_2 = 31.9988$ g.

1/4 mole of $O_2 = (1/4)(31.9988) = 8.00$ g ∎

■ EXAMPLE 187. The element vanadium can react with O_2 to form V_2O_5. What weight of V would be needed to react with 1 equivalent of oxygen to form this product?

SOLUTION: It takes 1 equivalent of reducing agent to react with 1 equivalent of oxidizing agent.

Since we are given 1 equivalent of oxygen, we need to take 1 equivalent of vanadium.

In this reaction, V goes from 0 to +5.
Each V atom gives up five electrons.
One mole of V = 5 equivalents.
One equivalent = 1/5 mole of V = (1/5)(50.942) = 10.188 g. ∎

■ EXAMPLE 188. You have an element X that reacts with O_2 to form an oxide, X_2O_3. If it takes 4.445 g of X to react with 0.150 equivalent of oxygen, what is the atomic weight of X?

SOLUTION: To match 0.150 equivalent of oxygen, we need 0.150 equivalent of reducing agent X.

This means 4.445 g of X contains 0.150 equivalent.

$$\frac{4.445 \text{ g}}{0.150 \text{ equivalent}} = 29.6 \text{ g/equivalent}$$

In going from X to X_2O_3, each atom X changes oxidation number from 0 to +3. So, each atom of X must give up three electrons. Consequently, 1 mole of X (or 1 gram-atom) is 3 equivalents.

$$(3 \text{ equivalents})\left(29.6 \ \frac{\text{g}}{\text{equivalent}}\right) = 88.8 \text{ g}$$

One mole of X = 88.8 g.
Therefore, the atomic weight of X is 88.8 amu. ∎

■ EXAMPLE 189. When MnO_4^- gets reduced in neutral solution, the product is likely to be MnO_2. How many equivalents are there per mole of MnO_4^- in such a reaction?

SOLUTION: Mn goes from $+7$ in MnO_4^- to $+4$ in MnO_2.
This means each Mn atom picks up three electrons.
Each mole of MnO_4^- picks up 3 moles of electrons.
By definition, this is 3 equivalents of oxidizing agent, so 1 mole of MnO_4^- is 3 equivalents. ∎

■ EXAMPLE 190.   When $Cr_2O_7^{2-}$ is reduced in acidic solution, the chromium is converted to Cr^{3+}. How many equivalents of reducing agent would you need to reduce 1 mole of $Cr_2O_7^{2-}$ in such a reaction?

SOLUTION: Cr goes from $+6$ in $Cr_2O_7^{2-}$ to $+3$ in Cr^{3+}.
Each Cr atom picks up three electrons.
But note there are two Cr atoms per $Cr_2O_7^{2-}$.
So, each $Cr_2O_7^{2-}$ needs to pick up six electrons.
One mole of $Cr_2O_7^{2-}$ picks up 6 moles of electrons.
By definition, this is 6 equivalents of oxidizing agent.
Each equivalent of oxidizing agent requires 1 equivalent of reducing agent.
So, 6 equivalents of $Cr_2O_7^{2-}$ require 6 equivalents of reducing agent. ∎

■ EXAMPLE 191.   How many equivalents of $Cr_2O_7^{2-}$ would it take to oxidize 0.136 equivalent of $N_2H_5^+$ by the reaction $N_2H_5^+ + Cr_2O_7^{2-} \rightarrow N_2 + Cr^{3+}$?

SOLUTION: One equivalent of any reducing agent requires 1 equivalent of any oxidizing agent.
If we have 0.136 equivalent of $N_2H_5^+$, we must take 0.136 equivalent of $Cr_2O_7^{2-}$. ∎

■ EXAMPLE 192.   How many moles of $Cr_2O_7^{2-}$ would it take to oxidize 0.136 equivalent of $N_2H_5^+$ by the reaction $N_2H_5^+ + Cr_2O_7^{2-} \rightarrow N_2 + Cr^{+3}$?

SOLUTION: It takes 0.136 equivalent of $Cr_2O_7^{2-}$ to do the job, because each equivalent of reducing agent needs 1 equivalent of oxidizing agent. In the reaction, Cr of $Cr_2O_7^{2-}$ goes from $+6$ to $+3$.
This means three electrons per Cr atom, or six electrons per $Cr_2O_7^{2-}$ ion.
One mole of $Cr_2O_7^{2-}$ equals 6 equivalents.

$$(0.136 \text{ equivalent})\left(\frac{1 \text{ mole}}{6 \text{ equivalents}}\right) = 0.0227 \text{ mole} \quad ∎$$

■ EXAMPLE 193.   Suppose it takes 1.40×10^{-3} mole of an unknown reducing agent to reduce 8.40×10^{-4} mole of MnO_4^- to Mn^{2+}. How many electrons would each formula-unit of the reducing agent need to furnish?

SOLUTION: Mn goes from $+7$ in MnO_4^- to $+2$ in Mn^{2+}.
Each Mn picks up five electrons.
Therefore, 1 mole of MnO_4^- is 5 equivalents.

$$(8.40 \times 10^{-4} \text{ mole of } MnO_4^-)\left(\frac{5 \text{ equivalents}}{\text{mole}}\right) = 4.20 \times 10^{-3} \text{ equivalent}$$

Oxidizing and reducing agents must match in equivalents. There must be 4.20×10^{-3} equivalent of reducing agent in the 1.40×10^{-3} mole of unknown. This means

$$\frac{4.20 \times 10^{-3} \text{ equivalent}}{1.40 \times 10^{-3} \text{ mole}} = 3 \text{ equivalents per mole of unknown}$$

Each mole of unknown supplies 3 moles of electrons.
Each formula-unit of unknown supplies 3 electrons. ∎

Exercises

*6.1. How many equivalents of acid are there in 6.85 g of H_2SO_4 (full neutralization)?

*6.2. How many equivalents of base are there in 6.85 g of $Ca(OH)_2$ (full neutralization)?

*6.3. What is the weight of 1 equivalent of each of the following, assuming complete neutralization: HF; $HClO_4$; H_2S; H_3AsO_4; $Mg(OH)_2$; KOH? (*Answer*: 20.00; 100.46; 17.04; 47.32; 29.16; 56.11 g.)

*6.4. How many equivalents of H_3AsO_4 are required to neutralize 2.50 moles of $Mg(OH)_2$?

*6.5. How many equivalents of H_3AsO_4 are required to neutralize 2.50 equivalents of base X?

**6.6. If the weight of 0.378 equivalent of $M(OH)_2$ is 26.79 g, what is the atomic weight of M? (*Answer*: 107.73 amu.)

**6.7. It takes 2.56×10^{-3} equivalent of KOH to neutralize 0.187 g of H_3XO_4. What is the atomic weight of X?

*6.8. Assuming maximum neutralization, which reagent and how much will be left unreacted when 1.50 equivalents of H_2SO_4 are mixed with 2.95 equivalents of NaOH?

**6.9. How many grams of which reagent will remain unneutralized upon mixing 5.03 g of H_3PO_4 and 0.0682 equivalent of $Ca(OH)_2$? (*Answer*: 2.80 g.)

**6.10. Suppose that it takes 55 g of X or 35 g of Y to give 1 equivalent of acid. A 1.50-g sample composed of a mixture of X and Y requires 1.50 g of NaOH for neutralization. What is the composition of the original sample?

6.11. Amphoteric material Q can neutralize acid and can also neutralize base. For 1.00 g of HCl, it takes 2.01 g of Q; for 1.00 g of NaOH, it takes 0.613 g of Q. What is the weight of 1 equivalent of Q?

*6.12. When the element tantalum (Ta) is allowed to react with an excess of oxygen, the equivalent weight of tantalum is observed to be 36.19 g. What is the formula of the oxide formed? (*Answer*: Ta_2O_5.)

*6.13. Under certain conditions the element manganese (Mn) can be converted to the oxide Mn_2O_7. What weight of manganese would be needed to furnish 0.35 equivalent for this reaction?

*6.14. How many grams of PbO_2 would it take to furnish 1 equivalent of PbO_2 for reactions in which it is reduced to each of the following: (a) Pb_3O_4; (b) PbO; (c) Pb?

*6.15. Suppose you are told to supply 0.375 equivalent of reducing agent for a particular reaction. If you furnish 6.37 g of V^{2+}, then to what oxidation state must the vanadium go?

6.16. If it takes 0.150 mole of ClO^- (going to Cl^-) to oxidize 12.6 g of an unknown chromium oxide to $Cr_2O_7^{2-}$, then what is the simplest formula of the unknown oxide? (*Answer*: CrO_2.)

6.17. If, in being oxidized to CO_2 and H_2O, it takes 39.02 g of $C_3H_4O_x$ to furnish 1 equivalent, then what is the value of x?

6.18. Suppose you are given a mixture of Mn_2O_3 and MnO_2. On reaction with HCl, the products are Cl_2 and Mn^{2+}. If 4.68 g of the mixture yield 0.0812 equivalent of chlorine, then what is the composition of the original mixture?

6.19. Given a complex organic acid of molecular formula $C_6H_xO_6$. If it requires 0.0154 equivalent of an oxidizing agent to oxidize completely 1.00 g of this compound to CO_2 and H_2O and if it takes 0.0170 equivalent of a base to neutralize completely 1.00 g of this compound, then how many hydrogen atoms are there in the molecule and how many of these are neutralizable by base?

6.20. Suppose you are given a 30.0-mg crystal of the nonstoichiometric compound $WO_{2.90}$. How many electrons would be released in the conversion of this crystal to WO_3? (*Answer*: 1.57×10^{19} electrons.)

7

GASES

TO DESCRIBE a sample of gas, the quantities that may be useful are the following:

P, the pressure, usually measured in atmospheres or millimeters of mercury (mm Hg). A relatively recent development is to use the unit "torricelli," written "Torr," instead of "millimeter of mercury."

V, the volume, usually measured in cubic centimeters (cm³), milliliters (ml), or liters. There used to be a great deal of fuss about the difference between a milliliter and a cubic centimeter when 1 cm³ used to be 0.99997 ml, but the milliliter was redefined in 1964 so that the cubic centimeter and the milliliter are now identical and can be used interchangeably.

n, the number of moles, calculable as the weight of sample divided by its molecular or molar weight. The number of moles can also be expressed as the number of molecules divided by the Avogadro number, 6.02×10^{23}.

T, the temperature, usually measured in degrees absolute (°A) or degrees Kelvin (°K). Referred to Celsius or centigrade temperature (which is usually denoted by t and is measured in °C), the absolute temperature is obtained by adding 273.15°.

$$T°K = t°C + 273.15°$$

People are trying to stamp out the Fahrenheit temperature (°F) but it refuses to disappear. If you need to convert, the appropriate relation is

$$t°C = \frac{5}{9}[t°F - 32°]$$

For an ideal gas (i.e., one whose behavior is exactly describable by the gas laws discussed below), the product PV divided by nT comes out to be a constant, the universal gas constant, usually denoted by R. The value of R

depends on what units are used for P, V, n, and T. We will say more about this general gas law later.

7.1 Boyle's law

Boyle's law states that for an ideal gas, the volume of a fixed weight at a fixed temperature varies inversely with the pressure exerted on it.

■ EXAMPLE 194.   If 3.0 g of gas at 25°C occupies 3.60 liters at a pressure of 1.00 atm, what will be its volume at a pressure of 2.50 atm? Assume ideal behavior.

SOLUTION: The pressure increases from 1.00 atm of 2.50 atm.
The pressure, therefore, has increased to 2.50/1.00 of its original value.
The volume changes inversely, so it must go to 1.00/2.50 of its original value.

$$\text{Final volume} = \left(\frac{1.00}{2.50}\right)(3.60 \text{ liters}) = 1.44 \text{ liters} \blacksquare$$

■ EXAMPLE 195.   A given sample of ideal gas occupies a volume of 11.2 liters at 0.863 atm. If you keep the temperature constant, to what pressure will you have to go to change the volume to 15.0 liters?

SOLUTION: You want the volume to go from 11.2 liters to 15.0 liters. Thus, you want the volume to increase to 15.0/11.2 of its original value. The pressure will have to change inversely.
So, you need the pressure to go to 11.2/15.0 of its original value.

$$\text{Final pressure} = \left(\frac{11.2}{15.0}\right)(0.863 \text{ atm}) = 0.644 \text{ atm} \blacksquare$$

■ EXAMPLE 196.   A sample of gas is trapped at a pressure of 1.55×10^{-6} mm Hg in a volume of 250.0 cm³. Assuming ideal behavior, what volume would this sample occupy if compressed to 1 atm pressure at the same temperature?

SOLUTION: One atm pressure is equivalent to 760 mm Hg.
Thus, you want the pressure to go from 1.55×10^{-6} mm Hg to 760 mm Hg; in other words, to $760/(1.55 \times 10^{-6})$ of its original value.
The volume would change inversely, or to $(1.55 \times 10^{-6})/760$ of its original value.

$$\text{Final volume} = \left(\frac{1.55 \times 10^{-6}}{760}\right)(250.0 \text{ cm}^3) = 5.10 \times 10^{-7} \text{ cm}^3 \blacksquare$$

■ EXAMPLE 197. The gas in outer space is at a pressure of about 5×10^{-14} Torr. Assuming constant temperature and ideal behavior, how much outer space could you compress into a 1-cm^3 box at 1 atm?

SOLUTION: One atm is 760 mm Hg, or 760 Torr.

You are going to compress from 5×10^{-14} Torr to 760 Torr—that is, to $760/(5 \times 10^{-14})$ of the original.

Therefore, the final volume will have to be $(5 \times 10^{-14})/760$ of the original.

$$\text{Final volume} = 1\text{cm}^3 = \left(\frac{5 \times 10^{-14}}{760}\right)(\text{original volume})$$

$$\text{Original volume} = \left(\frac{760}{5 \times 10^{-14}}\right)(1 \text{ cm}^3) = 2 \times 10^{16} \text{ cm}^3$$

[Note: Only one significant figure allowed here.] ■

In the foregoing problems, the solutions have been worked out by what is called the logical approach. The reasoning requires visualization of the compression or expansion process. An alternate way of setting up these Boyle's law problems is to express the inverse proportionality as an algebraic equation:

$$\frac{V \text{ initial}}{V \text{ final}} = \frac{P \text{ final}}{P \text{ initial}}$$

If any three of these quantities are specified, the fourth can be calculated.

■ EXAMPLE 198.   A lungful of air (350 cm^3) is exhaled into an evacuated chamber as big as an average room (75 m^3). Assuming constant temperature and ideal behavior, what would be the final pressure in the chamber if we start with 750 mm Hg in the lung?

SOLUTION:

V initial $= 350$ cm^3

V final $= 75$ m$^3 = (75)(100 \text{ cm})^3 = 75 \times 10^6$ cm^3

P initial $= 750$ mm Hg

P final $= ? ? ?$

From the equation $(V \text{ initial}/V \text{ final}) = (P \text{ final}/P \text{ initial})$ we can deduce, by rearranging the terms, that

$$P \text{ final} = P \text{ initial} \left(\frac{V \text{ initial}}{V \text{ final}}\right) = (750 \text{ mm Hg})\left(\frac{350 \text{ cm}^3}{75 \times 10^6 \text{ cm}^3}\right)$$

$$= 3.5 \times 10^{-3} \text{ mm Hg } ■$$

7.2   Charles' law

Charles' law states that given an ideal gas, its volume is directly proportional to the absolute temperature, provided that the pressure stays constant and provided we are working with a gas sample of fixed weight.

■ EXAMPLE 199.   Two grams of gas occupy 1.56 liters at 25°C and 1.00 atm pressure. What will be the volume of this sample if the gas is heated to 35°C with the pressure staying constant? Assume ideal behavior.

SOLUTION: Celsius temperature goes from 25°C to 35°C.
Absolute temperature goes from $(273 + 25)$ to $(273 + 35)$, or from 298°K to 308°K.
The absolute temperature has increased to 308/298 of the original.
The volume must also change proportionally.
Volume goes to 308/298 of the original.

$$\text{Final volume} = \left(\frac{308}{298}\right)(1.56 \text{ liters}) = 1.61 \text{ liters} \blacksquare$$

■ EXAMPLE 200.   A 268-cm^3 sample of ideal gas at 18°C and 748 Torr pressure is placed in an evacuated container of volume 648 cm^3. To what centigrade temperature must the assembly be heated so that the gas will fill the whole chamber at 748 Torr?

SOLUTION: The volume of the sample has to go from 268 to 648 cm^3, or to 648/268 of the original.
Absolute temperature must change proportionally.
Absolute temperature is to go to 648/268 of the original.
Original temperature is 18°C, or $18 + 273 = 291$°K.

$$\text{Final temperature} = \left(\frac{648}{268}\right)(291°\text{K}) = 704°\text{K}$$

Final Celsius temperature $= 704° - 273° = 431°\text{C} \blacksquare$

■ EXAMPLE 201.   Suppose you have a flask fitted with a capillary neck in which a glob of mercury rides so as to trap in the flask a certain sample of gas at atmospheric pressure. You put the flask in an ice bath (0°C) and you arbitrarily mark the mercury-blob position as 273 " volume units" of trapped gas. Now you put the flask in a water bath at 18°C. The mercury blob moves out. What number for volume should you put on the new position?

SOLUTION: The Celsius temperature has gone from 0 to 18°C. Absolute temperature has gone from 273 to 291°K.

If absolute temperature has gone to 291/273 of the original, the volume must change proportionally.

$$\text{Final volume} = \left(\frac{291}{273}\right)(\text{initial volume}) = \left(\frac{291}{273}\right)(273 \text{ "volume units"})$$

$$= 291 \text{ "volume units"} \blacksquare$$

■ EXAMPLE 202.   A freely collapsible balloon contains a certain amount of hot gas at atmospheric pressure. The initial volume is 2.64×10^6 liters. When the balloon falls into the ocean (15°C), the volume goes to 2.04×10^6 liters. What must have been the temperature of the hot gas initially? Assume ideal behavior.

SOLUTION: Volume goes from 2.64×10^6 liters to 2.04×10^6 liters. Volume goes to $(2.04 \times 10^6)/(2.64 \times 10^6)$ of the original.

Absolute temperature must also go to 2.04/2.64 of the original. Final temperature is 15°C = 288°K.

288°K must be 2.04/2.64 of original hot gas temperature.

$$\text{Original temperature} = \left(\frac{2.64}{2.04}\right)(288°K) = 373°K, \text{ or } 100°C \blacksquare$$

7.3   Dalton's law of partial pressures

This law simply states that when more than one gas is in the same container, the total pressure exerted by the mixture is the sum of the pressures that would have been exerted if each gas were there alone. In other words, so far as pressure is concerned, each gas acts completely independently of other gases present in the same volume. This can be stated mathematically in the following way: First, we invent something we call a "partial pressure." This is defined as the individual pressure one component exerts, treating it as if it were all by itself in the container. Let us designate the partial pressures of components 1, 2, 3, ..., as $p_1, p_2, p_3, \ldots$. Dalton's law states that the observed pressure P will be the sum of the partial pressures, or

$$P = p_1 + p_2 + p_3 + \cdots$$

■ EXAMPLE 203.   Given a completely evacuated 1.00-liter box at 25°C. Enough hydrogen is pumped in to make its partial pressure equal to 0.463 atm. At the same time enough oxygen is pumped in to make its partial pressure equal to 0.432 atm. What is the total pressure in the box, assuming ideal behavior?

SOLUTION: Total pressure equals sum of partial pressures.

$$\text{Total pressure} = p_{\text{hydrogen}} + p_{\text{oxygen}} = 0.463 \text{ atm} + 0.432 \text{ atm}$$
$$= 0.895 \text{ atm} \blacksquare$$

■ EXAMPLE 204. A given gas mixture consists of helium, neon, and argon, all at the same partial pressure. If the total pressure of the sample is 746 mm Hg, what is the pressure exerted by the helium?

SOLUTION: Let p_{He}, p_{Ne}, p_{Ar} be the partial pressures of the three components.
Then, total pressure $P = p_{He} + p_{Ne} + p_{Ar}$.
But, since the partial pressures are equal, $p_{He} = p_{Ne} = p_{Ar}$.
We can, therefore, substitute p_{He} for p_{Ne} and for p_{Ar}.

$P = 3p_{He} = 746$ mm Hg

$p_{He} = (1/3)(746$ mm Hg$) = 249$ mm Hg ■

■ EXAMPLE 205. You are given three boxes, A, B, and C, all at the same temperature, with volumes, respectively, of 1.20 liters, 2.63 liters, and 3.05 liters. Box A contains 0.695 g of nitrogen gas at a pressure of 742 Torr; box B contains 1.10 g of argon gas at a pressure of 383 Torr; box C is completely empty at the start of the experiment. What will the pressure become in box C if the contents of A and B are completely transferred to C? Assume ideal behavior.

SOLUTION: Treat the gases independently.
For pumping A into C, the sample goes from 1.20 liters to 3.05 liters. This is an increase of volume to 3.05/1.20 of the original. By Boyle's law, the pressure should change inversely—that is, to 1.20/3.05 of the original. Therefore, sample A in box C gives a pressure of

$$\left(\frac{1.20}{3.05}\right)(742 \text{ Torr}) = 282 \text{ Torr}$$

For pumping B into C, the sample goes from 2.63 liters to 3.05 liters. This is an increase of volume to 3.05/2.63 of the original. By Boyle's law, the pressure should go to 2.63/3.05 of the original, or

$$\left(\frac{2.63}{3.05}\right)(383 \text{ Torr}) = 330 \text{ Torr}$$

The final total pressure in box C will be the partial pressure exerted by gas A plus the partial pressure exerted by gas B:

$P = p_A + p_B = 292$ Torr $+ 330$ Torr $= 622$ Torr ■

7.4   Wet gases

One of the most useful applications of Dalton's law of partial pressures is for calculating about gases collected over water. The application is a common one because it is so simple to collect a gas X by allowing it to bubble into the mouth of an inverted bottle filled with water. However, gas X collected by such water displacement is invariably wet—that is, it is contaminated with water vapor. For most purposes, this contamination by water vapor is not serious, but it must be taken into account in measurement of the pressure of the gas. Ordinarily, the pressure in the surrounding room can be read on a barometer (the so-called "barometric pressure"), and if the water levels inside the bottle and outside the bottle are identical, then the total pressure inside the bottle is just equal to the barometric pressure. However, if the gas is "wet," the pressure inside the bottle is partly due to water vapor; this has to be subtracted from the barometric reading in order to find the true partial pressure of gas X.

How much correction do we need to subtract? This depends on the temperature of the water through which gas X is bubbled. The hotter the water, the more volatile it is, and the "wetter" the gas gets. In general, we make the assumption that the gas sample is "saturated" with water vapor— i.e., it contains as much water vapor as it can hold under equilibrium conditions. If the gas sample is "saturated" with water vapor, then the partial pressure of the water vapor in the gas phase is fixed by the temperature of the water in contact with the gas. Appendix C shows the pressure of aqueous vapor over liquid water in millimeters of Hg (torricellis) at various temperatures. When you meet a gas problem in which gas X is collected over water, look in this table to find out what correction for water vapor you need to subtract from the observed total pressure. To do this, you need to know the temperature of the liquid water over which the gas is collected. In general, it is assumed that the liquid water and the gas above it are at the same temperature. For precise work a thermometer has to be put in the water to find out what its temperature really is.

■ EXAMPLE 206. A sample of nitrogen gas is collected over water at 18°C. If the barometric reading is 742 mm Hg, what is the actual pressure of nitrogen in the sample?

SOLUTION: At 18°C, the vapor pressure of water is 15.48 mm Hg. The total pressure of the sample $= p_{nitrogen} + p_{water\ vapor}$.

$p_{nitrogen} =$ total pressure $- p_{water\ vapor}$

$= 742$ mm Hg $- 15.48$ mm Hg $= 727$ mm Hg ■

■ EXAMPLE 207. You wish to collect a sample of oxygen gas in which the pressure of oxygen is 732 mm Hg. You want to do this by water displacement on a day when the barometric pressure in the lab is 742 mm Hg. How cold should the water be?

SOLUTION:

Total pressure = 742 mm Hg

$$= p_{\text{oxygen}} + p_{\text{water vapor}}$$

$$= 732 \text{ mm Hg} + p_{\text{water vapor}}$$

$$p_{\text{water vapor}} = 742 - 732 = 10 \text{ mm Hg}$$

Consult the table to see at what temperature the vapor pressure of water becomes equal to 10 mm Hg. You will find it occurs when the water temperature is 11.2°C. ■

■ EXAMPLE 208. A given sample of dry methane occupies a volume of 368 cm³ at a temperature of 21.0°C and a pressure of 752 Torr. Suppose this sample is now bubbled through water at 21.0°C until it gets saturated with water vapor. If the total pressure remains 752 Torr, what volume will the "wet" gas have to occupy? Assume ideal behavior.

SOLUTION: At 21.0°C the vapor pressure of water comes out to be 18.6 mm Hg, or 18.6 Torr.

If the total pressure of the "wet" gas is 752 Torr, and if 18.6 Torr is due to the water vapor, only the difference (752 − 18.6, or 733 Torr) is due to the methane.

In other words, the pressure of the methane has gone from 752 Torr (in the dry state) to 733 Torr (in the wet state).

According to Boyle's law, for the pressure to go to 733/752 of the original, the volume must change inversely—that is, go to 752/733 of its original value. Therefore,

$$\text{final volume} = \left(\frac{752}{733}\right)(368 \text{ cm}^3) = 378 \text{ cm}^3 \quad ■$$

[*Note*: This problem illustrates a very important point. The volume occupied by gas X depends on the *pressure of* X; it does not depend on what other gases are present. Once we know what the pressure of X is, we can ignore all the other gases present—provided, of course, that ideal behavior is obeyed.]

7.5   Combined Boyle's, Charles', and Dalton's law behavior

It is a fundamental principle that the state of a gas sample does not depend on how you got it there. Specifically, it makes no difference in what sequence you carry out operations such as changing pressure, changing temperature, or adding other gases. The end result will be the same, provided you end up doing all the same operations.

This principle has direct application to problems, because frequently a gas will be taken from one pressure and temperature (P_1, T_1) to another pressure and temperature (P_2, T_2). In calculating what happens to, say, the volume in this case, we can separate the change into two steps: one corresponding to changing pressure from P_1 to P_2 and the other to changing temperature from T_1 to T_2. It makes no difference which we do first. The same principle holds for changing other variables (e.g., adding other gases).

The foregoing means that, given a set of initial conditions for the gas and a set of final conditions, an unknown quantity can be calculated by applying each correction successively. This kind of combined calculation is frequently encountered when the problem calls for "reduction to standard conditions." Standard conditions, which are the reference conditions for comparison, are 0°C and 1 atm pressure. They are usually referred to as STP (standing for "standard temperature and pressure").

■ EXAMPLE 209.   Three grams of a gas occupies 0.963 liter at 22°C and 0.969 atm. What will be the volume of this sample if taken to standard conditions, assuming ideal behavior?

SOLUTION: To go to standard conditions, the temperature must go from 22°C (or 295°K) to 0°C (or 273°K); the pressure must go from 0.969 to 1 atm. In other words,

the temperature goes to $\dfrac{273}{295}$ of the original

the pressure goes to $\dfrac{1.000}{0.969}$ of the original

The temperature change produces a directly proportional effect on the volume, so from the temperature change alone the final volume must go to 273/295 of the original.

The pressure change produces an inversely proportional effect on volume, so from the pressure change alone the final volume must go to 0.969/1.000 of the original.

Applying both corrections, we would find that

$$\text{final volume} = (\text{initial volume})\left(\frac{273}{295}\right)\left(\frac{0.969}{1.000}\right)$$

$$= (0.963 \text{ liter})\left(\frac{273}{295}\right)\left(\frac{0.969}{1.000}\right) = 0.864 \text{ liter} \blacksquare$$

■ EXAMPLE 210.   Three grams of a dry gas at $-40°C$ and 742 mm Hg occupies a volume of 862 cm^3. Assuming ideal behavior, what volume would this sample occupy if it were saturated with water vapor at 19°C and held at a barometric pressure of 756 mm Hg?

SOLUTION: Temperature goes from $-40°C$ (233°K) to 19°C (292°K). Thus, temperature goes to 292/233 of its original value.

Owing to temperature increase alone, the volume should go to 292/233 of its original value.

In the final state, the *total* pressure is 756 mm Hg. This includes the vapor pressure of water at 19°C. From the tables, we find that the vapor pressure of water at 19°C equals 16.5 mm Hg. Subtracting this 16.5 mm Hg from 756 mm Hg leaves 740 mm Hg as the actual pressure of the gas. So, pressure of the gas has gone from 742 to 740 mm Hg.

Pressure has gone to 740/742 of its original value.

According to Boyle's law, the volume should change inversely, and due to pressure change alone it should go to 742/740 of its original value.

$$\text{Final volume} = (\text{initial volume})\left(\frac{292}{233}\right)\left(\frac{742}{740}\right)$$

where the first fraction takes care of the temperature change and the second fraction takes care of the pressure change.

$$\text{Final volume} = (862 \text{ cm}^3)\left(\frac{292}{233}\right)\left(\frac{742}{740}\right) = 1080 \text{ cm}^3 \blacksquare$$

7.6   Avogadro's principle and the molar volume

The Avogadro principle states that at the same pressure and temperature, equal volumes of gases contain equal numbers of particles. This means, for example, that provided both are at the same temperature and pressure, 1 liter of hydrogen gas contains as many molecules as 1 liter of any other gas showing ideal behavior. Another way of stating the Avogadro principle is to say that the number of molecules per cubic centimeter of gas depends only on the temperature and pressure of the gas and not on the identity of the gas. The basic reason for the Avogadro principle is that gases are mostly empty space, so their behavior does not depend on what the molecules are that populate this empty space.

One mole of any gas contains the Avogadro number of molecules— that is, 6.02×10^{23} molecules. At STP, the volume occupied by these 6.02×10^{23} molecules is equal to 22.4 liters (no matter what the molecules are, so long as they constitute a gas showing ideal behavior). The number 22.4 liters is called the molar volume of an ideal gas at STP.

■ EXAMPLE 211. What would be the molar volume of an ideal gas at 25°C and 742 Torr?

SOLUTION: The molar volume is 22.4 liters at 0°C(273°K) and 760 Torr. We change the temperature from 273 to 298°K—that is, to 298/273 of its original value.

We also change the pressure from 760 to 742 Torr, thereby changing it to 742/760 of its original value.

Owing to the increase in temperature, the volume will go up to 298/273 of its original value.

Owing to the decrease in pressure, the volume will increase to 760/742 of its original value.

$$\text{Final volume} = (\text{initial volume at STP}) \left(\frac{298}{273}\right) \left(\frac{760}{742}\right)$$

$$= (22.4 \text{ liters}) \left(\frac{298}{273}\right) \left(\frac{760}{742}\right) = 25.0 \text{ liters} ■$$

■ EXAMPLE 212. Assuming a gas behaves ideally, how many molecules per cubic centimeter will it contain at −33°C and 726 Torr?

SOLUTION: One mole STP contains 6.02×10^{23} molecules and occupies 22.4 liters.

Therefore, at STP, we have

$$\frac{6.02 \times 10^{23} \text{ molecules}}{22,400 \text{ cm}^3} = 2.69 \times 10^{19} \text{ molecules/cm}^3$$

Now we decrease the temperature from STP (273°K) to −33°C (240°K). This shrinks the gas to 240/273 of the original, thereby increasing the particles per cubic centimeter to 273/240 of the original.

We also release the pressure from STP (760 Torr) to 726 Torr. This reduction of pressure to 726/760 of the original is the same as an expansion to 760/726 of the original volume. If we expand the gas, there will be fewer molecules per cubic centimeter proportionately, so the pressure drop decreases the particles per cubic centimeter to 726/760 of the original.

Final number of molecules per cubic centimeter

$$= \text{(initial number of molecules per cubic centimeter at STP)} \left(\frac{273}{240}\right) \left(\frac{726}{760}\right)$$

$$= (2.69 \times 10^{19} \text{ molecules per cubic centimeter}) \left(\frac{273}{240}\right) \left(\frac{726}{760}\right)$$

$$= 2.92 \times 10^{19} \text{ molecules per cubic centimeter.} \ \blacksquare$$

■ EXAMPLE 213.   An unknown gas X shows a density of 2.39 g/liter at 100.0°C and 715 Torr. What is the molecular weight of this gas, assuming ideal behavior?

SOLUTION: (We take our 2.39-g sample, reduce it to STP to find its volume at STP, then find out how many such volumes there would be in 22.4 liters.)

We go from 100°C (or 373°K) to STP (273°K). Thus, the absolute temperature goes to 273/373 of its original value.

The volume of 2.39 g goes to 273/373 of its original value.

Then we go from 715 Torr to STP (760 Torr). Thus, the pressure goes to 760/715 of the original.

The volume goes to 715/760 of its starting value.

$$\text{Final volume of 2.39 g of gas at STP} = (1.00 \text{ liter}) \left(\frac{273}{373}\right) \left(\frac{715}{760}\right)$$

$$= 0.689 \text{ liter}$$

Molar volume at STP is 22.4 liters. If we divide this 22.4 liters by 0.689 liter we will find out how many 2.39-g portions we need to get a mole. Then if we multiply the number of such portions—actually 22.4/0.689—by 2.39 g, we shall have the weight of 1 mole:

$$\left(\frac{22.4 \text{ liter/mole}}{0.689 \text{ liter}}\right) (2.39 \text{ g}) = 77.7 \ \frac{\text{g}}{\text{mole}} \ \blacksquare$$

7.7   Equation of state for an ideal gas

Boyle's law, Charles' law, and the Avogadro principle are special cases of gas behavior. The most general way of summarizing the behavior of a gas is in what is called the "equation of state." An equation of state tells mathematically how the pressure, temperature, volume, and number of moles in a sample must be related. For an ideal gas the equation is

$$PV = nRT$$

where P is the pressure, V is the volume, n is the number of moles, T is the absolute temperature, and R is a constant, usually called the "universal gas constant." If P is in atmospheres, V in liters, T in degrees absolute, then R has the numerical value of 0.08206 liter-atm per degree per mole. So long as a sample of gas approximates ideal behavior, PV divided by nT must come out to be equal to R no matter what the gas is and no matter what condition it is in.

■ EXAMPLE 214.   Show that Boyle's law can be derived from the equation of state for an ideal gas.

SOLUTION: Boyle's law holds for a fixed weight of sample at a fixed temperature. In other words, n and T are fixed numbers as well as R. This means we can lump n, R, and T into one constant—call it c. So, under Boyle's law conditions

$$PV = nRT = c$$

Suppose P_1 and V_1 describe the sample in the initial state and P_2 and V_2 describe the sample in the final state. The product PV must stay constant, so we can write

$$P_1V_1 = c = P_2V_2$$

which can be rewritten as

$$\frac{V_1}{V_2} = \frac{P_2}{P_1}$$

showing the Boyle's law inverse proportionality. ■

The equation of state is extremely useful because it enables us to calculate one of the variables P, V, n, or T as soon as three of the others are specified.

In particular, we can calculate n if P, V, and T are given. This kind of calculation takes on added significance if you stop to think that number of moles is hard to measure but P, V, and T can usually be determined with readily available instruments.

■ EXAMPLE 215. A tube of volume 36.8 cm³ contains oxygen (O_2) at a pressure of 2000 pounds per square inch (lb/in²) and a temperature of 343°C. How many grams of oxygen are there in this sample?

SOLUTION: We can calculate the number of moles of oxygen from the relation $n = (PV)/(RT)$, provided we use the proper units.

$$V = \text{volume in liters} = \frac{36.8 \text{ cm}^3}{1000 \text{ cm}^3/\text{liter}} = 0.0368 \text{ liter}$$

$$P = \text{pressure in atmospheres} = \frac{2000 \text{ lb/in}^2}{14.7 \text{ lb/in}^2/\text{atm}} = 136 \text{ atm}$$

$R = 0.08206$ liter-atm/°C mole

$T = 343°C = 343 + 273$, or $616°K$

$$n = \frac{PV}{RT} = \frac{(136)(0.0368)}{(0.08206)(616)} = 0.0990 \text{ mole}$$

One mole of oxygen = 32.00 g.

$$0.0990 \text{ mole} = (0.0990 \text{ mole})\left(32.00 \frac{\text{g}}{\text{mole}}\right) = 3.17 \text{ g} \ ■$$

7.8   Kinetic theory

The ideal equation of state can be derived from the simple model that gases consist of independent molecules that exert pressure by making elastic collisons with the walls of the container. The principal assumption that has to be introduced is that the average kinetic energy of the gas molecules ($1/2 \ mv^2$, where m is the mass of the molecule and v is its average velocity) is proportional to the absolute temperature.

The implication of this assumption that average kinetic energy is proportional to temperature is that once the temperature is fixed, the product mv^2 is fixed. This means that, given two gases at the same temperature, $m_1 v_1^2 = m_2 v_2^2$ or $v_1/v_2 = \sqrt{m_2/m_1}$, or "velocities are inversely proportional to the square roots of the molecular masses." A direct outgrowth of this is Graham's law of diffusion, which states that the "rate of diffusion of a gas is inversely proportional to the square root of its molecular mass." The other implication is that, for a given gas, the product mv^2 is proportional to the temperature. Since m for a given gas is constant, this means that the square of the average velocity of a gas molecule increases proportionally to the absolute temperature.

■ EXAMPLE 216.   If the average velocity of an oxygen molecule (O_2) is 4.25×10^4 cm/sec at 0°C, what is the average velocity of a CO_2 molecule at the same temperature?

SOLUTION: At the same temperature, average kinetic energies are equal.

$$\frac{1}{2} m_{O_2} v_{O_2}^2 = \frac{1}{2} m_{CO_2} v_{CO_2}^2$$

$$v_{CO_2}^2 = v_{O_2}^2 \left(\frac{m_{O_2}}{m_{CO_2}} \right)$$

$$v_{CO_2} = v_{O_2} \sqrt{\frac{m_{O_2}}{m_{CO_2}}}$$

The ratio of the masses of the O_2 and CO_2 molecules will be the same as the ratio of their molecular weights, that is, 32.00 to 44.01.

$$v_{CO_2} = v_{O_2} \sqrt{\frac{32.00}{44.01}}$$

$$= \left(4.25 \times 10^4 \frac{cm}{sec} \right) (0.853) = 3.62 \times 10^4 \frac{cm}{sec} \ ■$$

■ EXAMPLE 217.   If the average velocity of an O_2 molecule is 4.25×10^4 cm/sec at 0°C, what would it be at 25°C?

SOLUTION: The kinetic energy changes proportionally to the absolute temperature.

The absolute temperature changes from 273°K (for 0°C) to 298°K (for 25°C). The kinetic energies will be in the same ratio as the absolute temperatures.

$$\frac{(1/2) m v_{0°C}^2}{(1/2) m v_{25C°}^2} = \frac{273}{298}$$

From this we deduce, canceling $(1/2)m$ in the numerator against $(1/2)m$ in the denominator, that

$$\frac{v_{0°C}}{v_{25°C}} = \sqrt{\frac{273}{298}}$$

from which

$$v_{25°C} = \sqrt{\frac{298}{273}} \, v_{0°C}$$

$$= (1.04)(4.25 \times 10^4) = 4.42 \times 10^4 \text{ cm/sec} \blacksquare$$

7.9   Reactions involving gases

Once you have a balanced chemical equation, you know directly from the coefficients the relative number of moles of each substance involved in the reaction. If a particular reactant or product is a gas, then we can (by introducing the equation of state) say something about volumes of reagents as well as weights and numbers of moles. For example, in the reaction

$$Zn(s) + 2H^+ \text{(solution)} \longrightarrow Zn^{2+} \text{(solution)} + H_2(g)$$

the product H_2 is gaseous, so we cannot only say we get 1 mole of H_2 in the reaction but alternatively we can say we get 22.4 liters at STP of H_2. In other words, it may sometimes be convenient to focus attention on a directly measurable aspect of a gaseous reagent, such as its volume.

■ EXAMPLE 218.   You are given 1.00 lb of zinc. How many liters of hydrogen, to be measured over water at 22°C and a barometric pressure of 743 mm Hg, can you generate from it?

SOLUTION: (Figure out how many moles of Zn you have. From the balanced equation figure out how many moles of H_2 this will give. Use the gas laws to find out what volume this occupies under the specified conditions.)

$$(1.00 \text{ lb of Zn}) \left(454 \, \frac{g}{lb}\right) = 454 \text{ g of Zn}$$

One mole (or 1 gram-atom) of Zn weighs 65.37 g.

$$\frac{454 \text{ g of Zn}}{65.37 \text{ g/mole}} = 6.95 \text{ moles of Zn}$$

The equation $Zn + 2H^+ \rightarrow Zn^{2+} + H_2$ says that 1 mole of Zn reacts to produce 1 mole of H_2. So, we conclude that 6.95 moles of Zn will produce 6.95 moles of H_2. This H_2 is to be measured under the following conditions:

Temperature = 22°C = 295°K

Pressure = (743 mm Hg) − (vapor pressure of H_2O at 22°C)

$$= 743 - 19.8 = 723 \text{ mm Hg}$$

$$= \frac{723 \text{ mm Hg}}{760 \text{ mm Hg/atm}} = 0.951 \text{ atm}$$

Number of moles = 6.95 moles

From the equation of state $PV = nRT$ we calculate

$$V = \frac{nRT}{P} = \frac{(6.95)(0.08206)(295)}{(0.951)} = 177 \text{ liters } \blacksquare$$

■ EXAMPLE 219. Given that a Bunsen burner consumes CH_4 at 5.0 liters/min (pressure 773 Torr and temperature 28°C), how many liters per minute of oxygen need to be supplied (pressure 742 Torr and temperature 29°C) if the CH_4 is to be oxidized by O_2 to CO_2 and H_2O?

SOLUTION: (Figure out how many moles of CH_4 are consumed per minute. Then calculate from the equation the required number of moles of O_2. Convert to volume of O_2 at specified conditions.)

Five liters of CH_4 at 773 Torr and 301°K can be reduced to standard conditions as

$$(5.0)\left(\frac{773}{760}\right)\left(\frac{273}{301}\right) = 4.6 \text{ liters}$$

At STP, 1 mole equals 22.4 liters, so 4.61 liters of gas STP is

$$\frac{4.6 \text{ liters}}{22.4 \text{ liters/mole}} = 0.21 \text{ mole}$$

The equation for the reaction is

$$CH_4 + 2O_2 \longrightarrow CO_2 + 2H_2O$$

and shows that for every mole of CH_4, we need 2 moles of O_2. Therefore, for 0.21 mole of CH_4 we need 0.42 mole of O_2.

At 742 Torr and 302°K, 0.42 mole is

$$(0.42 \text{ mole})\left(22.4 \frac{\text{liters}}{\text{mole}}\right)\left(\frac{760}{742}\right)\left(\frac{302}{273}\right) = 11 \text{ liters}$$

Thus, we need 11 liters of O_2 per minute. ■

Exercises

*7.1. To squeeze 30.0 liters of nitrogen gas at 1.00 atm pressure into a volume of 1.50 liters, how much pressure needs to be applied? Assume constant temperature and ideal behavior.

*7.2. Given a sealed can containing 0.500 liter of nitrogen gas at 0.960 atm and 25°C. If the can is thrown into a fire where the temperature is 600°C, what will the pressure in the can become? Assume constant volume and ideal behavior.

**7.3. A collapsed polyethylene bag of unknown capacity is first connected to a nitrogen tank and half inflated to 730 Torr at 25°C. Then the bag is connected to a CO_2 tank and inflated the rest of the way to give a final pressure of 730 Torr and a temperature of 25°C. If the weight of the final contents of the bag is 3.60 g, what is the final fully inflated volume? Assume ideal behavior. (*Answer*: 2.54 liters.)

**7.4. Suppose that a gas pump transfers a volume of 250 cm^3 of nitrogen gas at 1.00 atm and 25°C into a 150-cm^3 container filled with oxygen gas at 1.00 atm and 35°C. If the final temperature becomes 32°C, what will be the final pressure? Assume ideal behavior.

**7.5. On a day when the barometric pressure is 740 Torr you are assigned the job of collecting by water displacement at 35°C the exact equivalent of 460 cm^3 of oxygen gas at STP. What volume should you collect under the actual experimental conditions? To compensate for the fact that the gas may not be fully saturated with water, should you collect more or less than this volume? Justify your answer.

**7.6. A 3.00-liter mixture of CO_2 and O_2 gas, total pressure 740 Torr and temperature 50°C, is scrubbed with a water spray at 20°C so as to dissolve all the CO_2 and remove it from the gas phase. If the final observed gas volume at 740 Torr and 20°C is 2.50 liters, what fraction of the molecules in the original mixture must have been CO_2? (*Answer*: 10.4%.)

*7.7. On being driven, an automobile tire is observed to show an increase in pressure from 26.0 lb/in^2 above atmospheric to 30.0 lb/in^2. Assuming the atmospheric pressure is 14.7 lb/in^2 and the tire starts cold at 25°C, what must the final temperature have become? Assume no change in tire volume.

*7.8. The nitrogen in a 30.0-liter container at 740 Torr and 55°C and the hydrogen in a 20.0-liter container at 650 Torr and 15°C are pumped into a 25.0-liter container at 32°C. What will the final pressure be?

*7.9. You are given a 1.00-g sample of a dry gas that occupies 1.00 liter at a pressure of 545 Torr and a temperature of 25°C. The sample is bubbled through water at 35°C and collected over water so that the total pressure is 640 Torr and the temperature is 35°C. What will be the observed volume of the wet gas? (*Answer*: 0.942 liter.)

7.10. A sealed tube contains 10.0 cm³ of oxygen gas and 10.0 cm³ of liquid water at a temperature of 48°C and a pressure of 740 Torr. The whole assembly is then plunged into ice water so the temperature becomes 0°C. What will the total final pressure be? Assume the total volume of the tube remains at 20.0 cm³.

7.11. A 30.0-g chunk of Dry Ice (CO_2) is dropped into a bottle of 0.750-liter volume at an initial air pressure of 1.00 atm and a temperature of 25°C. The bottle is now tightly capped and the CO_2 allowed to evaporate. What will be the ultimate pressure in the bottle assuming the temperature stays at 25°C?

7.12. How big a volume of dry oxygen gas at STP would you need to take to get the same number of oxygen molecules as there are hydrogen molecules in 25.0 liters at 0.850 atm and 35°C? (*Answer*: 18.8 liters.)

7.13. You are given an unknown gas X, the molecular weight of which is to be determined. It requires 36.8 g of gas X to inflate a plastic bag to a volume of 35.0 liters at a pressure of 740 Torr and a temperature of 26°C. What is the molecular weight of X?

7.14. What is the maximum volume of H_2O gas at 0.940 atm and 440°C that can be made from 0.565 liter of H_2 gas at 0.980 atm and 98°C if you have available no more than 0.135 liter of O_2 gas at 0.960 atm and 72°C?

7.15. How many moles of gas would it take to fill an average man's lungs, total capacity of which is about 4.5 liters? Assume 1.00 atm pressure and 37.0°C? (*Answer*: 0.18 mole.)

7.16. To what temperature would you have to heat CO_2 gas to give its molecules the same average velocity as that shown by O_2 molecules at 0°C, namely, 4.25×10^4 cm/sec?

7.17. What is the ratio of molecular masses of gases X and Y if X molecules show an average velocity of 5.3×10^4 cm/sec at 300°C and Y molecules show an average velocity of 4.9×10^4 cm/sec at 277°C?

7.18. The quantity of air inhaled and exhaled in one breath by an average man at rest is about 0.5 liter. Assuming the intake at 1 atm pressure and 25°C is 20% oxygen, and assuming all this oxygen is used to oxidize sugar ($C_6H_{12}O_6$) to gaseous CO_2 and H_2O, calculate the partial pressure of CO_2 you expect in the exhalation. Assume the final temperature is 37°C. (*Answer*: 0.21 atm.)

7.19. A fuel gas consisting of 65% CH_4 and 34.5% C_2H_6 by volume is completely burned to CO_2 and H_2O. If all the products were trapped and weighed what ratio of CO_2 to H_2O by weight should you observe?

***7.20.** Gaseous phosgene, $COCl_2$, at sufficiently high temperature may be partially decomposed into gaseous CO and Cl_2. In a given experiment it is found that injection of 0.631 g of $COCl_2$ into a volume of 0.472 liter at 900°K produces a total pressure of 1.872 atm. What fraction of the phosgene must have decomposed? (*Answer*: 87.5%.)

8

THERMOCHEMISTRY

THE FIELD of thermochemistry includes two important types of problems: (1) What temperature change occurs when a certain amount of heat is added to a substance?; (2) What heat effects occur when substances undergo chemical change?

8.1 Specific heat

"Specific heat" is commonly defined as the amount of heat needed by 1 g of a substance to raise its temperature by 1°C. (If you want to be fussy, the foregoing is a definition of "thermal capacity," and you should define specific heat as a unitless quantity that tells the ratio of the thermal capacity of a substance to the thermal capacity of water. However, few chemists stick with the exact definition, and the term "specific" is now generally taken to mean per gram or per some other given unit. To take specific heat to mean "per gram" is acceptable.) We shall refer to specific heat in terms of "calories per gram per degree." The *calorie* is defined as the amount of heat needed to warm 1 g of H_2O from 14.5 to 15.5°C. In practice, it turns out that the specific heat of water is essentially constant at about 1.00 cal/g deg all the way from 0 to 100°C.

The specific heat of an element or a compound is a characteristic property of the element or compound. Its numerical value depends on how the added heat gets distributed among the various forms of kinetic energy, such as vibrational motion. This in turn depends on the masses of the atoms involved and their bonding. Many years ago, before much was known about the structure of materials, it was recognized as an experimental fact that, for many solid elements, especially at higher temperatures, the "specific heat" times the "atomic weight" is about equal to 6.3. This law is frequently referred to as the law of Dulong and Petit, and the value 6.3 as the Dulong and Petit constant.

150

■ EXAMPLE 220. If it takes 9.98 cal to heat a chunk of gold weighing 18.69 g from 10.0 to 27.0°C, what is the specific heat of gold?

SOLUTION: It takes 9.98 cal for 18.69 g.
Per gram the required heat is

$$\frac{9.98 \text{ cal}}{18.69 \text{ g}} = 0.534 \frac{\text{cal}}{\text{g}}$$

But this takes it from 10.0 to 27.0°C, a rise of 17.0°C.
Per degree, the rise would be

$$\frac{0.534 \text{ cal/g}}{17.0 \text{ deg}} = 0.0314 \text{ cal/g deg.}$$

[*Note*: Both the " gram " and the " degree " are in the denominator.] ■

■ EXAMPLE 221. If the specific heat of nickel is 0.1146 cal/g deg, how much heat would you need to take a 168-g slug of nickel from -18.6 to $+57.2$°C?

SOLUTION: The increase in temperature is 75.8°C.
The weight of the sample is 168 g.
Therefore,

$$\text{the heat required} = \left(0.1146 \frac{\text{cal}}{\text{g deg}}\right)(75.8 \text{ deg})(168 \text{ g}) = 1.46 \text{ kcal} ■$$

■ EXAMPLE 222. A stick of molybdenum weighing 237 g and starting at a temperature of 100.0°C is thrust into 244 g H_2O starting at 10.0°C. If the final observed temperature of the whole system is 15.3°C, what would be the specific heat of molybdenum?

SOLUTION: The heat gained by the water on warming up equals the heat lost by the molybdenum on cooling.

Heat gained by H_2O = (specific heat)(mass)(temperature change)

$$= \left(1.00 \frac{\text{cal}}{\text{g deg}}\right)(244 \text{ g})(15.3° - 10.0°) = 1300 \text{ cal}$$

Heat lost by Mo = (specific heat)(mass)(temperature change)
$$= (\text{specific heat of Mo})(237 \text{ g})(100.0° - 15.3°) = 1300 \text{ cal}$$

$$\text{Specific heat of Mo} = \frac{1300 \text{ cal}}{(237 \text{ g})(84.7 \text{ deg})} = 0.065 \frac{\text{cal}}{\text{g deg}} ■$$

■ EXAMPLE 223.   An unknown metal X has a measured specific heat of 0.0408 cal/g deg. What would you estimate for its atomic weight?

SOLUTION: The Dulong and Petit law says

(specific heat)(atomic weight) ≅ 6.3.

So,

$$\text{atomic weight} \cong \frac{6.3}{\text{specific heat}} = \frac{6.3}{0.0408} \cong 150. \ ■$$

8.2   Heat of fusion

When a substance melts, it takes a certain amount of heat to convert it from the solid state to the liquid state. This heat is generally specified at the normal melting point and is called the *heat of fusion* or the *heat of melting*. It can be calculated either per gram or, more meaningfully, per mole. If per mole, it is called the *molar heat of fusion*.

■ EXAMPLE 224.   It takes 79.71 cal to melt 1 g of H_2O. What is the molar heat of fusion of H_2O?

SOLUTION: One mole of H_2O weighs 18.015 g.
If it takes 79.71 cal to melt 1 g, it will take

$$\left(79.71 \ \frac{\text{cal}}{\text{g}}\right)\left(18.015 \ \frac{\text{g}}{\text{mole}}\right)$$

or 1436 cal to melt 18.015 g.
Thus, the molar heat of fusion of H_2O is 1.436 kcal/mole. ■

The amount of heat required to melt 1 mole of material is equal numerically to the amount of heat liberated when 1 mole of material is frozen or solidified. Thus, we sometimes talk about *heat of freezing* or *heat of solidification*. Because heat of fusion needs to be *added* to the system, whereas heat of freezing comes *out of* the system to the surroundings, we say that these are of opposite sign. However, they are of equal magnitude.

■ EXAMPLE 225.   If the heat of fusion of H_2O is +1.436 kcal/mole, what is the molar heat of freezing of H_2O?

SOLUTION: To go from 1 mole of solid H_2O to 1 mole of liquid H_2O requires 1.436 kcal from the surroundings.

If we reverse the process and go from 1 mole of liquid H_2O to 1 mole of solid H_2O, we give up to the surroundings 1.436 kcal. Thus, if we assign a plus sign to the take-up of heat, we give a minus sign for the evolution of heat.

Therefore, the molar heat of freezing of H_2O is -1.436 kcal/mole. ∎

The question of sign ($+$ or $-$) for a heat effect is apt to be very confusing because there are two major ways of indicating heat changes, one by writing it as a chemical change and the other by writing it as a change in a property. For example, the fact that 1 mole of solid H_2O absorbs 1.436 kcal of heat in being converted to 1 mole of liquid H_2O can be represented as

$$H_2O(s) + 1.436 \text{ kcal} \longrightarrow H_2O(\ell)$$

where the 1.436 kcal is treated like a chemical reagent that is used up in the change from left to right. (We do not need to say 1.436 kcal "per mole" because we are reading the whole equation in terms of moles.) But anything appearing on the left side of a chemical equation can be moved over to the right side provided the algebraic sign is reversed. In other words, the reaction above can also be written

$$H_2O(s) \longrightarrow H_2O(\ell) - 1.436 \text{ kcal}$$

where the 1.436 kcal appears as a negative reagent on the right.

∎ EXAMPLE 226. Given that the molar heat of fusion of NaCl is 6.8 kcal/mole, write the fusion process as a chemical equation.

SOLUTION:

$$NaCl(s) + 6.8 \text{ kcal} \longrightarrow NaCl(\ell)$$

or

$$NaCl(s) \longrightarrow NaCl(\ell) - 6.8 \text{ kcal.} \quad ∎$$

This example indicates one obvious source of trouble: The heat shows up as a $+$ or $-$, depending on which side of the equation we put it on. Difficulties from this source can be minimized if we agree to try to keep the heat on the side where it has the plus sign. In other words, the first way of writing the equation

$$NaCl(s) + 6.8 \text{ kcal} \longrightarrow NaCl(\ell)$$

is preferred. If we reverse the process, we put the heat on the other side of the equation.

■ EXAMPLE 227. Including the heat effect, write the chemical equation for the process 1 mole of liquid NaCl going to 1 mole of solid NaCl.

SOLUTION: $NaCl(\ell) \rightarrow NaCl(s) + 6.8$ kcal. ■

Note there is no ambiguity here. When we go from left to right, 6.8 kcal is produced as a product and gets liberated by the system to the surroundings. The ambiguity comes in if we do not write the equation and if we do not consider the heat as a reagent. Then we have to decide how we are going to distinguish liberation of heat from consumption of heat. The problem becomes crucial when, as frequently done, the chemical change is written as a chemical equation but the heat effect is indicated separately to the side. In such cases, you need a strict rule to follow and it is this: "A heat effect is considered positive if heat is absorbed *by the system*." The rationale for this is that there is a property called the "heat content" (or enthalpy), symbolized by H, which can be used to describe a system. If the system absorbs heat, its heat content increases. The *increase in heat content* is usually designated as ΔH, which you read "delta aitch," and frequently it is given alongside the chemical equation. For example,

$$NaCl(s) \longrightarrow NaCl(\ell); \Delta H = +6.8 \text{ kcal}$$

or

$$NaCl(\ell) \longrightarrow NaCl(s); \Delta H = -6.8 \text{ kcal}$$

Obviously, there may be troubles. To minimize them, remember a positive ΔH means heat is *absorbed by the system* and a negative ΔH means heat is *evolved to the surroundings*. ΔH is a molar quantity and applies if the chemical equation is read in moles.

8.3   Heat of evaporation

It takes heat to convert 1 mole of a substance from a condensed phase, such as liquid or solid, to the gas phase. For liquids, the quantity is called the molar *heat of evaporation* (or heat of vaporization) and is generally quoted at the normal boiling point of the liquid. For solids, the heat required to convert 1 mole to the vapor is frequently called *heat of sublimation*, although heat of evaporation is also acceptable. For solids, the heat of evaporation can be quoted at almost any temperature, so it is usual to specify at what pressure the gas formed will be. The amount of heat required is, of course, characteristic of the material, but also it depends on the pressure of the gas phase produced.

Heat effects associated with vaporization can be indicated by writing a chemical equation in which heat is a reactant or product, or by specifying the ΔH associated with the given change. For example, for H_2O, where it takes 9.717 kcal to convert 1 mole of liquid H_2O into 1 mole of gaseous H_2O at 1 atm pressure, we can write either

$$9.717 \text{ kcal} + H_2O(\ell) \longrightarrow H_2O(g, p = 1 \text{ atm})$$

or

$$H_2O(\ell) \longrightarrow H_2O(g, p = 1 \text{ atm}); \qquad \Delta H = +9.717 \text{ kcal}$$

■ EXAMPLE 228.   If it takes 9.717 kcal to evaporate 1 mole of liquid H_2O at the normal boiling point to gas at 1 atm, how much heat would be required to do this for 1.00 g of H_2O?

SOLUTION: One mole of H_2O weighs 18.015 g.
To evaporate 1 mole of H_2O requires 9.717 kcal.
Therefore, for 1.00 g it takes

$$\frac{9.717 \text{ kcal/mole}}{18.015 \text{ g/mole}} = 0.5394 \text{ kcal.} \ ■$$

■ EXAMPLE 229.   How many calories of heat would you have available if you took 1.00 liter of steam ($T = 373°K, p = 1.00$ atm) and condensed it to liquid H_2O at 100°C? Assume ideal behavior.

SOLUTION: (From Example 228 we know that under these conditions the heat of evaporation is 9.717 kcal/mole. All we have to do is to figure out how many moles are condensed under these conditions.)
The general gas law says $PV = nRT$.

$$n = \frac{PV}{RT} = \frac{(1.00)(1.00)}{(0.08206)(373)} = 0.0327 \text{ mole}$$

If it takes 9.717 kcal to evaporate 1 mole of H_2O, the same heat must be liberated when 1 mole condenses.
If we get 9.717 kcal liberated on condensation of 1 mole, then with 0.0327 mole we liberate

$$(0.0327 \text{ mole})\left(9.717 \frac{\text{kcal}}{\text{mole}}\right) = 0.318 \text{ kcal} = 318 \text{ cal} \ ■$$

8.4   The Clausius-Clapeyron equation

The smaller the attractive forces between molecules of a liquid, the more likely the liquid is to be volatile. We should expect then that there will be a connection between the energy required to separate liquid molecules from each other and the degree of volatility. A quantitative measure of the energy required to separate liquid molecules from each other is the heat of evaporation—that is, the heat required to convert 1 mole of the liquid to 1 mole of the vapor. A quantitative measure of the degree of volatility is, of course, the vapor pressure of the liquid. So, we look for a relation between the heat of evaporation of a liquid and its vapor pressure. The relation will also need to involve the temperature because the temperature measures the average kinetic energy of the molecules, which after all is the source of the energy that overcomes the attractive forces in the evaporation process.

Such a relation, deducible from the first principles of thermodynamics, is the Clausius–Clapeyron equation. If ideal behavior is assumed in the gas phase, it has the form

$$\log p = \frac{-\Delta H}{4.576T} + C$$

where p is the vapor pressure of the liquid at absolute temperature T, ΔH is the heat required to evaporate 1 mole, and C is a constant that varies from liquid to liquid, depending also on the units used for pressure. For p we shall use millimeters of Hg (or torricellis). ΔH is expressed in calories per mole. If we know the numerical values of ΔH and C, we should be able to calculate p at any temperature T. Numerical values of ΔH and C can be found in tables in the handbooks; they can also be determined by knowing the vapor pressure p at two different temperatures.

■ EXAMPLE 230.   For liquid water, the vapor pressure is 17.535 Torr at 20.0°C. and 55.423 Torr at 40.0°C. Calculate ΔH and C from these data.

SOLUTION: (1) One set of conditions is $T = 20.0 + 273.15 = 293.2°K$, where $p = 17.535$ Torr. Substitute these numbers in the Clapeyron equation and get

$$\log p = -\frac{\Delta H}{4.576T} + C$$

$$\log (17.535) = -\frac{\Delta H}{(4.576)(293.2)} + C$$

$$1.2439 = -\frac{\Delta H}{1342} + C$$

(2) The other set of conditions is $T = 40.0 + 273.15 = 313.2°K$, where $p = 55.423$ Torr. Substitute these in the equation and get

$$\log (55.324) = -\frac{\Delta H}{(4.576)(313.2)} + C$$

$$1.7429 = -\frac{\Delta H}{1433} + C$$

We have two simultaneous equations, (1) and (2), which we can solve by solving for C in each case and setting the results equal to each other:

$$1.2439 + \frac{\Delta H}{1342} = 1.7429 + \frac{\Delta H}{1433}$$

In this way we get $\Delta H = 10,500$ cal/mole Substituting this value of ΔH in either (1) or (2) gives $C = 9.07$. ■

■ EXAMPLE 231. Using the constants calculated in the previous example, what vapor pressure would you expect for liquid water at 30.0°C?

SOLUTION: For H_2O the Clapeyron equation would have the form

$$\log p = \frac{-10,500}{4.576T} + 9.07$$

Substitute $T = 30°C = 30 + 273 = 303°K$.
Get $p = 31.4$ Torr. ■

■ EXAMPLE 232. Acetone, CH_3COCH_3, shows a measured vapor pressure of 100 Torr at 7.7°C and 400 Torr at 39.5°C. At what temperature would you predict that its normal boiling point will occur? Assume ΔH to be a constant over the temperature range required and that behavior is ideal.

SOLUTION: Substituting the two sets of data into the Clapeyron equation, we solve for ΔH and C. We get 7600 cal/mole and 7.91, respectively. Then we feed these constants into the Clapeyron equation, getting

$$\log p = \frac{-7600}{4.576T} + 7.92$$

The normal boiling point is defined as the temperature at which the vapor pressure of the liquid reaches 1 atm (760 Torr). So, we solve the foregoing equation for T after substituting $p = 760$ Torr. We find $T = 330°K$ (or 57°C), which compares to the observed normal boiling point of 56.5°C. ■

8.5 Heat of reaction

The amount of heat produced or consumed in a chemical reaction is called the "heat of reaction," and it can be shown either as a heat term in the chemical equation or by giving the ΔH for the chemical change as written. Because the heat change associated with a reaction is affected by the state of the materials—whether solid, liquid, or gas—you should be careful in writing the chemical equations to indicate by (s), (ℓ), or (g) what the state is for each material. The heat of reaction generally quoted is for molar quantities and is applicable when the accompanying chemical equation is read off in moles. For example,

$$2H_2(g) + O_2(g) \longrightarrow 2H_2O(g); \quad \Delta H = -115.60 \text{ kcal}$$

means that 2 moles of gaseous hydrogen react with 1 mole of gaseous oxygen to produce 2 moles of gaseous water and liberate 115.60 kcal in the process. (Recall that ΔH refers to the increase in the heat content of the system. If the ΔH is negative, the heat content of the system has decreased and heat has been liberated to the surroundings.)

■ EXAMPLE 233.   Using the data just given, calculate how much heat will be liberated when 1.00 g of hydrogen reacts as in the equation above.

SOLUTION: The equation tells us that when 2 moles of H_2 are used up, 115.60 kcal of heat is liberated.
The heat liberated is

$$\frac{115.60 \text{ kcal}}{2 \text{ moles of } H_2} = 57.80 \text{ kcal/mole of } H_2$$

One mole of H_2 weighs 2.016 g.

$$\frac{1.00 \text{ g of } H_2}{2.016 \text{ g/mole}} = 0.496 \text{ mole of } H_2$$

If we are to use up 0.496 mole of H_2 and the heat liberated is 57.80 kcal/mole of H_2, the total heat liberated will be the product of these two:

$$(0.496 \text{ mole})\left(57.80 \frac{\text{kcal}}{\text{mole}}\right) = 28.7 \text{ kcal} \; ■$$

Although the term "heat of reaction" is generally applicable to any change for which the chemical equation is specified, there are some special types of reaction for which heats take on special names. For example, the heat liberated by the reaction A + B → AB, where AB is a compound formed from the elements A and B, is called the *heat of formation*. In fact, if we are told that the heat of formation of compound X is so much, then we understand it to mean that we form X from the component elements in their standard (or commonly referred to) states and we liberate the heat specified. Thus, if we are told that the heat of formation of $CaCO_3(s)$ is 288.45 kcal/mole, we can refer this to the elemental states $Ca(s)$, $C(s)$, $O_2(g)$ from which $CaCO_3$ can be considered to be derived and can write an appropriate equation, which would be

$$Ca(s) + C(s) + \frac{3}{2}O_2(g) \longrightarrow CaCO_3(s); \qquad \Delta H = -288.45 \text{ kcal}$$

If you want to get rid of the fraction, you can double the whole equation, but then you must double ΔH also.

■ EXAMPLE 234. The heat of formation of solid $KClO_3$ is 93.50 kcal/mole. The standard states for the component elements are $K(s)$, $Cl_2(g)$, and $O_2(g)$. Write an appropriate equation and ΔH for forming $KClO_3$.

SOLUTION:

$$K(s) + \frac{1}{2}Cl_2(g) + \frac{3}{2}O_2(g) \longrightarrow KClO_3(s);$$

$$\Delta H = -93.50 \text{ kcal},$$

or

$$2K(s) + Cl_2(g) + 3O_2(g) \longrightarrow 2KClO_3(s); \quad \Delta H = -187.0 \text{ kcal.} \quad ■$$

Another special kind of reaction is the combustion reaction in which a compound, generally a hydrocarbon or derivative, is oxidized by O_2 to form $CO_2(g)$, $H_2O(g)$, and perhaps other products (e.g., $N_2(g)$ if N is one of the elements in the compound). Again, there is a special name for the associated heat of reaction; it is called *heat of combustion*. There are in the handbooks extended tables that list heats of combustion, giving the formula of the compound, its state, and the number of kilocalories in the combustion reaction. For example, you might be told that the heat of combustion of acetaldehyde, CH_3CHO, in its liquid state is 279.0 kcal. Immediately, you can write the chemical equation for the combustion reaction, since it is assumed the products are $CO_2(g)$ and $H_2O(g)$ [and $N_2(g)$ if necessary]. You supply the $O_2(g)$ and the coefficients for the equation $CH_3CHO(\ell)$ + $5/2O_2(g) → 2CO_2(g) + 2H_2O(g); \Delta H = -279.0$ kcal.

■ EXAMPLE 235. How much heat would be liberated by burning 1.00 g of liquid CH_3CHO as in the equation just given?

SOLUTION: The equation tells us 279.0 kcal is liberated when 1 mole of CH_3CHO is burned.
One mole of CH_3CHO weighs 44.05 g.

$$\frac{1.00 \text{ g of } CH_3CHO}{44.05 \text{ g/mole}} = 0.0227 \text{ mole}$$

If burning 1 mole liberates 279.0 kcal, then burning 0.0227 mole will liberate

$$(0.0227 \text{ mole})\left(279.0 \frac{\text{kcal}}{\text{mole}}\right) = 6.33 \text{ kcal.} \quad ■$$

8.6 Hess's law

This law summarizes the observation that the amount of heat liberated or absorbed in a chemical change is the same whether the change occurs in one step or in several consecutive steps. In other words, once the initial state and final state are specified, the heat change is fixed and does not depend on how we go from the initial state to the final state. In practice, Hess's law is important, because it means that we can evaluate indirectly heats of reaction that perhaps are difficult to measure in a direct experiment. We can do this because, just as chemical equations can be added together (or subtracted), their heat changes can also be added (or subtracted).

The following shows how the Hess law works. We have been told that the heat of formation of $CaCO_3(s)$ is 288.45 kcal. This means we can write the chemical equation, including the heat effect, as follows:

$$Ca(s) + C(s) + \frac{3}{2}O_2(g) \longrightarrow CaCO_3(s) + 288.45 \text{ kcal}$$

However, the reaction as written is very hard to study directly, and it is much more practical to examine the following set of reactions:

$$Ca(s) + \frac{1}{2}O_2(s) \longrightarrow CaO(s) + 151.85 \text{ kcal}$$

$$C(s) + O_2(g) \longrightarrow CO_2(g) + 94.05 \text{ kcal}$$

$$CaO(s) + CO_2(g) \longrightarrow CaCO_3(s) + 42.55 \text{ kcal}$$

If we just add up these equations, keeping to the left and to the right of the arrow each of the molar quantities as it appears, then we get

$$\left(\begin{array}{c} Ca(s) + \dfrac{3}{2}O_2(g) + C(s) \\ + \ CaO(s) + CO_2(g) \end{array}\right) \longrightarrow \left(\begin{array}{c} CaO(s) + CO_2(g) + CaCO_3(s) \\ + 151.85 \text{ kcal} + 94.05 \text{ kcal} \\ + 42.55 \text{ kcal} \end{array}\right)$$

Canceling those items that are duplicated on left and right, we get

$$Ca(s) + \frac{3}{2}O_2(g) + C(s) \longrightarrow CaCO_3(s) + 288.45 \text{ kcal}$$

which is the overall reaction we wish to study.

The foregoing sequence is not unique in leading to the final equation. For example, another possible set, which adds up to the same overall equation, is the following:

$$Ca(s) + 2C(s) \longrightarrow CaC_2(s) + 15.0 \text{ kcal}$$

$$CaC_2(s) + \frac{5}{2}O_2(g) \longrightarrow CaCO_3(s) + CO_2(g) + 367.5 \text{ kcal}$$

$$94.05 \text{ kcal} + CO_2(g) \longrightarrow C(s) + O_2(g)$$

$$\overline{\rule{0pt}{0pt}\hspace{10em}}$$

$$Ca(s) + C(s) + \frac{3}{2}O_2(g) \longrightarrow CaCO_3(s) + 288.4 \text{ kcal}$$

Note that in this sequence the formation of CO_2 occurs in the reverse sense as the third step.

■ EXAMPLE 236.   Calculate the heat evolved in the formation of 1 mole of $PbSO_4(s)$ from its elements, given the following:

$$Pb(s) + S(s) \longrightarrow PbS(s) + 22.54 \text{ kcal}$$

$$PbS(s) + 2O_2(g) \longrightarrow PbSO_4(s) + 196.96 \text{ kcal}$$

SOLUTION: Add the two equations, cancel the $PbS(s)$, which will appear on both sides, and the result is

$$Pb(s) + S(s) + 2O_2(g) \longrightarrow PbSO_4(s) + 219.50 \text{ kcal}$$

So, the heat evolved in the formation of 1 mole $PbSO_4(s)$ is 219.50 kcal. ■

■ EXAMPLE 237. Given that ΔH of formation of $CO_2(g)$ is -94.05 kcal/mole and the ΔH of formation of $CO(g)$ is -26.41 kcal/mole, calculate the ΔH for $CO(g) + \frac{1}{2} O_2(g) \rightarrow CO_2(g)$.

SOLUTION: Write the chemical equations for formation of CO_2 and of CO from the elements and subtract the CO equation from the CO_2 equation.

$$C(s) + O_2(g) \longrightarrow CO_2(g); \quad \Delta H = -94.05 \text{ kcal}$$

$$-\left(C(s) + \frac{1}{2} O_2(g) \longrightarrow CO(g); \quad \Delta H = -26.41 \text{ kcal} \right)$$

$$\frac{1}{2} O_2(g) \longrightarrow CO_2(g) - CO(g); \quad \Delta H = -94.05 - (-26.41)$$
$$= -67.64 \text{ kcal}$$

Note that we subtract ΔH's just like everything else.

Note also that $CO(g)$ occurs with a minus sign to the right of the arrow; it can be transferred to the left of the arrow if we change sign.

We get finally

$$CO(g) + \frac{1}{2} O_2(g) \longrightarrow CO_2(g); \quad \Delta H = -67.64 \text{ kcal} ■$$

■ EXAMPLE 238. Given that the ΔH of formation of $FeO(s)$ is -64.04 kcal/mole and the ΔH of formation of $Fe_2O_3(s)$ is -196.5 kcal/mole, calculate the ΔH for the reaction

$$2FeO(s) + \frac{1}{2} O_2(g) \longrightarrow Fe_2O_3(s)$$

SOLUTION:

$$Fe(s) + \frac{1}{2} O_2(g) \longrightarrow FeO(s); \quad \Delta H = -64.04 \text{ kcal}$$

$$2Fe(s) + \frac{3}{2} O_2(g) \longrightarrow Fe_2O_3(s); \quad \Delta H = -196.5 \text{ kcal}$$

In this case we need to double the first equation all the way across and then subtract it from the second equation.

The result

$$\frac{3}{2} O_2(g) - O_2(g) \quad\longrightarrow\quad Fe_2O_3(s) - 2FeO(s);$$

$$\Delta H = -196.5 - 2(-64.04)$$

can be rewritten as

$$2FeO(s) + \frac{1}{2} O_2(g) \quad\longrightarrow\quad Fe_2O_3(s); \quad \Delta H = -68.4 \text{ kcal } \blacksquare$$

8.7 Calorimetry

Calorimetry refers to the experimental determination of the heat of chemical reaction. The requirements are a well-insulated container in which to carry out the reaction, knowledge of the weight of each reactant and product, and some way to determine the heat change that accompanies the conversion from reactants to products. For the latter purpose, one of the simplest ways is to have the reaction vessel immersed in a known weight of water, for which the temperature change is carefully measured as the reaction occurs. Knowing the specific heat of water and its temperature change, we can calculate the number of calories gained (or lost) by the water. Assuming no heat leaks to the surroundings, the heat change in the water must be matched by the heat produced (or used up) in the reaction.

■ EXAMPLE 239.   A solid sample of benzoic acid, C_6H_5COOH, weighing 1.89 g is placed with excess oxygen in a sealed bomb in a calorimeter containing 18.94 kg of H_2O at 25.00°C. The reaction of complete oxidation is set off and there is observed a temperature rise in the H_2O of 0.632°C. What is the molar heat of combustion of benzoic acid? At 25.00°C, the specific heat of water is 0.99828 cal/g deg.

SOLUTION:

Amount of heat absorbed by the water

$$= \text{(mass of water)(specific heat)(temperature rise)}$$

$$= (18,940 \text{ g})\left(0.99828 \, \frac{\text{cal}}{\text{g deg}}\right)(0.632 \text{ deg}) = 11,900 \text{ cal}$$

$$= \text{amount of heat liberated by the combustion}$$

If the combustion of 1.89 g of benzoic acid liberates 11,900 cal, then for 1 mole of benzoic acid (which weighs 122.13 g) the amount of heat liberated would be

$$\left(122.13 \, \frac{\text{g}}{\text{mole}}\right)\left(\frac{11,900 \text{ cal}}{1.89 \text{ g}}\right) = 769 \, \frac{\text{kcal}}{\text{mole}}$$

Thus, the molar heat of combustion of C_6H_5COOH is 769 kcal. ∎

Another kind of calorimeter, called the *ice calorimeter*, measures heat changes by the amount of ice that is melted or frozen by the heat of the reaction. Since the heat of fusion of ice is well known at 1.436 kcal/mole, the product of moles-of-ice-melted times 1.436 kcal/mole tells us how much heat has been given up by the chemical change under study.

∎ EXAMPLE 240.   A bomb containing 5.40 g of Al and 15.97 g of Fe_2O_3 is placed in an ice calorimeter containing initially 8.000 kg of ice and 8.000 kg of liquid water. The reaction $2Al(s) + Fe_2O_3(s) \rightarrow Al_2O_3(s) + 2Fe(s)$ is set off by remote control and it is then observed the calorimeter contains 7.746 kg of ice and 8.254 kg of water. What is the ΔH for the foregoing reaction as written?

SOLUTION: The decrease in the amount of ice is from 8.000 to 7.746 kg, amounting to 254 g. So, 254 g of H_2O has been melted. This is

$$\frac{254 \text{ g}}{18.015 \text{ g/mole}} = 14.1 \text{ mole of } H_2O$$

One mole of H_2O requires 1.436 kcal for melting.

$$14.1 \text{ mole of } H_2O \text{ requires } (14.1 \text{ mole})\left(1.436 \frac{\text{kcal}}{\text{mole}}\right) = 20.2 \text{ kcal}$$

This amount of heat, 20.2 kcal, has been liberated by the reaction of 5.40 g of Al and 15.97 g of Fe_2O_3.

$$\frac{5.40 \text{ g of Al}}{26.98 \text{ g/mole}} = 0.200 \text{ mole of Al}$$

$$\frac{15.97 \text{ g of } Fe_2O_3}{159.7 \text{ g/mole}} = 0.100 \text{ mole of } Fe_2O_3$$

Since the balanced equation shows 1 mole Fe_2O_3, we work with that reagent: 20.2 kcal of heat per 0.100 mole of Fe_2O_3 means 202 kcal of heat per 1 mole of Fe_2O_3.

So, when the reaction $2Al(s) + Fe_2O_3(s) \rightarrow 2Fe(s) + Al_2O_3(s)$ occurs as written, 202 kcal is liberated to the surroundings. The heat content of the chemical system goes down by 202 kcal, so we write $\Delta H = -202$ kcal for the process. ∎

Exercises

*8.1. If the specific heat of silver is 0.0573 cal/g deg, how much heat would it take to raise the temperature of 1 kg of silver from $-50°C$ to $+150°C$?

*8.2. A 50-g chunk of unknown metal X is heated to 98.5°C and then dropped into 450 g of water initially at 25.00°C. The water temperature is observed to rise to 26.47°C. Calculate the specific heat of X.

**8.3. Given that the specific heat of aluminum is 0.226 cal/g deg, what final temperature would you expect to see if a 35-g chunk of aluminum initially at 100°C were dropped into 320 g of water starting at 23.75°C? (*Answer*: 25.59°C.)

*8.4. The newly discovered element hahnium is reported to have an atomic weight of 260. What value would you predict for its specific heat?

*8.5. To melt 1 g of mercury requires as much heat as to warm 1 g of liquid H_2O from 15.00°C to 17.82°C. Calculate the molar ΔH for the process $Hg(s) \rightarrow Hg(\ell)$.

**8.6. How much heat would it take to convert 5.0 g of solid H_2O at 0°C into gaseous H_2O at 1.00 atm pressure and 100°C? (*Answer*: 3600 cal.)

**8.7. If the heat of evaporation for ethyl alcohol, C_2H_5OH, is 9200 cal/mole at its normal boiling point, how many grams of ethyl alcohol could you evaporate by condensing 25 liters of steam ($p = 1$ atm) to liquid water at 100°C?

**8.8. Some boiling water (100°C) is poured on a hot stove so that it gets converted to steam at $p = 1$ atm. How much water would it take to cool a 15-kg mass of iron from 600°C to 100°C by this process? The specific heat of iron is 0.13 cal/g deg.

**8.9. If a bullet weighing 15.0 g and traveling 9000 cm/sec is fired into a bucket of water containing 500 g of H_2O at 25.00°C, what temperature rise should be observed in the H_2O, assuming all the kinetic energy of the bullet is converted to heat? Note that 1 g $cm^2/sec^2 = 2.39 \times 10^{-8}$ cal. (*Answer*: 0.029°C.)

**8.10. The vapor pressure of ethyl alcohol is 400 Torr at 63.5°C and 760 Torr at 78.4°C. Calculate the molar heat of vaporization in this temperature range.

**8.11. Using the data given in Exercise 8.10, what vapor pressure would you expect for ethyl alcohol at 70.0°C?

**8.12. The normal boiling point of liquid ammonia is $-33.4°C$; its heat of vaporization is 5560 cal/mole. What would be its boiling point at a pressure of 700 Torr? (*Answer*: $-35.0°C$.)

**8.13. Liquid nitrogen is a very useful refrigerant for low-temperature experiments. Its normal boiling point is $-195.8°C$; its heat of vaporization is 1330 cal/mole. You can cool liquid nitrogen by pumping on it. If you regulate the pump to maintain a pressure of 100 Torr, what temperature should you get?

**8.14. Liquid helium has a normal boiling point of $-268.9°C$. At this temperature it takes 6.0 cal to evaporate a gram of helium from the liquid state to the gaseous state. Suppose you want to use a liquid helium bath to produce a temperature of $2.0°$ above absolute zero. To what pressure would you need to keep evacuated the space above the liquid helium?

**8.15. Liquid oxygen doubles its vapor pressure in going from $-183.1°C$ to $-176.0°C$; liquid nitrogen doubles its vapor pressure in going from $-195.8°C$ to $-189.2°C$. What is the ratio of their heats of vaporization? (*Answer*: 1.3.)

*8.16. Given that $\Delta H = -115.6$ kcal for the reaction $2H_2(g) + O_2(g) \rightarrow 2H_2O(g)$, how much heat would be liberated on formation of 1.00 g of water vapor from the elements?

*8.17. The heat evolved on formation of liquid HNO_3 from $H_2(g)$, $N_2(g)$, and $O_2(g)$ is 657 cal per gram of HNO_3. Write an appropriate chemical equation for the reaction and indicate its ΔH.

**8.18. The heat absorbed on formation of $NO(g)$ is 720 cal/g; the heat absorbed on formation of $NO_2(g)$ is 176 cal/g. Calculate ΔH for the reaction $2NO(g) + O_2(g) \rightarrow 2NO_2(g)$. (*Answer*: -27.0 kcal.)

**8.19. The heat evolved on forming $MgO(s)$ from the elements is 3570 cal/g; the heat evolved on forming $Al_2O_3(s)$ from the elements is 3910 cal/g. Calculate ΔH for the reaction $2Al(s) + 3MgO(s) \rightarrow Al_2O_3(s) + 3Mg(s)$.

**8.20. The heat of formation of $CaO(s)$ is 151.9 kcal/mole; that of $CO_2(g)$ is 94.05 kcal/mole. Calculate the ΔH for the reaction of burning 1 mole of $CaC_2(s)$ in $O_2(g)$ to form $CaO(s)$ and $CO_2(g)$.

***8.21. The heat of combustion of ethyl alcohol, $C_2H_5OH(\ell)$, is 327.6 kcal/mole; that of dimethyl ether, $CH_3OCH_3(g)$, is 347.6 kcal/mole. Given that the vapor pressure of dimethyl ether is 760 Torr at $-23.7°C$ and 400 Torr at $-37.8°C$, calculate the molar ΔH that accompanies the rearrangement of atoms in $C_2H_5OH(\ell)$ to give $CH_3OCH_3(\ell)$. (*Answer*: 14.7 kcal.)

*8.22. An inverted U-tube containing $HNO_3(\ell)$ in one arm and $NaOH(s)$ in the other is placed in an ice-water mixture. On inverting the U-tube so that reaction occurs, it is found that 39.6 g of the ice melts. If the starting charge of reactants was 0.10 mole each, what must be the ΔH for the following reaction?

$$HNO_3(\ell) + NaOH(s) \longrightarrow NaNO_3(aq) + H_2O(\ell)$$

**8.23. The melting point of mercury is $-38.87°C$. Its heat of fusion is 2.82 cal/g. On the average the specific heat of solid mercury is 0.0306 cal/g; that of liquid mercury is 0.0336 cal/g. If a 10-g chunk of solid mercury is taken from a liquid nitrogen bath $(77°K)$ and dropped into an ice-water calorimeter containing 500 g of ice and 500 g of liquid water, what will be the final ratio of ice to water?

8.24. Given the following data:

$K(s) \longrightarrow K(g)$		$\Delta H = 21.5 \text{ kcal}$
$K(g) \longrightarrow K^+(g) + e^-$		$\Delta H = 100.1 \text{ kcal}$
$F_2(g) \longrightarrow 2F(g)$		$\Delta H = 36 \text{ kcal}$
$F^-(g) \longrightarrow F(g) + e^-$		$\Delta H = 79.6 \text{ kcal}$
$KF(s) \longrightarrow K^+(g) + F^-(g)$		$\Delta H = 192 \text{ kcal}$

calculate ΔH for the reaction $K(s) + \frac{1}{2} F_2(g) \longrightarrow KF(s)$.

(*Answer*: -132 kcal.)

8.25. The heat of formation of $MgO(s)$ is 144 kcal/mole. Given a photo flashbulb containing 1.00 g of magnesium ribbon in excess oxygen; if the bulb is fired in a calorimeter containing 18.94 kg of H_2O at 25.00°C what temperature rise of the water should be observed, assuming that all the magnesium converts to $MgO(s)$?

9

ELECTROCHEMISTRY

SO FAR AS we are concerned, at the introductory level, the important quantitative aspects of electrochemistry are units, the relation between chemical change and electrical current, and the relation between tendency for chemical reaction to occur and voltage.

9.1 Units

The practical unit of electrical charge is the *coulomb* (C), the exact definition of which is so complicated that very few people know exactly what it means. For us, it is enough to say that 1 C equals 2.998×10^9 electrostatic units. The electrostatic unit (esu) is the amount of charge that repels an identical charge 1 cm away in a vacuum with unit force (1 dyne, or 1 gram-cm/sec^2).

■ EXAMPLE 241. The charge on an electron is -1.602×10^{-19} C. How many electrostatic units is this?

SOLUTION. One $C = 2.998 \times 10^9$ esu

$$(1.602 \times 10^{-19} \text{ C}) \times \left(2.998 \times 10^9 \frac{\text{esu}}{\text{C}}\right) = 4.803 \times 10^{-10} \text{ esu}$$

Electronic charge $= -4.803 \times 10^{-10}$ esu ■

■ EXAMPLE 242. The electric charge on the aluminum ion is usually referred to as $+3$. How much is this in coulombs?

SOLUTION: Plus 3 means 3 units of positive charge on the scale where the charge of the electron is -1. We saw in Example 241 that the electronic charge is -1.602×10^{-19} C. For the Al^{3+} ion we need three such charges but of opposite sign, namely, $3(+1.602 \times 10^{-19}) = 4.806 \times 10^{-19}$ C. ■

The practical unit of electrical current is the *ampere*, (amp) which is the amount of current flowing when 1 C passes a given point in 1 sec. Frequently, an ampere is referred to as a "coulomb per second." It should be obvious from the dimensions that "current" multiplied by "time" gives charge, if current is in amperes (or coulombs per second) and if time is in seconds.

■ EXAMPLE 243.   The current in a given wire is 1.80 amp. How many coulombs will pass a given point on the wire in 1.36 min?

SOLUTION:

$$1.36 \text{ min is } (1.36 \text{ min})\left(60 \ \frac{\text{sec}}{\text{min}}\right) = 81.6 \text{ sec.}$$

1.80 amp = 1.80 C/sec
For 1.80 amp flowing for 81.6 sec, the charge transferred is

$$\left(1.80 \ \frac{C}{\text{sec}}\right)(81.6 \text{ sec}) = 147 \text{ C } ■$$

■ EXAMPLE 244.   Given that the electronic charge is 1.60×10^{-19} C, how much current would be flowing in a wire in which 1.27×10^{18} electrons are being transferred per minute?

SOLUTION:

Number of electrons transferred per second

$$= \left(1.27 \times 10^{18} \ \frac{\text{electrons}}{\text{min}}\right)\left(\frac{1 \text{ min}}{60 \text{ sec}}\right) = 2.12 \times 10^{16} \ \frac{\text{electrons}}{\text{sec}}$$

$$\text{Charge transferred per second} = \left(2.12 \times 10^{16} \ \frac{\text{electrons}}{\text{sec}}\right)$$

$$\times \left(1.60 \times 10^{-19} \ \frac{C}{\text{electron}}\right) = 3.39 \times 10^{-3} \ \frac{C}{\text{sec}}$$

or 3.39×10^{-3} amp. ■

Although the coulomb is the usual unit for measuring charge, the chemist finds (for reasons soon to be discussed) that a more convenient unit is the *faraday*. It corresponds to the charge carried by a mole of electrons and amounts to 96,487 C. A common step in electrochemical computations is to convert information given in coulombs, or amperes and seconds, into faradays, or vice versa.

■ EXAMPLE 245.   If a current of 80.0 microamp is drawn from a solar cell for 100 days, how many faradays are involved?

SOLUTION:

$$\text{Time in seconds} = (100 \text{ days})\left(24 \frac{\text{hours}}{\text{day}}\right)\left(60 \frac{\text{min}}{\text{hr}}\right)\left(60 \frac{\text{sec}}{\text{min}}\right)$$

$$= 8.64 \times 10^6 \text{sec}.$$

Current $= 80.0$ microamp $= 80.0 \times 10^{-6}$ amp

$$\text{Charge} = \text{current} \times \text{time} = \left(80.0 \times 10^{-6} \frac{C}{\text{sec}}\right)(8.64 \times 10^6 \text{ sec})$$

$$= 691 \text{ C}$$

Since 1 faraday equals 96,500 C, 691 C must equal

$$\frac{691 \text{ C}}{96,500 \text{ C/faraday}} = 7.16 \times 10^{-3} \text{ faraday} \quad ■$$

9.2   Electrolysis

The term electrolysis is applied to the process in which chemical change is brought about by electric current. To make computations for electrolysis, the main requirement is to know the half-reaction that occurs at each electrode. For example, when an aqueous solution of sodium chloride is electrolyzed, the changes that occur are the following: *at the anode* (where oxidation occurs):

$$2 \text{ Cl}^- \longrightarrow \text{Cl}_2(g) + 2e^-$$

at the cathode (where reduction occurs):

$$2\text{H}_2\text{O} + 2e^- \longrightarrow \text{H}_2(g) + 2\text{OH}^-$$

[Note that the anode is defined as the "electrode where oxidation occurs" and the cathode as the "electrode where reduction occurs." Some people define the cathode as the negative electrode and the anode as the positive electrode, as done with vacuum radio tubes, but there is then risk of confusion as to whether internal or external circuits are being used as the basis of the sign definition. In chemistry, it is better to link "cathode" with "reduction" and "anode" with "oxidation."]

Once the half-reaction is written, it can be read off in terms of ions, molecules, and electrons, or *moles* of ions, molecules, and electrons. As indicated in Section 9.1, 1 mole of electrons is 1 faraday (96,500C), so when half-reactions are read off in moles, the coefficient of e^- gives directly the number of faradays involved in the reaction.

■ EXAMPLE 246. When aqueous NaCl is electrolyzed, how many faradays need to be transferred at the anode to release 0.015 mole of Cl_2 gas?

SOLUTION: The half-reaction $2Cl^- \rightarrow Cl_2 + 2e^-$ tells us that at the anode, for every 2 moles of Cl^- ion used up, 1 mole of Cl_2 is formed along with 2 faradays of electrical charge that get transferred from the solution to the electrode and out the external circuit. Thus, we can say we get 1 mole of Cl_2 per 2 faradays transferred. In order to get 0.015 mole of Cl_2, we need

$$0.015 \text{ mole of } Cl_2 \left(\frac{2 \text{ faradays}}{1 \text{ mole of } Cl_2}\right) = 0.030 \text{ faraday} \quad ■$$

■ EXAMPLE 247. When aqueous NaCl is electrolyzed, how long must a current of 0.010 amp run in order to liberate 0.015 mole of H_2 at the cathode?

SOLUTION: The cathode half-reaction is $2H_2O + 2e^- \rightarrow H_2(g) + 2OH^-$. It states that for every 2 faradays consumed, 1 mole of H_2 is liberated. To get 0.015 mole of H_2 we need to use

$$(0.015 \text{ mole of } H_2)\left(\frac{2 \text{ faradays}}{1 \text{ mole of } H_2}\right) = 0.030 \text{ faraday}$$

$$(0.030 \text{ faraday} \left(96,500 \frac{C}{\text{faraday}}\right) = 2900 \text{ C}$$

The current is 0.010 amp (or 0.010 C/sec).

$$\frac{2900 \text{ C}}{0.010 \text{ C/sec}} = 2.9 \times 10^5 \text{ sec} \quad ■$$

■ EXAMPLE 248. In the electrolysis of molten zinc chloride, $ZnCl_2$, how many grams of Zn metal can be deposited at the cathode by passage of 0.010 amp for 1 hour?

SOLUTION: 0.010 amp for 1 hour is

$$\left(0.010 \frac{C}{\text{sec}}\right)\left(60 \frac{\text{min}}{\text{hour}}\right)\left(60 \frac{\text{sec}}{\text{min}}\right) = 36 \text{ C}$$

$$\frac{36 \text{ C}}{96,500 \text{ C/faraday}} = 3.7 \times 10^{-4} \text{ faraday}$$

In molten zinc chloride, the cathode reaction is

$$Zn^{2+} + 2e^- \longrightarrow Zn$$

which states that for every 2 faradays of electricity used up, 1 mole of Zn forms. Since we use 3.7×10^{-4} faraday, we must be getting

$$(3.7 \times 10^{-4} \text{ faraday})\left(\frac{1 \text{ mole of Zn}}{2 \text{ faradays}}\right) = 1.8 \times 10^{-4} \text{ mole of Zn}$$

One mole Zn is 65.37 g.

$$(1.8 \times 10^{-4} \text{ mole of Zn})\left(65.37 \frac{g}{\text{mole}}\right) = 0.012 \text{ g} \blacksquare$$

■ **EXAMPLE 249.** A simple way to measure the amount of charge going through an electrolytic cell is to set up in series with the studied cell one in which an aqueous solution of $AgNO_3$ is electrolyzed. ("In series" means to hook up two cells so that the cathode of one is wired directly to the anode of the other. The result will be that any electricity that goes through one cell has to come out of the other.) Suppose you have an aqueous $AgNO_3$ cell in series with one in which aqueous NaCl is being electrolyzed. If 0.0198 g of Ag is plated out on the cathode of the first cell, how many moles of H_2 will be liberated at the cathode of the second cell?

SOLUTION: *In the first cell:* The cathode half-reaction is $Ag^+ + e^- \rightarrow Ag(s)$. This shows that 1 mole of Ag is plated out for every faraday passed.

$$\frac{0.0198 \text{ g of Ag}}{107.870 \text{ g/mole}} = 1.84 \times 10^{-4} \text{ mole of Ag}$$

$$(1.84 \times 10^{-4} \text{ mole of Ag}) \times \left(1 \frac{\text{faraday}}{\text{mole of Ag}}\right) = 1.84 \times 10^{-4} \text{ faraday}$$

In the second cell: The number of faradays passed in the second cell is the same number (1.84×10^{-4}) as in the first cell, because all the electricity fed through the first cell must go into the second cell.

The cathode half-reaction is $2H_2O + 2e^- \rightarrow H_2 + 2OH^-$, which states that 1 mole of H_2 is formed for every 2 faradays passed through. Therefore, if we pass 1.84×10^{-4} faradays through the second cell, we must get

$$(1.84 \times 10^{-4} \text{ faraday})\left(\frac{1 \text{ mole of } H_2}{2 \text{ faradays}}\right) = 9.20 \times 10^{-5} \text{ mole of } H_2 \blacksquare$$

■ EXAMPLE 250. Suppose the half-reaction $6H_2O \rightarrow O_2 + 4H_3O^+ + 4e^-$ is occurring at an anode. You observe the formation of 36.6 ml of wet oxygen at barometric pressure 743 Torr and 25°C. How many moles of H_3O^+ have been set free at the anode?

SOLUTION: (Figure out how many moles of O_2 you have. Then use the half-reaction to calculate H_3O^+ formed.)

At 25°C, the vapor pressure of water is 23.8 Torr.

Therefore, the pressure of the oxygen is $743 - 23.8$, or 719 Torr. This corresponds to $719/760$ atm.

Using $PV = nRT$, we get

$$n = \frac{PV}{RT} = \frac{(719/760)(0.0365)}{(0.08206)(273 + 25)} = 1.41 \times 10^{-3} \text{ mole of } O_2$$

From the half-reaction $6H_2O \rightarrow O_2 + 4H_3O^+ + 4e^-$ we see there are 4 moles of H_3O^+ formed per mole of O_2. So, if we get 1.41×10^{-3} mole of O_2, we must form

$$(1.41 \times 10^{-3} \text{ mole of } O_2)\left(\frac{4 \text{ moles of } H_3O^+}{1 \text{ mole of } O_2}\right) = 5.64 \times 10^{-3} \text{ mole of } H_3O^+. \ ■$$

9.3 Galvanic cells

In *electrolysis*, current is forced by some external means to pass through a cell so as to produce chemical changes at the electrodes—specifically, an oxidation half-reaction at the anode and a reduction half-reaction at the cathode. A *galvanic cell* (sometimes also called a *voltaic* cell) is the direct opposite of this; that is, it has reactions occurring spontaneously at electrodes, the result of which is to produce an electric current in an external circuit. In making computations for galvanic cells, the most important thing again is to be able to specify the half-reactions. Once the half-reaction is written, it is an easy matter to figure out how many faradays are pushed out from the anode or sucked into the cell at the cathode.

In principle, any oxidation–reduction reaction can be used as a basis for a galvanic cell. The first step is to split the whole reaction into half-reactions (as we have done in Section 5.4 for balancing equations). Then set up a cell, divided in two by a porous partition, so that the components taking part in the oxidation half-reaction are in one compartment and the components taking part in the reduction half-reaction are in the other compartment. Appropriate electrodes, either inert (e.g., platinum) or one of the conducting components (e.g., zinc), serve to carry electrons out of the anode compartment or into the cathode compartment. The number of faradays coming out of the anode must just equal the number going into the cathode; this number is fixed by the amount of chemical change occurring at the electrodes.

■ EXAMPLE 251.   Given a galvanic cell in which the anode reaction is $Zn(s) \rightarrow Zn^{2+} + 2e^-$. How many faradays will you get out of the cell if you consume 1.00 g of Zn by this reaction?

SOLUTION: The half-reaction indicates that, for every 1 mole of Zn consumed, 2 moles of electrons (or 2 faradays of electrical charge) is set free.
One mole of Zn weighs 65.37 g.

$$\frac{1.00 \text{ g of Zn}}{65.37 \text{ g/mole}} = 0.0153 \text{ mole}$$

If 1 mole of Zn can liberate 2 faradays, then from 0.0153 mole of Zn we can get

$$(0.0153 \text{ mole of Zn})\left(\frac{2 \text{ faradays}}{1 \text{ mole of Zn}}\right) = 0.0306 \text{ faraday} \blacksquare$$

■ EXAMPLE 252.   Given a galvanic cell in which the cathode reaction is $Cl_2(g) + 2e^- \rightarrow 2Cl^-$. If 1.00 g of chlorine is consumed in this reaction, how many faradays will be pulled into the cell from outside?

SOLUTION: One mole of Cl_2 weighs 70.906 g.

$$\frac{1.00 \text{ g of Cl}_2}{70.906 \text{ g/mole}} = 0.0141 \text{ mole}$$

The half-reaction says that each time 1 mole of Cl_2 disappears, 2 faradays have to go with it. Therefore, if 0.0141 mole of Cl_2 is used up, the required electric charge is

$$(0.0141 \text{ mole of Cl}_2)\left(\frac{2 \text{ faradays}}{1 \text{ mole of Cl}_2}\right) = 0.0282 \text{ faraday} \blacksquare$$

■ EXAMPLE 253.   Given a galvanic cell in which the overall reaction is $Zn(s) + Cl_2(g) \rightarrow Zn^{2+} + 2Cl^-$. Using up 1.50 g of Zn by the reaction, how long could such a cell deliver 0.10 amp to the outside?

SOLUTION: We need consider only one half-reaction, since the other must occur in equivalent amount.

The zinc half-reaction is $Zn(s) \rightarrow Zn^{2+} + 2e^-$. It states that 2 faradays are released when 1 mole of Zn is used.

$$\frac{1.50 \text{ g of Zn}}{65.37 \text{ g/mole}} = 0.0229 \text{ mole of Zn}$$

$$(0.0229 \text{ mole of Zn})\left(\frac{2 \text{ faradays}}{1 \text{ mole of Zn}}\right) = 0.0458 \text{ faraday}$$

$$(0.0458 \text{ faraday})\left(96,500 \frac{C}{\text{faraday}}\right) = 4420 \text{ C}$$

The desired current is 0.10 amp, or 0.10 C/sec.

$$\frac{4420 \text{ C}}{0.10 \text{ C/sec}} = 4.4 \times 10^4 \text{ sec} \blacksquare$$

■ EXAMPLE 254. In the lead storage battery, the anode reaction is $Pb(s) + HSO_4^- + H_2O \rightarrow PbSO_4(s) + H_3O^+ + 2e^-$. A typical battery would be rated "100 ampere-hours," which means it has the chemical capacity to deliver 100 amp for 1 hour or 1 amp for 100 hours. How many grams of Pb would be used up at the foregoing anode to accomplish this?

SOLUTION: (Figure out how many faradays are involved. Then use the half-reaction to relate the faradays to the chemical change.)
One ampere for 100 hours is

$$\left(1 \frac{C}{\text{sec}}\right)(100 \text{ hours})\left(60 \frac{\text{min}}{\text{hour}}\right)\left(60 \frac{\text{sec}}{\text{min}}\right) = 3.6 \times 10^5 \text{ C}$$

This can be converted to faradays by dividing by the number of coulombs per faraday.

$$\frac{360,000 \text{ C}}{96,500 \text{ C/faraday}} = 3.73 \text{ faradays}$$

The half-reaction says that 1 mole of Pb disappears for each 2 faradays liberated. If we liberate 3.73 faradays, we use

$$(3.73 \text{ faradays})\left(\frac{1 \text{ mole of Pb}}{2 \text{ faradays}}\right) = 1.86 \text{ mole of Pb}$$

One mole of Pb is 207.19 g.

$$(1.86 \text{ moles of Pb})\left(207.19 \frac{g}{\text{mole}}\right) = 385 \text{ g.} \blacksquare$$

■ EXAMPLE 255.   In a flashlight cell the cathode reaction can be written $2MnO_2(s) + Zn^{2+} + 2e^- \rightarrow ZnMn_2O_4(s)$. If your flashlight cell is to give out a current of 4.6 milliamp, how long could it do this if we start with 3.50 g of MnO_2?

SOLUTION: One mole of MnO_2 weighs 86.94 g.

$$\frac{3.50 \text{ g of } MnO_2}{86.94 \text{ g/mole}} = 0.0403 \text{ mole}$$

The foregoing half-reaction indicates 2 faradays will be transferred each time 2 moles of MnO_2 get used up. Therefore, if we use up 0.0403 mole of MnO_2, we must transfer

$$(0.0403 \text{ mole of } MnO_2)\left(\frac{2 \text{ faradays}}{2 \text{ moles of } MnO_2}\right) = 0.0403 \text{ faraday}$$

$$(0.0403 \text{ faraday})\left(96,500 \frac{C}{\text{faraday}}\right) = 3890 \text{ C}$$

The rate is 4.6 milliamp, or $0.0046 \dfrac{C}{\text{sec}}$,

$$\frac{3890 \text{ C}}{0.0046 \text{ C/sec}} = 8.5 \times 10^5 \text{ sec}$$

(which is about 10 days). ■

9.4   Oxidation potentials

In setting up a galvanic cell, we take an oxidation–reduction reaction, separate it into the two half-reactions, and set these half-reactions up to occur individually in the two compartments of the cell (the so-called *half-cells*). In the preceding section we considered how much current can be fed to the outside by such a cell. But now, suppose we raise the question "What is the situation if there is no external connection between the anode and the cathode?" There is still a drive tending to push electrons out of the anode relative to the cathode, but the electrons have no place to go. The only thing that can happen is that a difference in voltage, also called a difference in potential, establishes itself between the anode and the cathode. We can measure this difference in voltage by hooking up a gadget, called a potentiometer, between the anode and cathode. Such a potentiometer can measure the voltage difference without itself drawing any electric current from the cell.

The observed voltage difference between two electrodes of a galvanic cell depends on the materials present in each compartment—what they are and at what concentration. For reference purposes, it has been agreed to consider the hydrogen half-cell based on the reaction $H_2(g) \rightarrow 2H^+ + 2e^-$, better written as $H_2(g) + 2H_2O \rightarrow 2H_3O^+ + 2e^-$, as a standard. We arbitrarily assign it a value of zero; that is, we say that it makes zero contribution to the observed voltage of a cell in which the hydrogen half-cell is part of the assembly. This assignment of zero contribution holds only if the H_2 and the H_3O^+ are in what are called their standard states. Strictly speaking, "standard state" means 25°C and "unit activity," but for our purposes it will be sufficient to interpret unit activity for a gas component as 1 atm pressure and unit activity for a dissolved species as 1 molal (conventionally written 1 m) concentration. (Concentrations are discussed in detail in Chapter 10.)

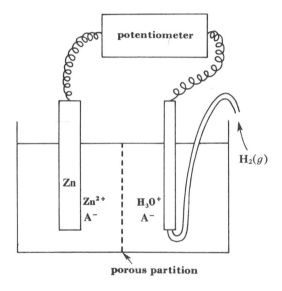

Fig. 24. Galvanic cell for the reaction $Zn(s) + 2H_3O^+ \rightarrow H_2(g) + Zn^{2+} + 2H_2O$

Figure 24 shows a galvanic cell set up to measure the voltage difference between the anode and cathode of a cell designed to use the reaction

$$Zn(s) + 2H_3O^+ \longrightarrow H_2(g) + Zn^{2+} + 2H_2O$$

The half-reaction $Zn(s) \rightarrow Zn^{2+} + 2e^-$ is set up in the left-hand compartment by putting a zinc bar into a solution containing Zn^{2+} and some anion A^-. The half-reaction $2H_3O^+ + 2e^- \rightarrow H_2(g) + 2H_2O$ is set up in the right-hand compartment by dipping an inert conductor such as carbon or platinum into a solution containing H_3O^+ and A^- and then allowing H_2 gas to bubble over the electrode surface. If the pressure of the hydrogen gas is 1 atm and everything else is at unit concentration, then the potentiometer will show a reading of 0.76 V. *We assume that all of this* 0.75 V *can be attributed to the Zn electrode*, since we decided to regard the hydrogen contribution as zero. Such voltages, referred to hydrogen, are called *oxidation potentials*, and they are used to give a quantitative measure of the tendency of a half-reaction to give off electrons.

In Appendix C some common half-reactions are listed along with their oxidation potentials. The numerical values hold strictly only when each species appearing in the half-reaction is in its standard state. Departures from unit pressure for gases and from unit molality for dissolved species are considered in Section 9.5.

When every species is in its standard state, the oxidation potential is usually designated E^0 (read "ee zero"). The following problems illustrate how E^0's are used. An E^0 for a single half-reaction gives a quantitative measure of the tendency of that half-reaction to occur, but also E^0's for two half-reactions can be combined to tell us something about the voltage to be expected for a corresponding galvanic cell and whether its overall chemical reaction tends to go spontaneously as written. If the overall E^0 is positive, the reaction tends to go spontaneously; if negative, it does not. "Spontaneously" means without having to put on an external voltage to force the reaction to go opposite its natural tendency.

■ EXAMPLE 256.   If the E^0 for $H_2(g) + 2H_2O \rightarrow 2H_3O^+ + 2e^-$ is zero, what is the E^0 for $2H_3O^+ + 2e^- \rightarrow H_2(g) + 2H_2O$?

SOLUTION: E^0 is zero in both directions.

If we say that a half-reaction is arbitrarily labeled as having zero tendency to go from left to right, then it must have zero tendency to go from right to left. ■

■ EXAMPLE 257.   If the E^0 for $Zn(s) \rightarrow Zn^{2+} + 2e^-$ is $+0.76$ V, what is the E^0 for $2e^- + Zn^{2+} \rightarrow Zn(s)$?

SOLUTION: We are really asking: "If a half-reaction has a big tendency to go to the right, compared to some standard reaction, what can you say about its tendency to go to the left, compared to that same standard?" The answer seems obvious: big tendency to go to the right means small tendency to go to the left. Specifically, Zn has a bigger tendency than H_2 to give off electrons by 0.76 V. Therefore, the species formed from the Zn (namely, Zn^{2+}) must have a smaller tendency to pick up electrons than the species formed from H_2 (namely, H_3O^+). If we use zero for the hydrogen half-reaction in either direction, then reversing the direction of $Zn \to Zn^{2+} + 2e^-$ means reversing the sign of its E^0. In display form, this looks as follows:

$$Zn(s) \to Zn^{2+} + 2e^- \qquad\qquad E^0 = +0.76 \text{ V}$$
$$H_2(g) + 2H_2O \to 2H_3O^+ + 2e^- \qquad E^0 = 0.00000 \text{ V}$$
$$2e^- + 2H_3O^+ \to H_2(g) + 2H_2O \qquad E^0 = 0.00000 \text{ V}$$
$$2e^- + Zn^{2+} \to Zn(s) \qquad\qquad E^0 = -0.76 \text{ V } \blacksquare$$

■ EXAMPLE 258.   If the E^0 for $Sn^{2+} \to Sn^{4+} + 2e^-$ is -0.15 V, what is the E^0 for $Sn^{4+} + 2e^- \to Sn^{2+}$?

ANSWER: $E^0 = +0.15$ V. ■

■ EXAMPLE 259.   If the E^0 for $Mn^{2+} + 12H_2O \to MnO_4^- + 8H_3O^+ + 5e^-$ is -1.51 V, what is the E^0 for MnO_4^- acting as an oxidizing agent?

ANSWER: $E^0 = +1.51$ V. ■

■ EXAMPLE 260.   The E^0 for $Cr_2O_7^{2-}$ going to Cr^{3+} in acidic solution is $+1.33$ V. What would be the E^0 for Cr^{3+} acting as a reducing agent and going to $Cr_2O_7^{2-}$ in acidic solution? Write the appropriate half-reaction.

ANSWER: $2Cr^{3+} + 21H_2O \to Cr_2O_7^{2-} + 14H_3O^+ + 6e^-$;
$E^0 = -1.33$ V. ■

In most tabulations of E^0's, the half-reactions are written so that the electrons appear on the right. This means that the value given applies when the species shown on the left of the half-reaction acts as a *reducing* agent. If we are interested in the reverse half-reaction, in which the species shown on the right of the half-reaction acts as an *oxidizing* agent, then we need to reverse the arrow and change the sign on E^0.

For deducing an E^0 for an overall reaction, we can add the E^0's for the two half-reactions involved so long as the half-reactions have been written to oppose each other. In other words, one half-reaction must show electrons on its right and the other half-reaction must show electrons on its left. When addings E^0's in this way, we ignore any disparity in number of electrons on right and left, since E^0 is a per-electron property.

■ EXAMPLE 261.   What is the E^0 for the reaction $Zn(s) + 2H_3O^+ \to Zn^{2+} + H_2(g) + 2H_2O$?

SOLUTION: (Write the two half-reactions with corresponding E^0's, and add them.)

$$Zn(s) \longrightarrow Zn^{2+} + 2e^- \qquad\qquad E^0 = +0.76 \text{ V}$$
$$2e + 2H_3O^+ \longrightarrow H_2(g) + 2H_2O \qquad E^0 = 0.00 \text{ V}$$
$$\overline{Zn(s) + 2H_3O^+ \longrightarrow Zn^{2+} + H_2(g) + 2H_2O \qquad E^0 = +0.76 \text{ V } \blacksquare}$$

■ EXAMPLE 262.   What is the E^0 for the reaction $2H_2O + H_2(g) + Zn^{2+}$ $\rightarrow Zn(s) + 2H_3O^+$?

SOLUTION:

$2H_2O + H_2(g) \longrightarrow 2H_3O^+ + 2e^-$		$E^0 = 0.00$ V
$2e^- + Zn^{2+} \longrightarrow Zn(s)$		$E^0 = -0.76$ V

$2H_2O + H_2(g) + Zn^{2+} \longrightarrow 2H_3O^+ + Zn(s)$ $E^0 = -0.76$ V ■

■ EXAMPLE 263.   What is the E^0 for the reaction $Zn(s) + Cl_2(g) \rightarrow Zn^{2+} + 2Cl^-$?

SOLUTION:

$Zn(s) \longrightarrow Zn^{2+} + 2e^-$		$E^0 = +0.76$ V
$Cl_2(g) + 2e^- \longrightarrow 2Cl^-$		$E^0 = +1.36$ V

$Zn(s) + Cl_2(g) \longrightarrow Zn^{2+} + 2Cl^-$ $E^0 = +2.12$ V

[Note that the E^0 for the Zn half-reaction is as given in Appendix C. But for the chlorine half-reaction, we find in the table $2Cl^- \rightarrow Cl_2 + 2e^-$ with $E^0 = -1.36$ V. This is the reverse half-reaction of what we need, so we flip the half-reaction over and change sign on E^0.] ■

■ EXAMPLE 264.   Under standard conditions, what will be the E^0 for the oxidation of I^- by H_3AsO_4 to I_2 and $HAsO_2$ in acidic solution? Should the reaction occur spontaneously?

SOLUTION: Write the two half-reactions

$2I^- \longrightarrow I_2 + 2e^-$

$2e^- + 2H_3O^+ + H_3AsO_4 \longrightarrow HAsO_2 + 4H_2O$

Get the E^0's from Appendix C. Add.

$2I^- \longrightarrow I_2 + 2e^-$	$E^0 = -0.54$ V
$2e^- + 2H_3O^+ + H_3AsO_4 \longrightarrow HAsO_2 + 4H_2O$	$E^0 = +0.56$ V

$H_3AsO_4 + 2H_3O^+ + 2I^-$
$\longrightarrow HAsO_2 + I_2 + 4H_2O$ $E^0 = +0.02$ V

Since the E^0 for the overall reaction is greater than zero, the reaction *ought to occur* spontaneously from left to right. ■

■ EXAMPLE 265. Write the equation for the oxidation of Cr^{3+} by Br_2 in acidic solution to form $Cr_2O_7^{2-}$ and Br^-. From the overall E^0 decide whether this oxidation should occur spontaneously.

SOLUTION:

$$2Cr^{3+} + 21H_2O \longrightarrow Cr_2O_7^{2-} + 14H_3O^+ + 6e^- \qquad E^0 = -1.33 \text{ V}$$
$$Br_2 + 2e^- \longrightarrow 2Br^- \qquad E^0 = +1.07 \text{ V}$$

$$3Br_2 + 2Cr^{3+} + 21H_2O$$
$$\longrightarrow Cr_2O_7^{2-} + 14H_3O^+ + 6Br^- \qquad E^0 = -0.26 \text{ V}$$

[Note that in getting the balanced equation we need to multiply the second half-reaction by three before adding. However, when we combine the E^0's we add them directly, since they do not depend on how much reaction occurs. The fact that E^0 for the final reaction is less than zero means that the change written will *not occur* spontaneously from left to right. Its natural tendency is to go in the opposite direction.] ■

9.5 Nernst equation

In Section 9.4 special note was taken of the fact that the oxidation potentials listed in the tabulations hold only for standard conditions, which generally mean 1 atm pressure and 1 m concentration. What happens if we change these conditions? Properly speaking, we should delay this discussion until after Chapter 10, which treats the topic of concentration in solution. However, the Nernst equation gives us a very simple way of expressing how the voltage of a half-reaction or of an overall reaction varies with concentration, so it really needs to be considered as part of electrochemistry. For anyone unfamiliar with methods of expressing concentration in solution, reference to Chapter 10 is advisable.

As expected, increasing the concentration of a reagent on the left side of a reaction (or half-reaction) or decreasing a concentration on the right gives an increased tendency of the reaction to go from left to right. Since the voltage expresses the tendency to go to the right, any influence that increases this tendency should make the voltage more positive. For example, the half-reaction

$$H_2(g) + 2H_2O \longrightarrow 2H_3O^+ + 2e^-$$

has a certain tendency to go to the right, which we are labeling zero. If we raise the pressure of $H_2(g)$ or decrease the concentration of H_3O^+, we increase the tendency to go to the right; that is, we make the voltage become higher than zero. Alternatively, if we decrease the pressure of $H_2(g)$ to less than 1 atm or if we raise the concentration of H_3O^+ to more than 1 m, we make the voltage of the reaction less than zero, or negative. If we let E^0 represent the voltage of the reaction when everything is in its standard state and E the voltage in a nonstandard state, we can relate one to the other by the Nernst equation:

$$E = E^0 - \frac{0.0592}{n} \log \left[\frac{\text{concentrations of species on right}}{\text{concentrations of species on left}} \right]$$

Here n stands for the number of electrons appearing in the half-reaction, or for the number of electrons transferred in an overall reaction. The expression in square brackets is made up by multiplying together the concentration of each species appearing in the equation raised to a power given by the coefficient of the species. (Any species whose concentration is not variable—e.g., H_2O in dilute aqueous solutions, solids that exist as separate phases—are ignored.) Specifically, for $H_2(g) + 2H_2O \rightarrow 2H_3O^+ + 2e^-$ we write

$$E = E^0 - \frac{0.0592}{2} \log \frac{[H_3O^+]^2}{[H_2(g)]}$$

Note first that the H_2O appearing on the left side of the half-reaction is ignored. Note also that H_3O^+ is taken to the square power because there is a coefficient 2 before H_3O^+ in the half-reaction; H_2 is taken to the first power because there is a coefficient 1 before H_2 in the half-reaction. $[H_3O^+]^2$ appears in the numerator because it comes from the right-hand side of the half-reaction; $[H_2]$ appears in the denominator because it appears on the left side of the reaction. (The $2e^-$ that appears in the half-reaction does not come into the logarithm expression; it is effectively taken care of by $n = 2$, dividing the 0.0592.)

How to express $[H_3O^+]$ and $[H_2(g)]$? For gases, we use pressure, so in place of $[H_2(g)]$ we can put p_{H_2}. For dissolved species, we use molality (normally indicated by m). In dilute solutions, the numerical values of molality are almost the same as those for molarity, but we shall be precise and stick with molality. For $[H_3O^+]$ we thus use $m_{H_3O}{}^+$. So, the Nernst equation says for the hydrogen electrode

$$E = E^0 - \frac{0.0592}{2} \log \frac{m_{H_3O^+}^2}{p_{H_2}}$$

■ EXAMPLE 266. What does the Nernst equation give for the hydrogen electrode when H_3O^+ is at 1 m concentration and H_2 is at 1 atm pressure?

SOLUTION: For $H_2(g) + 2H_2O \rightarrow 2H_3O^+ + 2e^-$; $E^0 = 0$

$$E = E^0 - \frac{0.0592}{2} \log \frac{m^2_{H_3O^+}}{p_{H_2}} = 0 - 0.0296 \log \frac{(1)^2}{1} = 0$$

Having substituted $m_{H_3O^+} = 1$ and $p_{H_2} = 1$, we get the logarithm of one, which turns out to be zero. So, not surprisingly, we find that $E = E^0 = 0$ in the standard state for this hydrogen half-reaction. ■

■ EXAMPLE 267. What happens to the voltage of the hydrogen electrode when the pressure of hydrogen is increased to 100 atm, assuming the concentration of H_3O^+ stays at 1 m?

SOLUTION: $H_2(g) + 2H_2O \rightarrow 2H_3O^+ + 2e^-$; $E^0 = 0$.

$$E = E^0 - \frac{0.0592}{2} \log \frac{m^2_{H_3O^+}}{p_{H_2}}$$

where $m_{H_3O^+} = 1$ and $p_{H_2} = 100$.

$$E = 0 - 0.0296 \log \frac{1}{100} = (-0.0296)(\log 10^{-2}) = (-0.0296)(-2)$$

$$= +0.0592 \text{ V}.$$

Thus, not surprisingly, the E for $H_2(100 \text{ atm}) + 2H_2O \rightarrow 2H_3O^+(1 \ m)$ $+ 2e^-$ is more positive than for $H_2(1 \text{ atm}) + 2H_2O \rightarrow 2H_3O^+(1 \ m) + 2e^-$. Increasing the pressure of H_2 makes for a greater tendency to go to the right. ■

■ EXAMPLE 268. What happens to the voltage of the hydrogen electrode when the molality of H_3O^+ is cut to 10^{-2} m, assuming the hydrogen pressure stays at 1 atm?

SOLUTION: $H_2(g) + 2H_2O \rightarrow 2H_3O^+ + 2e^-$; $E^0 = 0$.

$$E = E^0 - \frac{0.0592}{2} \log \frac{m^2_{H_3O^+}}{p_{H_2}}$$

where $m_{H_3O^+} = 10^{-2}$ and $p_{H_2} = 1$.

$$E = 0 - 0.0296 \log \frac{(10^{-2})^2}{1} = -0.0296 \log (10^{-4}) = +0.118 \text{ V}. ■$$

■ EXAMPLE 269.   What E do you expect for $Pb(s) \rightarrow Pb^{2+} + 2e^-$ in $0.015m$ Pb^{2+} solution? The E^0 is $+0.13$ V.

SOLUTION: $Pb(s) \rightarrow Pb^{2+} + 2e^-$; $E^0 = +0.13$ V.

$$E = E^0 - \frac{0.0592}{2} \log m_{Pb^{2+}} \quad \text{where} \quad m_{Pb^{2+}} = 0.015$$

$E = +0.13 - 0.0296 \log (0.015) \quad$ where $\quad \log (0.015) = -1.824$

$E = +0.18$

[Note: This problem illustrates the point that for a solid, such as $Pb(s)$, there is no effective way to change the concentration from the standard-state value. It is assigned unit activity, which means we can ignore it in working with the Nernst equation.] ■

■ EXAMPLE 270.   What will be the E for the half-reaction $4H_2O + HAsO_2 \rightarrow H_3AsO_4 + 2H_3O^+ + 2e^-$ under conditions where $HAsO_2$ is 0.10 m; H_3AsO_4 is 0.050 m; H_3O^+ is 10^{-6} m? The E^0 for this half-reaction is -0.56 V.

SOLUTION: $4H_2O + HAsO_2 \rightarrow H_3AsO_4 + 2H_3O^+ + 2e^-$; $E^0 = -0.56$ V.

$$E = E^0 - \frac{0.0592}{2} \log \frac{(m_{H_3AsO_4})(m_{H_3O^+})^2}{m_{HAsO_2}}$$

$E = -0.56 - 0.0296 \log \dfrac{(0.050)(10^{-6})^2}{0.10} = -0.20$ V

[Note: This problem illustrates two points. One is that H_2O is ignored, again on the basis that its activity does not change perceptibly from the standard state. The other point to note is that when there are several species showing up on one side of a half-reaction, their concentrations (taken to the appropriate powers) are multiplied together in setting up the logarithm term.] ■

■ EXAMPLE 271.   What will be the E for the reaction $Pb(s) + 2H_3O^+ \rightarrow Pb^{2+} + H_2(g) + 2H_2O$ under conditions where H_3O^+ is 0.010 m, Pb^{2+} is 0.10 m, and p_{H_2} is 10^{-6} atm?

SOLUTION: (Use the Nernst equation with the same procedure as above, putting all the species on the right of the chemical equation in the numerator of the log term and all the species on the left in the denominator. For n, note that two electrons are being transferred from Pb to $2H_3O^+$. This could be figured out either from noting that one Pb goes to one Pb^{2+} plus two electrons or that two H_3O^+ pick up two electrons to form one H_2 plus $2H_2O$.). But first we need to figure out the E^0 from the values in Appendix C.

$Pb(s) \longrightarrow Pb^{2+} + 2e^-$		$E^0 = +0.13$ V
$2H_3O^+ + 2e^- \longrightarrow H_2(g) + 2H_2O$		$E^0 = 0$

$$Pb(s) + 2H_3O^+ \longrightarrow Pb^{2+} + H_2(g) + 2H_2O \qquad E^0 = +0.13 \text{ V}$$

$$E = E^0 - \frac{0.0592}{2} \log \frac{(m_{Pb^{2+}})(p_{H_2})}{(m_{H_3O^+})^2}$$

$$= +0.13 - 0.0296 \log \frac{(0.10)(10^{-6})}{(0.010)^2} = 0.22 \text{ V } \blacksquare$$

■ EXAMPLE 272.   What will be the E for the reaction $2Al(s) + 3I_2(s)$ $\rightarrow 2Al^{3+} + 6I^-$ under conditions where Al^{3+} is 0.10 m and I^- is 0.010 m?

SOLUTION:

$Al(s) \longrightarrow Al^{3+} + 3e^-$		$E^0 = +1.66$ V
$I_2(s) + 2e^- \longrightarrow 2I^-$		$E^0 = +0.54$ V

$$2Al(s) + 3I_2(s) \longrightarrow 2Al^{3+} + 6I^- \qquad E^0 = +2.20 \text{ V}$$

Recall that the E^0 for a total reaction is just the sum of the E^0's for the individual half-reactions, provided the sign is consistent with the direction of reaction as it actually occurs. However, to get the net balanced equation we need to double the first half-reaction and triple the second half-reaction before adding the half-reactions up.

To get n for the total reaction $2Al(s) + 3I_2(s) \rightarrow 2Al^{3+} + 6I^-$ we can note that $2Al(s)$ must give up six electrons and $3I_2(s)$ must pick up six electrons to accomplish the process. Therefore, $n = 6$.

$$E = E^0 - \frac{0.0592}{6} \log (m_{Al^{3+}})^2 (m_{I^-})^6$$

Note that, since there are only solid phases on the left of the net chemical equation, no species appear in the denominator of the logarithm term:

$$m_{Al^{3+}} = 0.10 \text{ } m; \text{ } m_{I^-} = 0.010 \text{ } m$$

$$E = 2.20 - 0.00987 \log (0.10)^2 (0.010)^6 = +2.34 \text{ V } \blacksquare$$

▪ EXAMPLE 273.   Calculate the E for the reaction $2H_2(g) + O_2(g)$ $\rightarrow 2H_2O$, when $p_{H_2} = 5.0$ atm, $p_{O_2} = 2.5$ atm, and $m_{H_3O^+} = 0.60$ m.

SOLUTION:

$$H_2(g) + 2H_2O \longrightarrow 2H_3O^+ + 2e^- \qquad E^0 = 0$$

$$O_2(g) + 4H_3O^+ + 4e^- \longrightarrow 6H_2O \qquad E^0 = +1.23 \text{ V}$$

$$2H_2(g) + O_2(g) \longrightarrow 2H_2O \qquad\qquad E^0 = +1.23 \text{ V}$$

$$E = E^0 - \frac{0.0592}{n} \log \frac{1}{(p_{H_2})^2(p_{O_2})}, \qquad \text{where} \quad n = 4$$

$$E = +1.23 - \frac{0.0592}{4} \log \frac{1}{(5.0)^2(2.5)} = +1.26 \text{ V} ▪$$

Exercises

*9.1. What is the charge in coulombs of a dinegative sulfide ion?

*9.2. What would be the current in amperes in a synchrotron-type particle accelerator corresponding to the movement of 2.9×10^{16} protons per second?

*9.3. In a given cathode-ray tube the electron beam sweeps across the face of the tube with a beam current of 6.50×10^{-6} amp. What would be the total number of electrons deposited on the face in 10 min? (*Answer:* 2.43×10^{16}.)

*9.4. What is the total electric charge in electrostatic units of 1 mole of electrons?

*9.5. If starting the engine of an automobile drains the battery of a current of 180 amp for 2.4 sec, how many faradays have been transferred?

**9.6. Given an electron moving in a circular orbit at a velocity of 3.0×10^8 cm/sec. If the radius of the circle is 2.0 Å, what is the ampere current in the orbit? (Answer: 0.38 milliamp.)

**9.7. In the electrolysis of molten NaOH, the anode reaction is $4OH^- \rightarrow O_2(g) + 2H_2O + 4e^-$. Calculate the volume of oxygen gas produced at 1.10 atm and 800°C by the passage of 1.50 amp for 1 hour through such a melt.

*9.8. Small controlled amounts of H_3O^+ can be generated by passing current through an appropriate cell. Given the anode reaction $6H_2O \rightarrow O_2 + 4H_3O^+ + 4e^-$, how much current would you need to pass for 1 min to generate 2.0×10^{-4} mole of H_3O^+?

*9.9. At a price of 0.010 cent per ampere-hour, what would be the cost of generating a kilogram of aluminum by the reaction $Al^{3+} + 3e^- \rightarrow Al(s)$? (*Answer:* 30 cents.)

9.10. In the electrolysis of aqueous $AgNO_3$, the anode reaction is $6H_2O \rightarrow O_2(g) + 4H_3O^+ + 4e^-$, while the cathode reaction is $Ag^+ + e^- \rightarrow Ag(s)$. During the time it takes to plate out 0.060 g of Ag(s) on the cathode, what volume of O_2 (STP) should form at the anode?

9.11. Given a cell in which the cathode reaction is $Cu^{2+} + 2e^- \rightarrow Cu(s)$ and the anode reaction is $6H_2O \rightarrow O_2(g) + 4H_3O^+ + 4e^-$, calculate the volume of wet oxygen gas at barometric pressure 740 Torr and temperature 25°C you expect to collect at the anode to accompany deposition of 0.696 g of copper at the cathode.

9.12. Two cells, one containing aqueous $AgNO_3$ and the other containing aqueous $CuSO_4$, are set up in series. In a given electrolysis that results in depositing 1.25 g of silver in the first cell, how much copper should deposit simultaneously in the second cell? (*Answer*: 0.369 g.)

9.13. When dilute aqueous Na_2SO_4 is electrolyzed with copper electrodes, the reactions are $2H_2O + 2e^- \rightarrow H_2(g) + 2OH^-$ and $2Cu(s) + 3H_2O \rightarrow Cu_2O(s) + 2H_3O^+ + 2e^-$. In a given electrolysis it is observed that 22.5 ml of wet hydrogen gas (barometric pressure 765 Torr and temperature 28°C) forms at the cathode. What should be the accompanying weight increase at the anode?

9.14. Hydrogen peroxide, H_2O_2, can be made by electrolyzing cold $KHSO_4$ so as to give the reaction $2SO_4^{2-} \rightarrow S_2O_8^{2-} + 2e^-$ and following with distillation to give $2H_3O^+ + S_2O_8^{2-} \rightarrow H_2O_2 + 2H_2SO_4$. How long would you need to pass a current of 15.0 milliamp to form 3.45×10^{-3} mole of H_2O_2 by these reactions? Assume 100% yield.

9.15. In the electrolysis of dilute aqueous H_2SO_4 it is usually observed that the ratio of hydrogen to oxygen is greater than the precise value 2 expected from only the reactions $2H_2O + 2e^- \rightarrow H_2(g) + 2OH^-$ and $6H_2O \rightarrow O_2(g) + 4H_3O^+ + 4e^-$. The deviation is usually attributable to some formation of hydrogen peroxide at the anode by the reaction $4H_2O \rightarrow H_2O_2 + 2H_3O^+ + 2e^-$. In a given experiment we observe 49.82 ml of H_2 (wet, total pressure 738 Torr, temperature 26°C) and 23.65 ml of O_2 (wet, total pressure 736 Torr, temperature 26°C). What fraction of the anode current has gone to form H_2O_2? (*Answer*: 5.3%.)

9.16. In a galvanic cell using the reaction $Zn(s) + 2Ag^+ \rightarrow Zn^{2+} + 2Ag(s)$ what weight of zinc metal would be consumed in producing a current of 10.0 milliamp for 1 hour?

9.17. The overall cell reaction in the lead storage battery is $Pb(s) + PbO_2(s) + 2H_3O^+ + 2HSO_4^- \rightarrow 2PbSO_4(s) + 4H_2O$. During discharge, the anode Pb(s) converts to $PbSO_4(s)$ and the cathode $PbO_2(s)$ to $PbSO_4(s)$, also. What would be the weight gain of the anode and of the cathode when a lead storage cell generates a current of 150 amp for 5.0 sec?

9.18. Given that the overall reaction for the lead storage cell is $Pb(s) + PbO_2(s) + 2H_3O^+ + 2HSO_4^- \rightarrow 2PbSO_4(s) + 4H_2O$ and that for the

Edison cell is $Fe(s) + Ni_2O_3(s) + 3H_2O \rightarrow Fe(OH)_2(s) + 2Ni(OH)_2(s)$, calculate the relative efficiencies of these two cells in terms of coulombs produced per gram of reactants consumed. (*Answer*: 284 and 701 C/g.)

9.19. Given a fuel cell that uses the anode reaction $CH_4(g) + 10OH^-$ $\rightarrow CO_3^{2-} + 7H_2O + 8e^-$. You wish to employ this fuel cell to power an automobile with 80 amp for 1 hour. How many liters of $CH_4(g)$ (STP) would you need to provide? Assume 100% efficiency.

9.20. The density of liquid hydrogen is 0.070 g/ml, that of liquid oxygen is 1.149 g/ml. You are assigned the job of designing fuel tanks for the reaction $2H_2 + O_2 \rightarrow 2H_2O$ to power the fuel cells in a space capsule requiring 30 amp for 8 days. What volume tanks do you need? Assume 100% efficiency.

9.21. How long could you run a motor requiring 15 amp by the fuel-cell reaction $CH_4 + 2O_2 \rightarrow CO_2 + 2H_2O$ starting with a 50-liter tank of liquid methane (density 0.415 g/ml) and a 50-liter tank of liquid oxygen (density 1.149 g/ml)? Assume 100% efficiency. (*Answer*: 535 days.)

*9.22. What is the E^0 for the reaction $Zn(s) + Pb^{2+} \rightarrow Zn^{2+} + Pb(s)$?

*9.23. What is the E^0 for the reaction $5H_2O_2 + 2MnO_4^- + 6H_3O^+$ $\rightarrow 2Mn^{2+} + 5O_2 + 14H_2O$?

*9.24. Write the balanced equation for the oxidation of Br^- by NO_3^- in acidic solution to give Br_2 and NO and calculate its E^0. (*Answer*: $E^0 = -0.11$ V.)

*9.25. If you could make a galvanic cell in which the overall reaction is the oxidation of $Al(s)$ to Al^{3+} by MnO_4^- as it forms Mn^{2+}, what voltage would you expect under standard conditions?

9.26. Under standard conditions what voltage would you expect for a fuel cell using the reaction $2H_2(g) + O_2(g) \rightarrow 2H_2O$? What pressure of hydrogen would you need to make the voltage 1.00 V?

*9.27. What ratio of Pb^{2+} to Sn^{2+} concentration would you need to reverse the reaction $Sn(s) + Pb^{2+} \rightarrow Sn^{2+} + Pb(s)$? (*Answer*: 2.2.)

*9.28. Calculate the E^0 for the reaction $Au(s) + NO_3^- + 4Cl^- + 4H_3O^+$ $\rightarrow AuCl_4^- + NO + 6H_2O$. What would be the effect of changing chloride concentration from standard conditions to 10^{-2} m?

9.29. The E^0 for the reference half-reaction $H_2 + 2H_2O \rightarrow 2H_3O^+ + 2e^-$ is zero. What would the voltage become in pure water where the H_3O^+ concentration is 10^{-7} m, assuming H_2 pressure stays constant? To what pressure would the H_2 have to go to restore the voltage in pure water to zero?

***9.30.** It is observed that the voltage of a galvanic cell using the reaction $M(s) + xH_3O^+ \rightarrow M^{x+} + x/2H_2(g) + H_2O$ varies linearly with the log of the square root of the hydrogen pressure and the cube root of the M^{x+} concentration. What is the value of x?

10

SOLUTIONS

IN THIS CHAPTER we consider two general topics: (a) How do we express concentrations in solution? (b) What are some of the quantitative relations between the concentration of a solution and its properties (e.g., freezing point, boiling point, vapor pressure)?

For describing the concentration of a solution, the following are in common use: molarity, normality, molality, mole fraction, formality, weight per cent, and volume per cent.

10.1 Molarity

The molarity of a solution is defined as the number of moles of solute per liter of solution. It is usually designated by M. Unless otherwise specified, the solvent is assumed to be water; in cases where ambiguity may arise, we can specify *aqueous solutions*. Since we assume the solvent to be water, we need for the molarity only a number (which tells us how many moles of solute are used to make a liter of solution) and a formula to tell what the solute is. Thus, 0.15 molar HCl (written 0.15 M HCl) means a concentration corresponding to 0.15 mole of HCl per liter of aqueous solution.

Since the molarity only tells the strength of the solution (not how much we have of it), we generally need to specify how much solution we work with, so we can know how much reagent is available. The total number of moles of solute available is equal to (molarity) × (volume in liters). This is consistent dimensionally, (moles per liter) × (liters) = moles.

The most common source of difficulty in working molarity problems is failure to distinguish between moles and moles per liter.

■ EXAMPLE 274. A solution is made by dissolving 127 g of ethyl alcohol (C_2H_5OH) in enough water to make 1.35 liters of solution. What is the molarity of the solution?

189

SOLUTION: One mole of C_2H_5OH weighs 46.07 g.

$$\frac{127 \text{ g of } C_2H_5OH}{46.07 \text{ g/mole}} = 2.76 \text{ moles}$$

We have 2.76 moles of C_2H_5OH in 1.35 liters of solution. Therefore,

$$\text{the molarity} = \frac{\text{moles}}{\text{liters}} = \frac{2.76 \text{ moles}}{1.35 \text{ liters}} = 2.04 \ M.$$

[*Note:* At this point we reemphasize that the dividing line between a numerator and a denominator means " per." So, $\frac{\text{moles}}{\text{liters}}$ is " moles *per* liter ".] ∎

■ EXAMPLE 275.   You have a solution that is 0.693 *M* HCl. For a certain reaction you need to have 0.0525 mole of HCl. How much solution do you take?

SOLUTION: 0.693 *M* HCl means 0.693 mole of HCl per liter of solution. You need 0.0525 mole of HCl. Therefore, if you divide moles needed by moles per liter, you will have liters needed.

$$\frac{0.0525 \text{ mole of HCl}}{0.693 \text{ mole/liter}} = 0.0758 \text{ liter} \blacksquare$$

■ EXAMPLE 276.   Suppose you mix 3.65 liters of 0.105 *M* NaCl with 5.11 liters of 0.162 *M* NaCl. Assuming the volumes are additive—that is, that the total volume after mixing is 8.76 liters—what would be the concentration of the final solution?

SOLUTION: 3.65 liters of 0.105 *M* NaCl contributes

$$(3.65 \text{ liters})\left(0.105 \ \frac{\text{mole}}{\text{liter}}\right) = 0.383 \text{ mole NaCl}$$

5.11 liters of 0.162 *M* NaCl contributes

$$(5.11 \text{ liters})\left(0.162 \ \frac{\text{mole}}{\text{liter}}\right) = 0.828 \text{ mole NaCl}$$

The total moles of NaCl is $0.383 + 0.828 = 1.211$ moles.
The total volume of the solution is 8.76 liters.

Therefore, the final concentration is

$$\frac{1.211 \text{ moles}}{8.76 \text{ liters}} = 0.138 \ M. \ \blacksquare$$

■ EXAMPLE 277.   You have a supply of 0.100 M KMnO$_4$. You wish to make 40.0 ml of 1.95×10^{-3} M KMnO$_4$ by diluting with water. How do you proceed? Assume volumes are additive.

SOLUTION: You want 40.0 ml of 1.95×10^{-3} M KMnO$_4$. $1.95 \times 10^{-3}M$ means 1.95×10^{-3} mole of KMnO$_4$ per liter.
Therefore, in 40.0 ml, or 0.0400 liter, you will have

$$(0.0400 \text{ liter})\left(1.95 \times 10^{-3} \frac{\text{mole of KMnO}_4}{\text{liter}}\right) = 7.80 \times 10^{-5} \text{ mole}$$

To get this 7.80×10^{-5} mole of KMnO$_4$ you need to use 0.100 M KMnO$_4$, which contains 0.100 mole of KMnO$_4$ per liter. If you divide moles required by moles per liter, you will get liters of 0.100 M solution needed.

$$\frac{7.80 \times 10^{-5} \text{ mole}}{0.100 \text{ mole/liter}} = 7.80 \times 10^{-4} \text{ liter, or } 0.780 \text{ ml}$$

So, you take 0.780 ml of 0.100 M KMnO$_4$ solution and add enough water to bring the total volume to 40.0 ml. ■

■ EXAMPLE 278.   If you have a supply of 0.150 M NaOH solution and a supply of 0.250 M NaOH, in what ratio should you mix these solutions to prepare some 0.169 M NaOH? Assume additive volumes.

SOLUTION: Let x = liters of 0.150 M NaOH to be used and let y = liters of 0.250 M NaOH to be used.

Total moles of NaOH = $0.150x + 0.250y$

Total volume of solution = $x + y$.

Final concentration = $\dfrac{(0.150x + 0.250y) \text{ mole}}{(x + y) \text{ liters}} = 0.169 \ M$

Solve this equation for the ratio x/y.
Cross-multiply to get $0.150x + 0.250y = 0.169x + 0.169y$.
Simplify to $0.081y = 0.019x$.

Solve to

$$\frac{x}{y} = \frac{0.081}{0.019} = 4.3$$

Therefore, we need to take 4.3 times as much 0.150 M NaOH solution as we take of the 0.250 M NaOH solution. ∎

10.2 Normality

The normality of a solution is defined as the number of equivalents of solute per liter of solution; it usually is designated by a capital N preceded by a number. We read 6 N HCl as "six normal HCl"; it means a solution that contains 6 equivalents of HCl per liter of solution. Normality is a convenient concentration term for work with solutions of acids and bases and for work with solutions of oxidizing and reducing agents. As discussed in detail in Chapter 6, the concept of equivalent simplifies acid–base calculations and oxidation–reduction calculations. Since the bulk of the work with such reactions is done by mixing solutions, the importance of normality as a concentration unit cannot be slighted. You should review Chapter 6, albeit rapidly, at this point.

For acids and bases, the points to remember are these: One equivalent of acid furnishes 1 mole of H^+ (or H_3O^+). One equivalent of base uses up 1 mole of H^+ (or H_3O^+). In a neutralization reaction, 1 equivalent of any acid is required for stoichiometric reaction with 1 equivalent of any base. Since "normality" specifies the number of equivalents per liter, the product of normality times volume (in liters) gives the number of equivalents of the reagent. The requirement that the number of equivalents of acid be equal to the number of equivalents of base leads to the simple relation that the (normality of the acid solution)(volume of the acid solution) = (normality of the base solution)(volume of base solution).

∎ EXAMPLE 279. You are directed to make a solution of phosphoric acid, H_3PO_4, for use in a neutralization reaction where all three hydrogens are to be neutralized. How many grams of H_3PO_4 do you need to make 18.68 ml of 0.1079 N H_3PO_4?

SOLUTION: 0.1079 N means 0.1079 equivalent per liter. 18.68 ml of such solution contains

$$(0.01868 \text{ liter})\left(0.1079 \frac{\text{equivalent}}{\text{liter}}\right) = 0.002016 \text{ equivalent}$$

One mole of H_3PO_4 furnishes 3 moles of H^+, so 1 mole of H_3PO_4 is the same as 3 equivalents. One mole of H_3PO_4 weighs 97.995 g, so 1 equivalent is 1/3 of this, or 32.665 g.

To get 0.002016 equivalent of H_3PO_4 we need to take

$$(0.002016 \text{ equivalent})\left(32.665 \frac{g}{\text{equivalent}}\right) = 0.06585 \text{ g} \blacksquare$$

■ **EXAMPLE 280.** What is the normality of an H_2SO_4 solution, 23.67 ml of which is required to neutralize 26.73 ml of 0.0936 N NaOH solution?

SOLUTION: (Figure out how many equivalents of base you have. You will then know how many equivalents of acid you will need. Then calculate the concentration needed.)

0.0936 N NaOH means 0.0936 equivalent of NaOH per liter of solution. You have 0.02673 liter, or

$$(0.02673 \text{ liter})\left(0.0936 \frac{\text{equivalent}}{\text{liter}}\right) = 0.00250 \text{ equivalent NaOH}$$

You need an equal number of equivalents of acid, so you need 0.00250 equivalent of H_2SO_4, which you will find in the 23.67 ml of acid solution. Therefore, the normality of the acid solution must be

$$\frac{0.00250 \text{ equivalent}}{0.02367 \text{ liter}} = 0.106 \ N \blacksquare$$

■ **EXAMPLE 281.** You are given a solution that is $7.69 \times 10^{-3} M$ $Ca(OH)_2$. What would be its normality for complete neutralization?

SOLUTION: One mole of $Ca(OH)_2$ furnishes 2 moles of OH^- if completely neutralized. One mole of $Ca(OH)_2$ therefore equals 2 equivalents.

$7.69 \times 10^{-3} M$ means 7.69×10^{-3} mole/liter.

$$(7.69 \times 10^{-3} \text{ mole})\left(2 \frac{\text{equivalents}}{\text{mole}}\right) = 1.54 \times 10^{-2} \text{ equivalent of } Ca(OH)_2$$

7.69×10^{-3} mole/liter means 1.54×10^{-2} equivalent/liter.
So, $7.69 \times 10^{-3} M$ $Ca(OH)_2$ is $1.54 \times 10^{-2} N$ $Ca(OH)_2$. ■

■ **EXAMPLE 282.** You are given a solution that is $6.68 \times 10^{-3} M$ H_3PO_4. If you add 36.2 ml of water to 18.6 ml of this solution, what will be the normality of the final solution? Assume additive volumes.

SOLUTION: You start with 18.6 ml of 6.68×10^{-3} M H_3PO_4. This contains

$$(0.0186 \text{ liter})\left(6.68 \times 10^{-3} \frac{\text{mole}}{\text{liter}}\right) = 1.24 \times 10^{-4} \text{ mole of } H_3PO_4$$

When you add water, you do not change the number of moles of H_3PO_4 present.

One mole of H_3PO_4 equals 3 equivalents.

1.24×10^{-4} mole of $H_3PO_4 = 3.72 \times 10^{-4}$ equivalent of H_3PO_4.

The final volume of solution is $36.2 + 18.6 = 54.8$ ml. Therefore, the final normality is

$$\frac{3.72 \times 10^{-4} \text{ equivalent}}{0.0548 \text{ liter}} = 6.79 \times 10^{-3} \text{ } N \text{ } H_3PO_4 \blacksquare$$

For oxidizing and reducing agents, the points to remember are these: One equivalent of a reducing agent furnishes 1 mole of electrons. One equivalent of an oxidizing agent uses up 1 mole of electrons. In an oxidation–reduction reaction, 1 equivalent of any reducing agent is required for stoichiometric reaction with 1 equivalent of any oxidizing agent. Again, since normality gives equivalents per liter, the product (normality of oxidizing solution) (volume of oxidizing solution) gives the number of equivalents of oxidizing agent. This must equal the number of equivalents of reducing agent, which is given by (normality of reducing solution)(volume of reducing solution).

Because oxidizing and reducing agents can frequently form a variety of products for which the electron change differs, it is necessary to make clear what electron change is being considered when the normality of an oxidizing or reducing solution is defined. In other words, you are supposed to tell what the solution will be used for before you can really interpret a normality label on the bottle.

■ EXAMPLE 283.   You wish to make up 17.31 ml of 0.692 N $KMnO_4$ solution to be used in a reaction where MnO_4^- goes to Mn^{2+}. How many grams of $KMnO_4$ do you need?

SOLUTION: 0.692 N $KMnO_4$ means 0.692 equivalent of $KMnO_4$ per liter. In 17.31 ml of such a solution, you will have

$$(0.01731 \text{ liter})\left(0.692 \frac{\text{equivalent}}{\text{liter}}\right) \text{ or } 0.0120 \text{ equivalent of } KMnO_4$$

To be used where MnO_4^- goes to Mn^{2+} means a five-electron change. So we want 1 mole of $KMnO_4$ to pick up 5 moles of electrons—that is, 1 mole of $KMnO_4$ to be 5 equivalents.

One mole of $KMnO_4$ weighs 158.04 g.

One equivalent $= 1/5$ mole $= (1/5)(158.04) = 31.608$ g.

Since we need 0.0120 equivalent of $KMnO_4$, we will need to take

$$(0.0120 \text{ equivalent})\left(31.608 \ \frac{g}{\text{equivalent}}\right) \text{ or } 0.379 \text{ g}$$

■ EXAMPLE 284.   In acidic solution $Cr_2O_7^{2-}$ oxidizes Sn^{2+} to form Cr^{3+} and Sn^{4+}. How many milliliters of 0.1097 N $K_2Cr_2O_7$ will you need to react with 25.36 ml of 0.2153 N $SnCl_2$ in this way?

SOLUTION: (Figure out how many equivalents you have in the reducing solution. You will then know you need an equal number of equivalents of oxidizing agent.)

0.2153 N means 0.2153 equivalent of reducing agent per liter. In 25.36 ml you will have

$$(0.02536 \text{ liter})\left(0.2153 \ \frac{\text{equivalent}}{\text{liter}}\right)$$

$$= 0.005460 \text{ equivalent of reducing agent}$$

You need 0.005460 equivalent of oxidizing agent.

It comes from 0.1097 N $K_2Cr_2O_7$, which means 0.1097 equivalent of oxidizing agent per liter.

To get 0.005460 equivalent of oxidizing agent, you need to take

$$\frac{0.005460 \text{ equivalent}}{0.1097 \text{ equivalent/liter}} = 0.04977 \text{ liter or } 49.77 \text{ ml} \ ■$$

■ EXAMPLE 285.   Working under conditions where ClO_3^- will go to Cl^-, how many milliliters of 1.79×10^{-3} M $NaClO_3$ solution need to be taken to oxidize 36.92 ml of 1.50×10^{-4} N $NaHSO_3$ and 2.345×10^{-5} equivalent of solid $NaHSO_3$ to give HSO_4^-?

SOLUTION: (Figure out the total number of equivalents of reducing agent that need to be oxidized. Take enough oxidizing solution to provide the equivalents needed to do this.)

1.50×10^{-4} N $NaHSO_3$ means 1.50×10^{-4} equivalent/liter.

In 36.92 ml, there will be

$$(0.03692 \text{ liter})\left(1.50 \times 10^{-4}\,\frac{\text{equivalent}}{\text{liter}}\right) = 0.554 \times 10^{-5} \text{ equivalent}$$

This has to be added to the 2.34×10^{-5} equivalent to give a total of 2.89×10^{-5} equivalent of $NaHSO_3$ that needs to be oxidized.

Since 1 equivalent of reducing agent needs 1 equivalent of oxidizing agent, we conclude that 2.89×10^{-5} equivalent of $NaClO_3$ will be needed. Since the ClO_3^- will go to Cl^-, a six-electron change, 1 mole of $NaClO_3$ will act as 6 equivalents.

To get 2.89×10^{-5} equivalent of $NaClO_3$, we need to take

$$(2.89 \times 10^{-5} \text{ equivalent})\left(\frac{1 \text{ mole}}{6 \text{ equivalents}}\right) = 4.82 \times 10^{-6} \text{ mole}$$

The $NaClO_3$ solution is specified to be 1.79×10^{-3} M, so we need

$$\frac{4.82 \times 10^{-6} \text{ mole}}{1.79 \times 10^{-3} \text{ mole/liter}} = 2.69 \times 10^{-3} \text{ liter or } 2.69 \text{ ml} \ \blacksquare$$

10.3 Molality

A 1 molal solution (designated 1 m and pronounced with a decided accent on the "a" of molal) is a solution in which there is 1 mole of solute dissolved per kilogram of solvent. For aqueous solutions, this means 1000 g of H_2O plus 1 mole of solute. As an example, 1 m NaCl is read "one molal NaCl" and designates a solution composed of 1 mole of NaCl per 1000 g of H_2O. The big advantage of molality, one that appeals mightily to the physical chemists, is that the molality of a given solution does not change if the temperature of the solution is changed. The same cannot be said for molarity where, for example, the expansion usually brought about by raising the temperature changes the volume of the solution and therefore changes the moles per liter of solution. For this reason, the high-precision chemist affixes to his molarity designation the temperature at which the label applies—for example, 0.6843 M HCl (18°C). Fortunately, at least near room temperature, solutions do not expand much, so the change of molarity with temperature is likely to be small (less than 1% change for a 10°C rise). The other fortunate fact is that the density of water and of most dilute aqueous solutions is very close to 1 g/ml, which means that molarity will be almost equal to molality for *dilute* solutions. (Dilute here means less concentrated than 0.01 M or 0.01 m.)

To convert from molality to molarity, we need to know the density of the solution. Otherwise, we cannot figure out the volume of solution.

■ EXAMPLE 286. A solution is made by dissolving 1.68 g of NaCl in 869 g of H_2O. What is its molality (or molal concentration)?

SOLUTION: One mole of NaCl weighs 58.44 g.

$$\frac{1.68 \text{ g of NaCl}}{58.44 \text{ g/mole}} = 0.0287 \text{ mole of NaCl}$$

869 g of H_2O = 0.869 kg of H_2O

$$\text{Molality} = \frac{0.0287 \text{ mole NaCl}}{0.869 \text{ kg } H_2O} = 0.0330 \ m \text{ NaCl} ■$$

■ EXAMPLE 287.   How many grams of $Al_2(SO_4)_3$ do you need to make 87.62 g of 0.0162 m $Al_2(SO_4)_3$ solution?

SOLUTION: 0.0162 m $Al_2(SO_4)_3$ means 0.0162 mole of $Al_2(SO_4)_3$ per kilogram of H_2O.
One mole of $Al_2(SO_4)_3$ weighs 342.15 g.
0.0162 mole of $Al_2(SO_4)_3$ weighs

$$(0.0162 \text{ mole})\left(342.15 \ \frac{g}{\text{mole}}\right) = 5.54 \text{ g}$$

So, in this solution there are 5.54 g of $Al_2(SO_4)_3$ per 1000 g of H_2O. Stated another way, there are 5.54 g of $Al_2(SO_4)_3$ per 1005.54 g of solution, where we have simply added the weights of the two components.
If there are 5.54 g of $Al_2(SO_4)_3$ per 1005.54 g of solution, then given 87.62 g of the solution there will be

$$(87.62 \text{ g of solution})\left(\frac{5.54 \text{ g of } Al_2(SO_4)_3}{1005.54 \text{ g of solution}}\right) = 0.483 \text{ g of } Al_2(SO_4)_3 ■$$

■ EXAMPLE 288.   You have a solution that contains 410.3 g of H_2SO_4 per liter of solution at 20°C. If the density of the solution is 1.243 g/ml, what is the molality and molarity of the solution?

SOLUTION: The density of the solution is 1.243 g/ml.
Therefore, 1 liter of solution weighs

$$\left(1.243 \ \frac{g}{ml}\right)\left(1000 \ \frac{ml}{\text{liter}}\right) = 1243 \text{ g.}$$

We are told that 410.3 g of this is H_2SO_4, so the rest of the weight (1243 − 410 = 833 g) must be water.

$$\frac{410.3 \text{ g of } H_2SO_4}{98.078 \text{ g/mole}} = 4.183 \text{ moles}$$

833 g of $H_2O = 0.833$ kg of H_2O

$$\text{Molality} = \frac{4.183 \text{ moles of } H_2SO_4}{0.833 \text{ kg of } H_2O} = 5.02 \ m \ H_2SO_4$$

$$\text{Molarity} = \frac{4.183 \text{ moles of } H_2SO_4}{1 \text{ liter of solution}} = 4.183 \ M \ H_2SO_4 \ \blacksquare$$

10.4   Mole fraction

For some solutions, particularly gases and nonaqueous solutions, the most useful concentration expression is the one that tells what fraction of the molecules present are of a particular species. Thus, mole fraction is the number of moles of one substance in the solution divided by the total number of moles of all kinds of substances in the solution. Because mole fraction is computed as moles divided by moles, it has no units. Furthermore, if you know the mole fraction of all but one component in a solution, then you can calculate the mole fraction of that last component from the fact that in a given mixture the sum of the mole fractions equals unity.

■ EXAMPLE 289.   A box contains 3.89×10^{-3} mole of N_2 gas and 1.54×10^{-3} mole of H_2 gas. What is the mole fraction of each component present?

SOLUTION: You have 3.89×10^{-3} mole of N_2 plus 1.54×10^{-3} mole of H_2, which makes a total of 5.43×10^{-3} mole present.
Of the total 5.43×10^{-3} mole, N_2 accounts for 3.89×10^{-3} mole. Therefore, the mole fraction of N_2 is

$$\frac{3.89 \times 10^{-3} \text{ mole of } N_2}{5.43 \times 10^{-3} \text{ mole of total}} = 0.716$$

The mole fraction of H_2 must be $1 − 0.716 = 0.284$. Or, calculating it directly, the mole fraction of H_2 is

$$\frac{1.54 \times 10^{-3} \text{ mole of } H_2}{5.43 \times 10^{-3} \text{ mole total}} = 0.284 \ \blacksquare$$

■ EXAMPLE 290.   What is the mole fraction of C_2H_5OH in an aqueous solution that is simultaneously 3.86 m C_2H_5OH and 2.14 m CH_3OH?

SOLUTION: 3.86 m C_2H_5OH means 3.86 moles of C_2H_5OH per kilogram of H_2O. 2.14 m CH_3OH means 2.14 moles of CH_3OH per kilogram of H_2O.

In 1 kg of H_2O, we have

$$\frac{1000 \text{ g}}{18.015 \text{ g/mole}} = 55.51 \text{ moles of } H_2O$$

Total moles in the solution = 3.86 moles of C_2H_5OH + 2.14 moles of CH_3OH + 55.51 moles of H_2O = 61.51 moles

$$\text{Mole fraction of } C_2H_5OH = \frac{3.86 \text{ moles of } C_2H_5OH}{61.51 \text{ moles total}} = 0.0628 \text{ ■}$$

10.5 Formality

Formality was introduced as a concentration unit because some people felt that there was too much ambiguity in the term molarity, particularly when it was used to describe solutions in which the species placed in solution broke up or otherwise changed to create other species. In such cases, the only thing you could be sure of was what you put into the solution; only *formally* did you know what was there after the interaction with the solvent had occurred. There was also another cogent consideration. In many cases we know only the simplest formula of the solute, not its true molecular formula. In such cases, 1 mole of solute means nothing unless the precise formula is also written down for which the formula-weight can be unambiguously calculated. This makes it necessary to describe the solution concentration in terms of a specified formula, and in this sense the terms "formal" and "formality" are closely tied to the formula given.

A "one formal solution," designated 1 F, means a solution in which 1 gram-formula-weight of the specific formula-designated solute has been dissolved in enough water to make 1 liter of solution. As an example, 0.69 F HCl is read "zero point six nine formal HCl" and designates a solution wherein 0.69 gram-formula-weights of HCl have been dissolved per liter of solution. Since most people would call such a solution 0.69 M HCl, there are arguments for not bothering with formal concentration units at all. However, they have considerable use, particularly on the West Coast, and you should know what all the shouting is about. The case in favor of "formal" and "formality" is easier to defend once we are aware of the fact that 0.69 M HCl has *no* HCl molecules in solution, all of them having converted into H_3O^+ and Cl^-.

▪ EXAMPLE 291.   If you dissolve 1.89 g of Na_2SO_4 in enough water to make 0.086 liter of solution, what will be its formal concentration?

SOLUTION: The formula-weight of Na_2SO_4 is 142.04 amu.

$$\frac{1.89 \text{ g of } Na_2SO_4}{142.04 \text{ g/gram-formula-weight}} = 0.0133 \text{ gram-formula-weight}$$

Formality is defined as the number of gram-formula-weights per liter, so we divide 0.0133 by 0.086:

$$\frac{0.0133 \text{ gram-formula-weight}}{0.086 \text{ liter}} = 0.15 \ F \ ▪$$

▪ EXAMPLE 292.   You are given 0.692 liter of 0.106 F HCl. Knowing that HCl in water is 100% dissociated into H_3O^+ and Cl^-, how many moles of Cl^- are there in this sample?

SOLUTION: 0.106 F HCl means 0.106 gram-formula-weight of HCl per liter.
Since we have 0.692 liter of solution, we must have used

$$(0.692 \text{ liter})\left(0.106 \ \frac{\text{gram-formula-weight}}{\text{liter}}\right)$$

$$= 0.0734 \text{ gram-formula weight}$$

One gram-formula-weight of HCl furnishes 1 mole of H_3O^+ and 1 mole of Cl^- when it dissociates.
Therefore, 0.0734 gram-formula-weight of HCl dissociates to furnish 0.0734 mole of Cl^-. ▪

10.6   Per cent by weight and per cent by volume

Since many solutions are made up simply by weighing out the various constituents and mixing them up, it is a simple matter to calculate what per cent of the total weight of the solution is contributed by each component. Obviously, the sum of all the *per cents by weight* should add up to 100%.

▪ EXAMPLE 293.   The concentrated sulfuric acid that is peddled commercially is normally 95% H_2SO_4 by weight. If its density is 1.834 g/ml, what is its molarity?

SOLUTION: Take 1 liter of the solution. Since its density is given as 1.834 g/ml, you know that 1 liter must weigh

$$(1000 \text{ ml})\left(1.834 \frac{g}{ml}\right) = 1834 \text{ g}$$

Of this total weight, 95% is H_2SO_4.

$$(0.95)(1834 \text{ g}) = 1700 \text{ g of } H_2SO_4$$

One mole of H_2SO_4 weighs 98.08 g.
In 1700 g of H_2SO_4, we have

$$\frac{1700 \text{ g}}{98.08 \text{ g/mole}} = 17 \text{ moles}$$

Since we have 17 moles of H_2SO_4 per liter of solution, the molarity of the solution must be 17 M H_2SO_4. ∎

Per cent by weight as a concentration unit is perfectly straightforward, since every time we add x grams of substance A to y grams of substance B we always get $x + y$ grams of mixture. Per cent by volume has the added fillip that volumes do not behave this way. In fact, if we dissolve x ml of A in y ml of B we will generally not end up with $x + y$ ml of solution. *Per cent by volume* is defined as the per cent of the final volume that is represented by the initial volume of the component in question. As a specific example, if we mix 50.00 ml of C_2H_5OH with 50.00 ml of H_2O we get only 96.54 ml total of solution. If we compute the percent by volume of C_2H_5OH, we get

$$\frac{50.00 \text{ ml of } C_2H_5OH}{96.54 \text{ ml of total solution}} \times 100 = 51.79\%$$

If we compute the per cent by volume of H_2O, we get

$$\frac{50.00 \text{ ml of } H_2O}{96.54 \text{ ml of total solution}} \times 100 = 51.79\%$$

So, we have the interesting point that per cents by volume of a solution's components need not necessarily add up to a hundred.

■ EXAMPLE 294. A solution is made by adding enough water to 32.86 g of C_2H_5OH to make a total volume of 100.00 ml. If the density of pure C_2H_5OH is 0.7851 g/ml, what will be the concentration of the solution expressed as per cent of C_2H_5OH by volume?

SOLUTION: Initial weight of C_2H_5OH is 32.86 g.
We are given its density as 0.7851 g/ml.
Therefore, we know the initial volume of the C_2H_5OH to be

$$\frac{32.86 \text{ g}}{0.7851 \text{ g/ml}} = 41.85 \text{ ml}$$

We are told that the final volume of the solution is 100.00 ml. Therefore,

$$\text{per cent by volume} = \frac{41.85 \text{ ml}}{100.00 \text{ ml}} \times 100 = 41.85\% \blacksquare$$

10.7   Raoult's law

One of the most important observations made for solutions is that the vapor pressure of the solvent is lowered proportionally to the concentration of dissolved particles in the solution. This is the gist of Raoult's law, which states that the partial pressure p_1 of the solvent above the solution divided by the vapor pressure of the pure solvent p_1^0 is equal to the mole fraction of solvent present. If we designate by x_1 the mole fraction of solvent in the system, then Raoult's law can be written

$$\frac{p_1}{p_1^0} = x_1$$

For a two-component system—that is, one consisting of solvent 1 and solute 2—the mole fractions x_1 and x_2 must add up to unity. So, we can write $x_1 + x_2 = 1$ or $x_1 = 1 - x_2$. Substituting above, we get

$$\frac{p_1}{p_1^0} = 1 - x_2$$

Solving this for x_2, we get

$$x_2 = 1 - \frac{p_1}{p_1^0} = \frac{p_1^0 - p_1}{p_1^0}$$

which is another way of stating Raoult's law. This latter part is important because it states that the vapor-pressure lowering, $p_1^0 - p_1$, is directly proportional to the mole fraction of solute. (Recall that p_1^0 is a constant number characteristic of the pure solvent.) Experimentally, this is significant because it means that a measurement of the vapor-pressure lowering brought about by the addition of a solute can tell us what fraction of all the particles in the final solution are actually derived from the solute. This may be particularly useful to know when the solute molecules break up into smaller fragments on coming in contact with solvent.

▪ EXAMPLE 295. Assuming Raoult's law is followed, what happens to the vapor pressure of water if 43.68 g of sugar, $C_{12}H_{22}O_{11}$, is dissolved in 245.0 ml of water at 25°C? At 25°C the density of water is 0.9971 g/ml and its vapor pressure is 23.756 Torr. Sugar does not dissociate in water.

SOLUTION: (Figure out the mole fraction that is water. Then use Raoult's law to calculate the vapor pressure.)

$$(245.0 \text{ ml } H_2O)\left(0.9971 \frac{g}{ml}\right) = 244.3 \text{ g } H_2O$$

One mole of H_2O is 18.015 g.

$$\frac{244.3 \text{ g of } H_2O}{18.015 \text{ g/mole}} = 13.56 \text{ moles of } H_2O$$

One mole $C_{12}H_{22}O_{11}$ weights 342.30 g.

$$\frac{43.68 \text{ g of } C_{12}H_{22}O_{11}}{342.30 \text{ g/mole}} = 0.128 \text{ mole}$$

Total moles = 13.56 moles of H_2O + 0.128 mole of sugar = 13.69 moles

$$\text{Mole fraction } H_2O = \frac{13.56 \text{ moles of } H_2O}{13.69 \text{ moles total}} = 0.9905$$

According to Raoult's law $p_1 = x_1 p_1^0$:

x_1 = mole fraction of water = 0.9905
p_1^0 = vapor pressure of the pure water = 23.756 Torr
p_1 = vapor pressure of H_2O over the solution
p_1 = (0.9905)(23.756 Torr) = 23.53 Torr. ▪

▪ EXAMPLE 296. If 106.3 g of compound Z dissolves in 863.5 g of benzene (C_6H_6) to lower the vapor pressure of benzene from 98.6 to 86.7 Torr, what must be the molecular weight of Z? Assume no dissociation.

SOLUTION: (Use the vapor-pressure lowering to fix the mole fraction of Z in the solution. Then calculate the weight of 1 mole of Z.)

The vapor pressure of the benzene goes from 98.6 to 86.7 Torr. This represents a decrease of 11.9 Torr. According to Raoult's law, the drop in vapor pressure (11.9 Torr) divided by the original vapor pressure (98.6 Torr) gives the mole fraction of solute. So, in this solution the mole fraction of compound Z is 11.9/98.6, or 0.121.

$$\text{Mole fraction of Z} = \frac{\text{moles of Z}}{\text{moles of Z} + \text{moles of benzene}} = 0.121$$

We have 863.5 g of benzene. One mole of C_6H_6 weighs 78.11 g, so 863.5 g is

$$\frac{863.5 \text{ g}}{78.11 \text{ g/mole}} = 11.05 \text{ moles of benzene}$$

Substituting this above we get

$$\text{mole fraction of Z} = \frac{\text{moles of Z}}{\text{moles of Z} + 11.05} = 0.121$$

Solving for moles of Z, we get 1.52 moles.
The problem tells us what weight of Z we have, 106.3 g. So,

$$\text{molecular weight of Z} = \frac{106.3 \text{ g}}{1.52 \text{ moles}} = 69.9 \frac{\text{g}}{\text{mole}} \blacksquare$$

10.8 Henry's law

The more you squeeze on a gas, the more of it you dissolve in a solvent. Stated more elegantly, the concentration of a dissolved gas is proportional to the pressure of the gas, at least for dilute solutions and for cases where there is no chemical reaction between gas and solvent. This is called Henry's law. It is usually written

$$K = \frac{p}{x}$$

where p is the partial pressure of the gas in question in the gas phase over the solution (in millimeters of Hg, or torricellis) and x is the mole fraction of the dissolved gas in the liquid solution phase. K is called the Henry's law constant and is characteristic of the solute, the solvent, and the temperature.

■ EXAMPLE 297.   At 20°C, oxygen gas, O_2, dissolves in water to an extent consistent with a Henry's law constant of 2.95×10^7. Under ordinary atmospheric conditions, where p_{O_2} is about 0.21 atm, how many moles of O_2 gas will be dissolved in 1000 g of water?

SOLUTION:

$$p_{O_2} = 0.21 \text{ atm} = (0.21 \text{ atm})\left(760 \frac{\text{Torr}}{\text{atm}}\right) = 160 \text{ Torr}$$

We are given the Henry's law constant K as 2.95×10^7.

$$K = 2.95 \times 10^7 = \frac{p_{O_2}}{x_{O_2}} = \frac{160}{x_{O_2}}$$

From this, we calculate x_{O_2}, the mole fraction of O_2 in the solution, as $160/2.95 \times 10^7 = 5.4 \times 10^{-6}$.

$$\text{Mole fraction of } O_2 = \frac{\text{moles of } O_2}{\text{moles of } O_2 + \text{moles of } H_2O} = 5.4 \times 10^{-6}$$

This is such a very dilute soution that, in the denominator, moles of O_2 is negligible when added to moles of H_2O.

$$\text{Moles of } H_2O = \frac{1000 \text{ g}}{18.015 \text{ g/mole}} = 55.51 \text{ moles}$$

$$\text{Mole fraction of } O_2 = 5.4 \times 10^{-6} = \frac{\text{moles of } O_2}{55.51 \text{ moles total}}$$

$$\text{Moles of } O_2 = (55.51)(5.4 \times 10^{-6}) = 3.0 \times 10^{-4} ■$$

■ EXAMPLE 298.   At 20°C, 9.3 ml of helium gas (temperature 20°C and pressure 730 Torr) can be dissolved in 1000 g of water if the helium pressure over the solution is kept at 730 Torr. Calculate the Henry's law constant for helium. Assume ideal behavior.

SOLUTION: (Figure out how many moles of helium you have. Then calculate the mole fraction of helium in the solution. Then get K.)

From the ideal gas law $PV = nRT$ we get

$$n = \frac{PV}{RT} = \frac{(730/760)(0.0093)}{(0.08206)(293)} = 3.7 \times 10^{-4} \text{ mole of helium}$$

$$\frac{1000 \text{ g of } H_2O}{18.015 \text{ g/mole}} = 55.5 \text{ moles of } H_2O$$

$$\text{Mole fraction of helium} = \frac{3.7 \times 10^{-4}}{55.5 + 3.7 \times 10^{-4}} = 6.7 \times 10^{-6} \blacksquare$$

10.9   Elevation of boiling point

Because addition of a nonvolatile solute lowers the vapor pressure of a solution, a higher temperature will be needed to produce boiling and, therefore, there is raising of the boiling point (bp). The bp elevation is characteristic for each solvent; it is proportional to the molal concentration of solute particles dissolved therein. For H_2O as a solvent, 1 mole of dissolved particles in 1000 g of H_2O produces a bp elevation of 0.512°C. This is called the *molal boiling-point elevation constant for water*. Because it is the concentration of particles and not their identity that is important, observed boiling-point elevations can be used to deduce how many moles of particles there are in the solution per kilogram of solvent. This can be particularly useful in experimentally fixing the molecular weight of a dissolved species. However, in cases where the solute molecules dissociate into smaller particles, the number of moles of particles in the solution will exceed the apparent number put into the solution; this will have to be taken into account because each particle contributes equally to the boiling-point elevation.

■ EXAMPLE 299.   What will be the normal boiling point of a solution made by dissolving 4.26 g of glucose ($C_6H_{12}O_6$) in 87.9 g of H_2O? Glucose is a nondissociating molecular solute. The molal boiling-point elevation constant for H_2O is 0.512°C.

SOLUTION: One mole of $C_6H_{12}O_6$ is 180.16 g.

$$\frac{4.26 \text{ g of } C_6H_{12}O_6}{180.16 \text{g/mole}} = 0.0236 \text{ mole of } C_6H_{12}O_6$$

0.0236 mole of $C_6H_{12}O_6$ per 0.0879 kg of H_2O is the same as

$$\frac{0.0236 \text{ mole of } C_6H_{72}O_6}{0.0879 \text{ kg of } H_2O} = 0.268 \, m$$

For a 1 m solution, the bp elevation is 0.512°C.
For a 0.268 m solution, the bp is 0.268 as much, or

$$(0.268\ m)\left(0.512\ \frac{°C}{m}\right) = 0.137°C$$

If the original boiling point was 100°C (corresponding to pure water), it will now be 100.137°C. ∎

∎ EXAMPLE 300.   Compound X dissolves in water to the extent of 1.89 g of X per 85.00 ml of H_2O (density 0.998 g/ml). The boiling point under atmospheric pressure of this solution is 100.106°C. What is the apparent molecular weight of X?

SOLUTION: The boiling-point elevation observed is 0.106°C. For a 1 m solution, the bp elevation would have been 0.512°C. Therefore, we have a solution that is

$$\frac{0.106°C}{0.512°C/m}\quad \text{or}\quad 0.207\ m$$

The solution has 1.89 g of X per 85.00 ml of H_2O, which is the same as

$$1.89\ \text{g of X per } (85.00\ \text{ml})\left(0.998\ \frac{g}{ml}\right)\quad \text{or per 84.83 g of } H_2O$$

But 1.89 g of X per 84.83 g of H_2O corresponds to

$$\left(\frac{1.89\ \text{g of X}}{84.83\ \text{g of } H_2O}\right)\left(1000\ \frac{g}{kg}\right) = 22.3\ \text{g of X per kilogram of } H_2O$$

Equating the 0.207 m and the 22.3 g of X per kilogram of H_2O, we get

$$\frac{22.3\ \text{g of X}}{0.207\ \text{mole}} = 108\ \text{g of X per mole}$$

Therefore, the apparent molecular weight of X is 108. ∎

10.10   Depression of the freezing point

In the same way that dissolved particles raise the boiling point proportionally to their molal concentration, they depress the freezing point (fp) as well. The fp lowering is characteristic of the solvent and is directly proportional to the molal concentration of dissolved particles in the solution. For H_2O, 1 mole of any kind of dissolved particles in 1 kg of water depresses the freezing point by 1.86°C. This number is called the *molal freezing-point depression constant*. Because freezing points are generally easier to measure accurately than boiling points, freezing-point lowering is a common experimental technique for finding out molecular weights of dissolved species. Here again, however, you have to be on guard for the possibility of dissociation of the solute when it is put in the water. Any increase in the concentration of particles proportionally depresses the freezing point.

■ EXAMPLE 301.   Given that the molal-freezing-point depression constant of H_2O is 1.86°C, what freezing point do you expect for water in which there is dissolved 17.9 g of $C_{12}H_{22}O_{11}$ (a nondissociating solute) per 47.6 g of H_2O?

SOLUTION: One mole of $C_{12}H_{22}O_{11}$ is 342.30 g.

$$\frac{17.9 \text{ g of } C_{12}H_{22}O_{11}}{342.30 \text{ g/mole}} = 0.0523 \text{ mole}$$

0.0523 mole of $C_{12}H_{22}O_{11}$ per 47.6 g of H_2O is the same as

$$\left(\frac{0.0523 \text{ mole}}{47.6 \text{ g}}\right)\left(1000 \frac{\text{g}}{\text{kg}}\right) = 1.10 \text{ moles/kg of } H_2O$$

If 1 mole/kg of H_2O produces a fp lowering of 1.86°C, then 1.10 moles/kg of H_2O should produce a lowering that is 1.10 as great, or

$$(1.10 \, m)\left(1.86 \frac{°C}{m}\right) = 2.05°C$$

So, the freezing point of the solution ought to be $-2.05°C$. ■

■ EXAMPLE 302.   Substance X dissolves in water to give a solution that contains 4.68 g of X per 287 g of H_2O. If the measured freezing point of the solution is 0.153°C below zero, what is the apparent molecular weight of X?

SOLUTION: 4.68 g of X per 287 g of H_2O is the same as

$$\left(\frac{4.68 \text{ g of X}}{287 \text{ g of } H_2O}\right)\left(1000 \frac{\text{g of } H_2O}{\text{kg of } H_2O}\right) = 16.3 \frac{\text{g of X}}{\text{kg of } H_2O}$$

The observed fp is $-0.153°C$. For a 1 m solution, it would be $-1.86°C$.

Therefore, the concentration of the solution is

$$\frac{0.153°C}{1.86°C/m} = 0.0823 \ m$$

So, the 16.3 $\dfrac{\text{g of X}}{\text{kg of H}_2\text{O}}$ must represent 0.0823 m.

Thus, 16.3 g of X must be 0.0823 mole of X.

$$\frac{16.3 \text{ g}}{0.0823 \text{ mole}} = 198 \text{ g/mole of X} \ \blacksquare$$

■ EXAMPLE 303. Zinc sulfate, $ZnSO_4$, is expected as a salt to be dissociated into Zn^{2+} and SO_4^{2-} in solution. For a solution that is made up by dissolving 0.5811 g of $ZnSO_4$ in 0.180 kg of H_2O, the observed freezing point is −0.0530°C. Calculate the apparent per cent dissociation of $ZnSO_4$ into Zn^{2+} and SO_4^{2-}.

SOLUTION: One mole of $ZnSO_4$ weighs 161.43 g.

$$\frac{0.5811 \text{ g of ZnSO}_4}{161.43 \text{ g/mole}} = 0.00360 \text{ mole of ZnSO}_4$$

$$\frac{0.00360 \text{ mole of ZnSO}_4}{0.180 \text{ kg of H}_2\text{O}} = 0.0200 \text{ mole of ZnSO}_4 \text{ per kilogram of H}_2\text{O}$$

Let x = number of moles of $ZnSO_4$ apparently dissociated. This will produce x moles of Zn^{2+} and x moles of SO_4^{2-} and leave $0.0200 - x$ moles of undissociated $ZnSO_4$ per kilogram of water. The total number of moles of all kinds of particles will be x of Zn^{2+} + x of SO_4^{2-} + $(0.0200 - x)$ of $ZnSO_4$, or $(0.0200 + x)$ moles total per kilogram of H_2O.

The observed freezing-point depression is 0.0530°C, which corresponds to

$$\frac{0.0530°C}{1.86°C/m} = 0.0285 \ m$$

We equate the $0.0200 + x$ moles of particles per kilogram of H_2O to the 0.0285 m and solve for x:

$$0.0200 + x = 0.0285$$
$$x = 0.0085$$

The question asks for per cent dissociation.

$$\text{Per cent dissociation} = \frac{\text{moles dissociated}}{\text{moles originally available}} \times 100$$

$$= \frac{0.0085}{0.0200} \times 100 = 42\% \blacksquare$$

Exercises

*10.1. What would be the molarity of a solution made by dissolving 38.5 g of HCl in enough water to make 650 ml of solution?

*10.2. Given a solution that is labeled 0.50 M HCl, how many milliliters of it would you need to take to get 0.0026 mole of HCl?

*10.3. Assuming volumes are additive, what would be the molar concentration of the final mixture made by combining 385 ml of 0.725 M HCl and 525 ml of 0.325 M HCl? (*Answer*: 0.494 M.)

*10.4. Assuming volumes are additive, how much water would you have to add to 15.0 ml of 2.15 M HCl to make it 0.95 M HCl?

*10.5. What would be the final concentration of a mixture made by adding 1.69 × 10⁻³ mole of HCl and 1.50 ml of H_2O to 11.25 ml of 0.239 M HCl? Assume liquid volumes are additive and represent total volume of final solution.

**10.6. How much 0.370 M H_2SO_4 can you make from 35.0 ml of 1.80 M H_2SO_4 using no more than 50.0 ml of H_2O? Assume volumes are additive. (*Answer*: 62.9 ml.)

**10.7. You are given 0.100 liter of 0.250 M H_2SO_4, 0.200 liter of 0.380 M H_2SO_4, and 0.300 liter of 0.520 M H_2SO_4. Using these sources alone (i.e., no added water), what is the maximum volume of 0.325 M H_2SO_4 that you can make? Assume volumes are additive.

*10.8. For full neutralization how many equivalents of acid are there in each of the following: (a) 35.0 ml of 0.650 N H_2SO_4; (b) 25.0 ml of 0.455 M H_2SO_4; (c) 15.0 ml of 0.506 M H_3PO_4; (d) 5.0 ml of 4.55 M HCl?

*10.9. What volume of 0.678 M NaOH would you need just to neutralize a mixture of 0.109 equivalent of HCl with 435 ml of 0.150 N HCl? (*Answer*: 0.257 liter.)

*10.10. What is the normality of an H_2SO_4 solution, 63.81 ml of which is required to neutralize 0.0539 equivalent of base?

*10.11. What is the normality of a solution made by mixing 36.82 ml of 0.0728 N H_3PO_4, 21.87 ml of 0.0538 N H_3PO_4, and 30.00 ml of water? Assume volumes are additive.

**10.12. How much 2.50 × 10⁻³ M $Ca(OH)_2$ can you make from 1.00 liter of 0.0600 N $Ca(OH)_2$ and 1.00 liter of water? Assume additive volumes. (*Answer*: 1.09 liters.)

**10.13. To make up 17.31 ml of 0.692 N $KMnO_4$ solution for a reaction where MnO_4^- goes to Mn_2O_3, how many grams of $KMnO_4$ would be needed?

***10.14. A given sample of impure Fe_2O_3 is analyzed by reducing the iron to Fe^{2+} and then reoxidizing it with a $KMnO_4$ solution. If a 3.583-g sample of the unknown requires 37.69 ml of 0.850 N $KMnO_4$ for the reoxidation, what per cent of the original sample was Fe_2O_3? Assume MnO_4^- goes to Mn^{2+}.

***10.15. An unknown solution of $KMnO_4$ is standardized by reaction with a weighed amount of solid $Na_2C_2O_4$. Products of the reaction are Mn^{2+} and CO_2. If 1.689 g of $Na_2C_2O_4$ requires 37.69 ml of the $KMnO_4$ solution, what is the normality of $KMnO_4$ solution? (*Answer:* 0.669 N.)

***10.16. A given aqueous solution of hydrogen peroxide weighs 3.479 g. On treatment with KI and acid, the H_2O_2 is reduced to H_2O as the I^- goes to I_2. The I_2 formed, measured by the reaction $I_2 + 2S_2O_3^{2-} \rightarrow 2I^- + S_4O_6^{2-}$, corresponds to 26.87 ml of 0.1635 N $Na_2S_2O_3$. What is the per cent of H_2O_2 in the original sample?

*10.17. Given 13.55 g of $CuSO_4 \cdot 5H_2O$, how many grams of 0.150 m $CuSO_4$ solution can you make?

**10.18. What will be the molality of the solution made by mixing 85.3 g of 1.35 m HCl solution, 0.867 g of HCl, and 89.1 g of H_2O? (*Answer:* 0.784 m.)

**10.19. Given a sample of "concentrated sulfuric acid" that is 95.6% by weight H_2SO_4 and has a density of 1.8 g/ml. Calculate the molality of such a solution.

**10.20. Suppose you mix equal volumes of 5.0 M H_2SO_4 solution (density 1.286 g/ml) and 5.0 m H_2SO_4 solution (density 1.243 g/ml). What will be the molality of the resulting solution?

**10.21. Suppose you mix equal weights of 5.0 M H_2SO_4 solution (density 1.286 g/ml) and 5.0 m H_2SO_4 solution (density 1.243 g/ml). What will be the molality of the resulting solution? (*Answer:* 5.62 m.)

*10.22. A liter of methyl alcohol, CH_3OH (density 0.796 g/ml), is mixed with a liter of ethyl alcohol, C_2H_5OH (density 0.789 g/ml). What is the mole fraction of C_2H_5OH in the final mixture?

**10.23. If you add 10.0 g of CH_3OH to 10.0 g of a CH_3OH–C_2H_5OH mixture in which the mole fraction of CH_3OH is originally 0.500, what does the mole fraction of CH_3OH become in the new mixture?

**10.24. Given a solution of Na in liquid NH_3 where the mole fraction of Na is 0.0416; if the density of the solution is 0.65 g/ml, what is its molarity? (*Answer:* 1.6 M Na.)

*10.25. What would be the formality of a solution made by dissolving 12.5 g of $K_2Cr_2O_7$ in 100 ml of H_2O? The solution density is 1.08 g/ml.

**10.26. Given that the salt $FeNH_4(SO_4)_2 \cdot 6H_2O$ is totally dissociated in aqueous solution to give Fe^{2+}, NH_4^+, and SO_4^{2-}, calculate the formality of the final solution and its molarity with respect to SO_4^{2-} if we dissolve 60.0 g of $FeNH_4(SO_4)_2 \cdot 6H_2O$ in 0.750 liter of 1.35 F $FeNH_4(SO_4)_2$. The final volume of the solution is 0.775 liter.

*10.27. "Concentrated hydrochloric acid" as supplied by the manufacturer is likely to be an aqueous solution having a density of 1.198 g/ml and containing 40.0% HCl by weight. What is the formality of such a solution? (*Answer*: 13.1 F HCl.)

**10.28. An aqueous solution that is 50.0% by weight C_2H_5OH has a density of 0.9125 g/ml. If pure C_2H_5OH has a density of 0.7874 g/ml, what is the per cent by volume of alcohol in the aqueous solution? Calculate also the per cent by volume of water in the mixture.

**10.29. One recipe for making "denatured alcohol" is to mix 5 gallons of wood alcohol (CH_3OH) with 100 gallons of 95% (by volume) ethanol (C_2H_5OH). The density of wood alcohol is 0.7866 g/ml; that of 95% ethanol is 0.8100 g/ml; that of pure ethanol is 0.7874 g/ml. What is the mole fraction of CH_3OH in the denatured stuff?

**10.30. The so-called 95% alcohol is actually 94.9% by volume of C_2H_5OH. Its density is 0.810 g/ml at 25°C. Calculate the per cent by weight of C_2H_5OH in "95% alcohol," given that the density of pure C_2H_5OH is 0.7874 g/ml at 25°C. (*Answer*: 92.3%).

**10.31. Given a solution containing 20.5 g of solid I_2 dissolved in 100 ml of CH_3OH (density 0.787 g/ml). If the vapor pressure of pure CH_3OH is 100 Torr at 21.2°C, what would be the vapor pressure of the CH_3OH after the I_2 had dissolved in it?

**10.32. The vapor pressure of pure water is 23.756 Torr at 25°C. How much sugar ($C_6H_{12}O_6$) would you have to add to 100 g of H_2O to bring the vapor pressure down to 23.000 Torr?

**10.33. When 2.182 g of phenanthrene is dissolved in 100 g of CCl_4, the vapor pressure of CCl_4 drops from 85.513 Torr to 83.932 Torr. Deduce from these data the molecular weight of phenanthrene. (*Answer*: 178.2.)

*10.34. If 0.00178 g of N_2 dissolves in 100 g of H_2O when the nitrogen pressure is 737 Torr, how many g of N_2 would dissolve in the 100 g of H_2O when in contact with air at 737 Torr? Assume air is 78.03% N_2 by volume.

**10.35. At 20°C and 700 Torr, 653 ml of NH_3 gas will dissolve in 1 ml of water. Calculate how many grams of NH_3 can dissolve in a kilogram of water when the ammonia pressure is 760 Torr.

**10.36. At 20°C, the Henry's law constant for CO_2 is 1.08×10^6 Torr. If a 1-liter bottle of soda pop is opened so the CO_2 pressure falls from 2 atm to 1 atm, how many milliliters of CO_2 gas (20°C and 1 atm) should escape? (*Answer*: 940 ml.)

*10.37. When $CaCl_2$ dissolves in water, each mole of $CaCl_2$ forms 1 mole of Ca^{2+} and 2 moles of Cl^-. Assuming ideal behavior, what should be the normal boiling point of an aqueous solution that is 0.550 m $CaCl_2$?

10.38. The normal boiling point of chloroform, $CHCl_3$, is 61.26°C. Assuming I_2 is nonvolatile, how many grams of I_2 would you need to dissolve in 235 g of $CHCl_3$ to bring the boiling point down to 60.00°C? The normal boiling-point elevation constant of $CHCl_3$ is 3.63°C/m.

10.39. A solution is made by dissolving 20.0 g of glucose ($C_6H_{12}O_6$) and 30.0 g of sucrose ($C_{12}H_{22}O_{11}$) in 225 g of H_2O. Assuming that these sugars are nondissociated in water but assuming NaCl is 100% dissociated in aqueous solution, how many grams of NaCl would you have had to dissolve in 225 g of H_2O to yield the same boiling point? (*Answer:* 5.80 *g.*)

*10.40.** What would be the freezing point of 0.350 m C_2H_5OH solution in H_2O?

10.41. If you have the problem of lowering the freezing point of H_2O from 0°C to 0°F (-17.8°C), what volume of methyl alcohol (CH_3OH) should you add to 10.0 liters of H_2O? The density of CH_3OH is 0.787 g/ml.

10.42. The normal freezing point of benzene is 5.50°C; its molal freezing point depression constant is 4.90°C/m. If 1.00 g of the element arsenic is dissolved in 86.0 g of benzene, the observed freezing point is 5.31°C. Deduce from these data the formula of the arsenic molecule in benzene. (*Answer:* As_4.)

GASEOUS EQUILIBRIA

IN THIS CHAPTER we take up a general consideration of chemical equilibrium based on the behavior of gases. In succeeding chapters we consider the various special types of chemical equilibrium normally encountered in aqueous solutions.

11.1   Evaluation of K

The whole idea of chemical equilibrium is essentially this: All chemical reactions, such as $A + B \rightarrow C + D$, are reversible in the sense that the products of the reaction, C and D, can react to regenerate the starting materials, A and B. If allowed enough time, the system will come to rest (i.e., to equilibrium) where not only C and D but also A and B are present at characteristic concentrations that do not change with time. The condition required for equilibrium is that the concentrations of species on the right of the equation raised to powers conforming to the coefficients in the chemical equation divided by the concentrations of species on the left of the equation raised to the proper powers must equal a constant characteristic of the system. For example, for the system

$$N_2(g) + 3H_2(g) \rightleftharpoons 2NH_3(g)$$

where the double arrow sign emphasizes that the chemical equation is to be read in both directions, the condition for equilibrium is that

$$\frac{C_{NH_3}^2}{C_{N_2} C_{H_2}^3} = K$$

Here C_{NH_3} stands for the concentration of ammonia; C_{N_2} stands for the concentration of nitrogen; and C_{H_2} stands for the concentration of hydrogen in the system containing NH_3, N_2, and H_2 all together at equilibrium. K is a number that is fixed for the system consisting of NH_3, N_2, and H_2, although in general it will vary if the temperature is changed. It is called the *equilibrium constant* for the reaction.

■ **EXAMPLE 304.** You are given a box containing NH_3, N_2, and H_2 at equilibrium at 1000°K. Analysis of the contents shows that the concentration of NH_3 is 0.102 mole/liter, N_2 is 1.03 moles/liter, and H_2 is 1.62 moles/liter. Calculate K for the reaction $N_2(g) + 3H_2(g) \rightleftarrows 2NH_3(g)$ at 1000°K.

SOLUTION: Write the correct expression for K so that it corresponds to the reaction as written. Be careful to put the stuff from the right of the equation in the numerator and the stuff from the left of the equation in the denominator. Make sure that the exponents match the coefficients in the equation:

$$K = \frac{C_{NH_3}^2}{C_{N_2} C_{H_2}^3}$$

Then substitute $C_{NH_3} = 0.102$ mole/liter, $C_{N_2} = 1.03$ moles/liter, and $C_{H_2} = 1.62$ moles/liter:

$$K = \frac{(0.102)^2}{(1.03)(1.62)^3} = 2.37 \times 10^{-3} \ ■$$

Strictly speaking, we should put units on the K to correspond to the concentration units employed. In Example 304, for example, K would be in units of one-over-(moles per liter) squared. However, there are several good reasons for not bothering with the dimensional units. One is that they are easily supplied, if necessary, by looking at the exponents in the numerator and denominator. The units will cancel out if the number of gas molecules is the same on the two sides of the chemical equation, but not if they differ. The second point is that we almost always work with molar concentrations (i.e., moles per liter) when calculating with K. If other concentration units are used, it is customary to call attention to them; if nothing is said about concentration units, we assume we are dealing with K calculated in terms of moles per liter for each species. Third, in the sophisticated treatment of chemical equilibrium a very clever dodge has been invented whereby each species concentration is expressed as a ratio to what the concentration would be in its standard reference state. As a result, concentration units per se disappear and are replaced by unitless parameters called *activities*. All of this is just for the record. So far as your present work with equilibrium constants is concerned, calculate with moles per liter unless otherwise directed.

A more immediate caution is to note that K will take on a different numerical value depending on which way you choose to write the chemical equation. This is illustrated in the next problem. To be absolutely clear, it is strongly recommended that you specifically write the chemical equation for which your K is being defined.

■EXAMPLE 305.   (Compare Example 304.) You are given a box containing NH_3, N_2, and H_2 at equilibrium at 1000°K. Analysis of the contents shows that the concentration of NH_3 is 0.102 mole/liter, N_2 is 1.03 moles/liter, and H_2 is 1.62 moles/liter. Calculate K for the reaction $2NH_3(g) \rightleftarrows N_2(g) + 3H_2(g)$.

SOLUTION: Write the expression for K so that it matches the chemical equation.

$$K = \frac{C_{N_2} C_{H_2}^3}{C_{NH_3}^2}$$

Species from the right of the chemical equation go in the numerator; species from the left of the equation go in the denominator. Then substitute the equilibrium concentrations: $C_{N_2} = 1.03\,M$; $C_{H_2} = 1.62\,M$; $C_{NH_3} = 0.102\,M$.

$$K = \frac{(1.03)(1.62)^3}{(0.102)^2} = 4.21 \times 10^2$$

Note that the numerical value of this K is the reciprocal of that obtained in Example 304. The general point is that when you reverse the direction of the chemical equation, you invert the equilibrium constant. ■

■EXAMPLE 306.   (Compare with Example 305.) You are given a box containing NH_3, N_2, and H_2 at equilibrium at 1000°K. Analysis of the contents shows that the concentration of NH_3 is 0.102 mole/liter, N_2 is 1.03 moles/liter, and H_2 is 1.62 moles/liter. Calculate K for the reaction

$$NH_3(g) \;\rightleftarrows\; \tfrac{1}{2}N_2(g) + \tfrac{3}{2}H_2(g)$$

SOLUTION: Set up the K so that it matches the chemical equation.

$$K = \frac{(C_{N_2})^{1/2}(C_{H_2})^{3/2}}{C_{NH_3}}$$

The powers to which the concentrations are to be raised must match the coefficients in the chemical equation. Then substitute the equilibrium concentrations: $C_{N_2} = 1.03$ M; $C_{H_2} = 1.62$ M; $C_{NH_3} = 0.102$ M.

$$K = \frac{(1.03)^{1/2}(1.62)^{3/2}}{0.102} = 20.5$$

Note that this K is the square root of the K we got in Example 305. The general rule is that if you double a chemical equation throughout, the corresponding K will be the square of the original; if you triple the equation, the K will be the cube of the original value. ■

■ EXAMPLE 307.   Hydrogen gas, sulfur vapor, and hydrogen sulfide gas are in equilibrium with each other under the following conditions: 1.68 moles of H_2S, 1.37 moles of H_2, and 2.88×10^{-5} mole of S_2 are in a volume of 18.0 liters at 750°C. Calculate the K for the reaction $2H_2(g) + S_2(g) \rightleftarrows 2H_2S(g)$

SOLUTION: Set up the K to correspond with the chemical equation.

$$K = \frac{(C_{H_2S})^2}{(C_{H_2})^2(C_{S_2})}$$

Then calculate the moles per liter of each constituent.

$$C_{H_2S} = \frac{1.68 \text{ moles}}{18.0 \text{ liters}} = 0.0933 \ M$$

$$C_{H_2} = \frac{1.37 \text{ moles}}{18.0 \text{ liters}} = 0.0761 \ M$$

$$C_{S_2} = \frac{2.88 \times 10^{-5} \text{ mole}}{18.0 \text{ liters}} = 1.60 \times 10^{-6} \ M$$

Substitute these values in the expression for K.

$$K = \frac{(0.0933)^2}{(0.0761)^2(1.60 \times 10^{-6})} = 9.39 \times 10^5 \ ■$$

■ EXAMPLE 308.   At 500°K, PCl_5 decomposes rather extensively into PCl_3 and Cl_2. It is found, for example, that if you put 1.000 mole of PCl_5 in a 1-liter box at 500°K, 13.9% of it decomposes to PCl_3 and Cl_2. Calculate K for the decomposition reaction $PCl_5(g) \rightleftarrows PCl_3(g) + Cl_2(g)$.

SOLUTION: According to the equation $PCl_5 \rightarrow PCl_3 + Cl_2$, for every 1 mole of PCl_5 that decomposes, 1 mole of PCl_3 and 1 mole of Cl_2 appears.

You are told that 13.9% of 1.000 mole, or 0.139 mole, of PCl_5 decomposes. This will produce 0.139 mole of PCl_3 and 0.139 mole of Cl_2 and leave 1.000 − 0.139, or 0.861, mole of PCl_5 undecomposed. So, for concentrations at equilibrium, we have the following:

$C_{PCl_3} = 0.139$ mole/liter

$C_{Cl_2} = 0.139$ mole/liter

$C_{PCl_5} = 0.861$ mole/liter

where we have divided the moles present by the volume, 1 liter:

$$K = \frac{C_{PCl_3} C_{Cl_2}}{C_{PCl_5}} = \frac{(0.139)(0.139)}{0.861} = 0.0224 \ \blacksquare$$

11.2   Calculations from K

The numerical value of K for a given reaction can be obtained by experimental study of one equilibrium set of concentrations involving that reaction at a specified temperature. Once K is obtained, it can be used to describe any equilibrium system involving those same species at the same temperature. There are in special handbooks tabulations of K for various reactions at different temperatures, or else numbers from which K can be derived. So far as we are concerned, the K value will be given, and what we need to look at now is how to calculate required data from the given K.

Before we look at some specific problems, it might be noted that there is a common way of designating concentrations that you should be familiar with. In Section 11.1 we were careful to write, for example, C_{NH_3} for the concentration of NH_3 in moles per liter in the system. We did this to emphasize that K involves *concentrations*—that is, moles *per liter* and not just moles. However, now that you know what K is, we can introduce the common convention, which is to indicate the concentration in moles per liter of a species by putting its formula in square brackets. Thus, $[NH_3]$ stands for moles of NH_3 per liter and is the same thing we previously designated as C_{NH_3}. Similarly, instead of writing C_{H_2}, we can use $[H_2]$, meaning concentration of H_2 in moles per liter. To emphasize the equivalence of these symbols, we shall show the K's both ways for the first several problems and then gradually work over completely to the square-bracket notation.

■ EXAMPLE 309.   You are given that K for $N_2(g) + 3H_2(g) \rightleftarrows 2NH_3(g)$ is equal to 2.37×10^{-3} at $1000°K$. If you have a system containing these species at equilibrium at $1000°K$, what must be the concentration of NH_3 if you fix N_2 at 2.00 moles/liter and H_2 at 3.00 moles/liter?

SOLUTION: You are dealing with an equilibrium system, so the equilibrium condition

$$K = \frac{C_{NH_3}^2}{C_{N_2}C_{H_2}^3} = 2.37 \times 10^{-3} = \frac{[NH_3]^2}{[N_2][H_2]^3}$$

must be satisfied. You are told that the concentration of N_2 is 2.00 moles/liter; so, $C_{N_2} = 2.00$. You are also told that the concentration of H_2 is 3.00 moles/liter; therefore, $C_{H_2} = 3.00$. All we have to do is to substitute these values in the equation above and solve for the concentration of NH_3:

$$\frac{C_{NH_3}^2}{C_{N_2}C_{H_2}^3} = \frac{C_{NH_3}^2}{(2.00)(3.00)^3} = 2.37 \times 10^{-3}$$

$$C_{NH_3} = 0.358 \text{ mole/liter} \blacksquare$$

■ EXAMPLE 310.   If $K = 2.37 \times 10^{-3}$ at $1000°K$ for $N_2(g) + 3H_2(g) \rightleftarrows 2NH_3(g)$ what must be the equilibrium concentration of H_2 in a system known to contain 0.683 M N_2 and 1.05 M NH_3?

SOLUTION:

$$K = \frac{C_{NH_3}^2}{C_{N_2}C_{H_2}^3} = 2.37 \times 10^{-3}$$

$$C_{N_2} = 0.683 \ M; \qquad C_{NH_3} = 1.05 \ M; \qquad C_{H_2} = \ ?$$

Substitute these values and solve for the concentration of H_2:

$$\frac{C_{NH_3}^2}{C_{N_2}C_{H_2}^3} = \frac{(1.05)^2}{(0.683)(C_{H_2})^3} = 2.37 \times 10^{-3}$$

$$C_{H_2} = 8.80 \text{ moles/liter} \blacksquare$$

■ EXAMPLE 311.   Given $K = 1.03 \times 10^{-3}$ for $H_2S(g) \rightleftarrows H_2(g) + \frac{1}{2}S_2(g)$ at 750°C. How many moles of S_2 would you have at equilibrium in a 3.68-liter box in which there are 1.63 moles of H_2S and 0.864 mole of H_2?

SOLUTION: Remember that what goes into the constant is moles *per* *liter*. So, first we convert our given data into moles per liter:

$$C_{H_2S} = \frac{1.63 \text{ moles}}{3.68 \text{ liters}} = 0.443 \ M$$

$$C_{H_2} = \frac{0.864 \text{ mole}}{3.68 \text{ liters}} = 0.235 \ M$$

Then we substitute these in the expression for the equilibrium constant.

$$K = \frac{C_{H_2} C_{S_2}^{1/2}}{C_{H_2S}} = 1.03 \times 10^{-3} = \frac{(0.235)(C_{S_2})^{1/2}}{0.443}$$

Solving for C_{S_2}, we get 3.77×10^{-6} mole/liter.

But the question asks for how many moles there are in the box. We know the concentration of S_2 must be 3.77×10^{-6} mole/liter; we know the total volume of the box is 3.68 liters.

$$\text{Moles of } S_2 = \left(3.77 \times 10^{-6} \frac{\text{mole}}{\text{liter}}\right)(3.68 \text{ liters}) = 1.39 \times 10^{-5} \text{ mole} \blacksquare$$

■ EXAMPLE 312.   You are told that $K = 3.76 \times 10^{-5}$ for $I_2(g) \rightleftarrows 2I(g)$ at 1000°K. You start an experiment by injecting 1.00 mole of I_2 into a 2.00-liter box at 1000°K. Let it come to equilibrium. What will be the concentration of I_2 and I in the system at equilibrium?

SOLUTION: (Figure out how many moles per liter of I_2 you start with. Let x of these break up to form I to establish equilibrium. Express $[I_2]$ and $[I]$ in terms of x and calculate x by substituting in the equilibrium constant expression.)

Given 1.00 mole of I_2 in 2.00 liters.
Therefore,

$$\text{the initial concentration of } I_2 = \frac{1.00 \text{ mole}}{2.00 \text{ liters}} = 0.500 \ M$$

Let x moles per liter break up. This leaves $(0.500 - x)$ moles/liter of I_2 at equilibrium. The reaction of breakup, $I_2 \rightarrow 2I$, shows that for every 1 mole of I_2 that breaks up, 2 moles of I are formed. If we break up x moles of I_2, we must form $2x$ moles of I.

When equilibrium is finally established, there will be $(0.500 - x)$ moles/liter of I_2 and $2x$ moles/liter of I. We can write $[I_2] = 0.500 - x$ and $[I] = 2x$. For $I_2(g) \rightleftarrows 2I(g)$ the equilibrium condition is given by

$$K = \frac{[I]^2}{[I_2]} = \frac{(2x)^2}{0.500 - x} = 3.76 \times 10^{-5}$$

We have one unknown and one algebraic equation, so we should be able to solve it by the method for quadratic equations. It gives $x = 2.16 \times 10^{-3}$. [We can also get an approximate answer by noting that, since the right-hand side of the algebraic equation is small, the left-hand side must also be small.

This can be true only if x is small, particularly compared to 0.500. If we neglect x in the denominator, we have $(2x)^2/(0.500) = 3.76 \times 10^{-5}$, which solves to $x = 2.17 \times 10^{-3}$, close enough to the exact answer to be acceptable.]
We can say, finally, that the concentrations are,

$$[I_2] = 0.500 - x = 0.500 - 2.17 \times 10^{-3} = 0.498 \ M$$
$$[I] = 2x = (2)(2.17 \times 10^{-3}) = 4.34 \times 10^{-3} \ M \ \blacksquare$$

■ EXAMPLE 313. For the decomposition of PCl_5 at 760°K, the equilibrium constant is $K = 33.3$ for $PCl_5(g) \rightleftarrows PCl_3(g) + Cl_2(g)$. You have a sample tube of volume 36.3 ml into which you inject 1.50 g of PCl_5. What will be the concentration of PCl_5 in the tube when equilibrium is finally established? What will PCl_3 and Cl_2 be?

SOLUTION: (Figure out the initial concentration of PCl_5. Let x moles/ liter decompose to give x moles/liter of PCl_3 and x moles/liter of Cl_2. Substitute the equilibrium concentrations in terms of x into the expression for K and solve for x.)
One mole of PCl_5 weighs 208.24 g.

$$\frac{1.50 \text{ g of } PCl_5}{208.24 \text{ g/mole}} = 0.00720 \text{ mole}$$

The volume of the system is 36.3 ml, or 0.0363 liter.
Therefore, the initial concentration of PCl_5 is

$$\frac{0.00720 \text{ mole}}{0.0363 \text{ liter}} = 0.198 \text{ mole/liter}$$

Let x equal the moles/liter of PCl_5 that decompose. This will leave $(0.198 - x)$ mole/liter undecomposed, and it will produce x moles/liter of PCl_3 and x moles/liter of Cl_2.
At equilibrium:

$$[PCl_5] = 0.198 - x \qquad K = \frac{[PCl_3][Cl_2]}{[PCl_5]} = 33.3$$
$$[PCl_3] = x$$
$$[Cl_2] = x \qquad \frac{(x)(x)}{0.198 - x} = 33.3$$

Approximation methods are no good for solving this equation because the right-hand side can be large only if $(0.198 - x)$ is small. In other words, x must be of the order of 0.198, which makes for lots of trouble. You can solve the equation by the quadratic formula and get $x = 0.197$. However, this does not help much because substituting this $x = 0.197$ into the expression $[PCl_5] = 0.198 - x$ gives $[PCl_5] = 0.198 - 0.197 = 0.001$ mole/liter. By the rules of significant figures, we have only one digit in our answer and that, unfortunately, is a doubtful digit. ∎

Can we get a better answer? We can, if we use a trick chemists frequently resort to when faced by approximation-method difficulties and that is "to approach equilibrium from the other side." Since the final equilibrium state does not depend on how we get there, we can use an indirect approach, the essence of which is to assume first that *all* the PCl_5 decomposes and then some of the resulting PCl_3 and Cl_2 recombines.

Specifically, if we start with 0.198 mole per liter of PCl_5 injected into the box, we first assume that all 0.198 mole of PCl_5 decomposes to form 0.198 mole of PCl_3 plus 0.198 mole of Cl_2 consistent with the equation $PCl_5 \rightarrow PCl_3 + Cl_2$. Then we look at the "back-reaction" $PCl_5 \leftarrow PCl_3 + Cl_2$ and define a new unknown y to represent the moles per liter of PCl_5 "regenerated." Such regeneration of PCl_5, of course, can only come at the expense of the hypothetical PCl_3 and Cl_2 produced, so we need to reduce their hypothetical 0.198 M concentrations by y.

In terms of this new unknown y, we have at equilibrium:

$$[PCl_5] = y$$

$$[PCl_3] = 0.198 - y$$

$$[Cl_2] = 0.198 - y$$

$$K = \frac{[PCl_3][Cl_2]}{[PCl_5]} = 33.3$$

$$\frac{(0.198 - y)(0.198 - y)}{y} = 33.3$$

where we have substituted in the same equation for K, since the requirement for equilibrium has not changed. The final algebraic equation may look just as hopeless as before, except that now y is small compared to 0.198. We can solve approximately, by neglecting y in the numerator.

$$\frac{(0.198 - y)(0.198 - y)}{y} \cong \frac{(0.198)(0.198)}{y} \cong 33.3$$

The sign $\cong$ means "is approximately equal to." Solving the approximate equality on the right, we get $y \cong 0.00117$. Substituting this y in the expressions for $[PCl_5]$, $[PCl_3]$, and $[Cl_2]$, we get for the final equilibrium concentrations:

$$[PCl_5] = y = 0.00117 \ M$$

$$[PCl_3] = 0.198 - y = 0.198 - 0.00117 = 0.197 \ M$$

$$[Cl_2] = 0.198 - y = 0.198 - 0.00117 = 0.197 \ M$$

That these are really acceptable values can be checked by substituting them in $K = [PCl_3][Cl_2]/[PCl_5]$ to see if the value $K = 33.3$ is reproduced.

■ EXAMPLE 314.   For $NO_2(g) \rightleftarrows NO(g) + \frac{1}{2}O_2(g)$ at 500°K, the equilibrium constant K is 1.25×10^{-3}. Calculate the concentration of each species at equilibrium following the injection of 0.0683 mole of NO_2 into a volume of 0.769 liter at 500°K.

SOLUTION:

$$\text{Initial concentration of } NO_2 = \frac{0.0683 \text{ mole}}{0.769 \text{ liter}} = 0.0888 \ M$$

Let x = moles per liter of NO_2 that decompose. This leaves $(0.0888 - x)$ moles/liter of NO_2 at equilibrium. It will produce some NO and O_2 by the reaction $NO_2 \rightarrow NO + \frac{1}{2}O_2$. Since the disappearance of 1 mole of NO_2 results in formation of 1 mole of NO and $\frac{1}{2}$ mole of O_2, we conclude that the decomposition of x moles of NO_2 must result in formation of x moles of NO and $\frac{1}{2}x$ moles of O_2. At equilibrium, we will have

$[NO_2] = 0.0888 - x$

$[NO] = x$

$[O_2] = \frac{1}{2}x$

$$K = \frac{[NO][O_2]^{1/2}}{[NO_2]} = 1.25 \times 10^{-3}$$

$$\frac{(x)(\frac{1}{2}x)^{1/2}}{(0.0888 - x)} = 1.25 \times 10^{-3}$$

We could solve this equation exactly by squaring both sides and using a method for cubic equations. An easier method is by *successive approximation*. We note that x may be small compared to 0.0888, so, as a *first approximation*, we ignore the x in the denominator, assume $(0.0888 - x) \cong 0.0888$ and solve:

$$\frac{(x)(\frac{1}{2}x)^{1/2}}{(0.0888)} \cong 1.25 \times 10^{-3}$$

Multiplying both sides by 0.0888 gives

$$(x)(\tfrac{1}{2}x)^{1/2} \cong (1.25 \times 10^{-3})(0.0888) = 1.11 \times 10^{-4}$$

Squaring both sides gives

$$(x)^2(\tfrac{1}{2}x) \cong (1.11 \times 10^{-4})^2 = 1.23 \times 10^{-8}$$

Multiplying both sides by 2 gives

$$x^3 \cong 2.46 \times 10^{-8}$$

Taking the cube root of both sides gives

$$x \cong 2.91 \times 10^{-3}$$

This is almost but not quite negligible compared to 0.0888, so we try a *second approximation* where we estimate x in $(0.0888 - x)$ to be equal to the value just computed. Assume $(0.0888 - x) \cong (0.0888 - 2.91 \times 10^{-3}) = 0.0859$ and put this in the denominator of the original equation to get

$$\frac{(x)(\tfrac{1}{2}x)^{1/2}}{0.0859} \cong 1.25 \times 10^{-3}$$

Solving this equation for x gives $x = 2.85 \times 10^{-3}$. If $x = 2.85 \times 10^{-3}$, the denominator $(0.0888 - x)$ would be 0.0860, which is close enough to the value 0.0859 we estimated that we have some confidence in x. Using $x = 2.85 \times 10^{-3} M$, we get for the equilibrium concentrations:

$$[NO_2] = 0.0888 - x = 0.0888 - 2.85 \times 10^{-3} = 0.0860 \ M$$
$$[NO] = x = 2.85 \times 10^{-3} \ M$$
$$[O_2] = \tfrac{1}{2}x = \tfrac{1}{2}(2.85 \times 10^{-3}) = 1.42 \times 10^{-3} \ M$$

If we put these values back in $K = [NO][O_2]^{1/2}/[NO_2]$, we get

$$K = \frac{(2.85 \times 10^{-3})(1.42 \times 10^{-3})^{1/2}}{0.0860} = 1.25 \times 10^{-3}$$

which checks out with the given $K = 1.25 \times 10^{-3}$. ∎

■ EXAMPLE 315. Suppose that ammonia decomposes at 600°K to establish the equilibrium $NH_3(g) \rightleftarrows \tfrac{1}{2}N_2(g) + \tfrac{3}{2}H_2(g)$ for which $K = 0.395$. You have a box of volume 1.00 liter into which you inject 2.65 g of NH_3 at 600°K. What will be the final concentrations of NH_3, N_2, and H_2 after equilibrium is established?

SOLUTION: (This is an example of a problem that is likely to be quite difficult, because K is of the order of 1 and that means neither side of the chemical equation can be considered negligible with respect to the other in the final state. However, the troubles are usually of the pencil-pushing type, requiring lots of scratch paper; there is actually little difficulty in setting up the calculation.)

$$\text{Initial concentration of } NH_3 = \frac{2.65 \ g}{17.03 \ g/mole} = 0.156 \ M$$

Let x moles per liter of NH_3 decompose via $NH_3 \rightarrow \frac{1}{2}N_2 + \frac{3}{2}H_2$. This will diminish the NH_3 concentration by x and produce $\frac{1}{2}x$ moles of N_2 and $\frac{3}{2}x$ moles of H_2. So, in terms of x, at equilibrium we would have

$$[NH_3] = 0.156 - x; \quad [N_2] = \frac{1}{2}x; \quad [H_2] = \frac{3}{2}x$$

$$K = \frac{[N_2]^{1/2}[H_2]^{3/2}}{[NH_3]} = \frac{(\frac{1}{2}x)^{1/2}(\frac{3}{2}x)^{3/2}}{0.156 - x} = 0.395$$

This equation cannot be solved by the usual approximation methods because x is not negligible compared to 0.156. (Try to neglect x in the denominator and see what happens.) The alternative is to solve the equation exactly. In this case, we are fortunate, because the equation comes out to be a simple quadratic, although at first sight the exponents look more complicated. We simplify as follows:

$$\frac{(\frac{1}{2}x)^{1/2}(\frac{3}{2}x)^{3/2}}{(0.156 - x)} = \frac{(\frac{1}{2})^{1/2}(\frac{3}{2})^{3/2}(x^{1/2})(x^{3/2})}{(0.156 - x)} = \frac{(3)^{3/2}x^2}{(2)^2(0.156 - x)}$$

Here we have simply added up the exponents of the multiplied terms and split the fraction between numerator and denominator. $(3)^{3/2}$ is the square root of three "cubed," or $\sqrt{27}$, or 5.20. $(2)^2$ is, of course, two squared, or 4. Progressively simplifying, we get

$$\frac{5.20}{4} \frac{x^2}{(0.156 - x)} = 1.30 \frac{x^2}{(0.156 - x)} = 0.395$$

where the final number is the value of K. The final two terms

$$1.30 \frac{x^2}{(0.156 - x)} = 0.395$$

reduce to the quadratic equation as conventionally written: $1.30x^2 + 0.395x - 0.0616 = 0$. The solution to this equation from the quadratic formula gives $x = 0.114$. So, at equilibrium, we have

$$[NH_3] = 0.156 - x = 0.156 - 0.114 = 0.042 \ M$$
$$[N_2] = \frac{1}{2}x = \frac{1}{2}(0.114) = 0.0570 \ M$$
$$[H_2] = \frac{3}{2}x = \frac{3}{2}(0.114) = 0.171 \ M \ \blacksquare$$

In the preceding problems, decomposition has occurred to produce a material that was not there at the start of the experiment. In the more general case, some of the material may already be there initially, in which case it will have to be taken into account in setting up the expression for the concentrations.

■ EXAMPLE 316.   Given that $K = 3.76 \times 10^{-5}$ for $I_2(g) \rightleftarrows 2I(g)$ at $1000°K$. Suppose you inject 1.00 mole of I_2 into a 2.00-liter box that already contains 5.00×10^{-3} mole of I. What will be the concentrations of I_2 and I at equilibrium?

SOLUTION: (Compare this problem to Example 312):

$$\text{Injection concentration of } I_2 \text{ is } \frac{1.00 \text{ mole}}{2.00 \text{ liter}} = 0.500 \; M.$$

Let x mole/liter decompose. This will form $2x$ mole/liter of I in addition to the I already there, which is at a concentration of

$$\frac{5.00 \times 10^{-3} \text{ mole}}{2.00 \text{ liters}} = 2.50 \times 10^{-3} \text{ mole/liter}$$

Therefore, at equilibrium,

$$[I_2] = 0.500 - x; \; [I] = 2.50 \times 10^{-3} + 2x$$

where $[I_2]$ takes into account that x mole/liter of the injected 0.500 M has disappeared, and the $[I]$ takes into account that $2x$ mole/liter of I has formed to add to the 2.50×10^{-3} M originally present.

$$K = \frac{[I]^2}{[I_2]} = 3.76 \times 10^{-5} = \frac{(2.50 \times 10^{-3} + 2x)^2}{0.500 - x}$$

Since x is probably small compared to 0.500 in the denominator, we can approximate $(0.500 - x) \cong 0.500$. However, we cannot neglect the $2x$ compared to 2.50×10^{-3} in the numerator, since the two terms may be about the same size.

So, solve the equation

$$3.76 \times 10^{-5} \cong \frac{(2.50 \times 10^{-3} + 2x)^2}{0.500}$$

The quadratic formula gives us $x = 9.15 \times 10^{-4}$. Substitute this value of x in the expressions for $[I_2] = 0.500 - x$ and $[I] = 2.50 \times 10^{-3} + 2x$ to get the final equilibrium concentrations:

$$[I_2] = 0.500 - x = 0.500 - 9.15 \times 10^{-4} = 0.499 \ M$$
$$[I] = 2.50 \times 10^{-3} + 2x = 2.50 \times 10^{-3} + (2)(9.15 \times 10^{-4})$$
$$= 4.33 \times 10^{-3} \ M \ \blacksquare$$

■ EXAMPLE 317.   Given that $K = 33.3$ for $PCl_5(g) \rightleftarrows PCl_3(g) + Cl_2(g)$ at $760°K$. What will be the final equilibrium state of the system if 1.50 g of PCl_5 and 15.0 g of PCl_3 are simultaneously injected into a volume of 36.3 ml at $760°K$? (Compare Example 313.)

SOLUTION:

$$\frac{1.50 \text{ g of } PCl_5}{208.24 \text{ g/mole}} = 0.00720 \text{ mole}$$

Injection concentration of $PCl_5 = \dfrac{0.00720 \text{ mole}}{0.0363 \text{ liter}} = 0.198 \ M.$

$$\frac{15.0 \text{ g of } PCl_3}{137.33 \text{ g/mole}} = 0.109 \text{ mole}$$

Injection concentration of $PCl_3 = \dfrac{0.109 \text{ mole}}{0.0363 \text{ liter}} = 3.00 \ M.$

At the start of the experiment, we thus have: $[PCl_5] = 0.198 \ M$; $[PCl_3] = 3.00 \ M$; and $[Cl_2] = 0 \ M$.

Now let x moles/liter of PCl_5 decompose via the reaction $PCl_5 \rightarrow PCl_3 + Cl_2$. This will decrease the PCl_5 concentration by x, and raise that of PCl_3 and of Cl_2 by x.

At equilibrium,

$$[PCl_5] = 0.198 - x; \qquad [PCl_3] = 3.00 + x; \qquad [Cl_2] = x$$

$$K = \frac{[PCl_3][Cl_2]}{[PCl_5]} = \frac{(3.00 + x)(x)}{0.198 - x} = 33.3$$

Solve this equation for x, using the quadratic formula, since the usual approximation of negligible x will not hold in this case. It turns out that $x = 0.181 \ M$:

$$[PCl_5] = 0.198 - x = 0.198 - 0.181 = 0.017 \ M$$
$$[PCl_3] = 3.00 + x = 3.00 + 0.181 = 3.18 \ M$$
$$[Cl_2] = x = 0.181 \ M$$

[Note: Comparing with Example 313, we note that the effect of initially present PCl_3 is to impede somewhat the decomposition of PCl_5]. ■

■ EXAMPLE 318. For the equilibrium $H_2(g) + CO_2(g) \rightleftarrows H_2O(g) + CO(g)$ the K is 4.40 at 2000°K. Calculate the concentration of each species at equilibrium after 1.00 mole of H_2 and 1.00 mole of CO_2 are simultaneously put into a 4.68-liter box at 2000°K.

SOLUTION:

$$\text{Initial concentration of } H_2 = \frac{1.00 \text{ mole}}{4.68 \text{ liters}} = 0.214 \ M.$$

$$\text{Initial concentration of } CO_2 = \frac{1.00 \text{ mole}}{4.68 \text{ liters}} = 0.214 \ M$$

Initial concentrations of H_2O and of CO = zero.

Let x = moles per liter of H_2O (or CO) form in order to get equilibrium established. From the stoichiometry of the change $H_2 + CO_2 \rightarrow H_2O + CO$ we get x moles of H_2O and x moles of CO each time x moles of H_2 and x moles of CO_2 disappear. Therefore, at equilibrium,

$$[H_2] = 0.214 - x; \quad [CO_2] = 0.214 - x; \quad [H_2O] = x; \quad [CO] = x$$

$$K = \frac{[H_2O][CO]}{[H_2][CO_2]} = \frac{(x)(x)}{(0.214 - x)(0.214 - x)} = 4.40$$

This equation can be solved either by use of the quadratic formula or by noting that the left side is a perfect square:

If we take the square root of both sides of

$$\frac{(x)(x)}{(0.214 - x)(0.214 - x)} = 4.40$$

we get

$$\frac{x}{0.214 - x} = \sqrt{4.40} = 2.10$$

The result is a simple linear equation that solves to give $x = 0.145$. Substituting for x in the concentration expressions, we get the equilibrium values:

$$[H_2] = 0.214 - x = 0.214 - 0.145 = 0.069 \ M$$
$$[CO_2] = 0.214 - x = 0.214 - 0.145 = 0.069 \ M$$
$$[H_2O] = x = 0.145 \ M$$
$$[CO] = x = 0.145 \ M \ ■$$

■ EXAMPLE 319. For the equilibrium $H_2(g) + CO_2(g) \rightleftarrows H_2O(g) + CO(g)$ the K is 4.40 at 2000°K. What will be the composition of the final equilibrium system if 1.00 mole of H_2, 1.00 mole of CO_2 and 1.00 mole of H_2O are simultaneously put into a 4.68-liter box at 2000°K?

SOLUTION:

$$\text{Initial concentration of } H_2 = \frac{1.00 \text{ mole}}{4.68 \text{ liters}} = 0.214 \; M.$$

$$\text{Initial concentration of } CO_2 = \frac{1.00 \text{ mole}}{4.68 \text{ liters}} = 0.214 \; M.$$

$$\text{Initial concentration of } H_2O = \frac{1.00 \text{ mole}}{4.68 \text{ liters}} = 0.214 \; M.$$

Initially, there is none of the fourth ingredient CO.

Let $x =$ moles per liter of CO that will form by the net change $H_2 + CO_2$ → $H_2O + CO$. This will increase the concentration of H_2O by x moles/liter and decrease the concentrations of H_2 and of CO_2 by x moles/liter. At equilibrium, then, we have

$$[H_2] = 0.214 - x; \quad [CO_2] = 0.214 - x; \quad [H_2O] = 0.214 + x; \quad [CO] = x$$

$$K = \frac{[H_2O][CO]}{[H_2][CO_2]} = \frac{(0.214 + x)(x)}{(0.214 - x)(0.214 - x)} = 4.40$$

Solve this by the quadratic formula and get $x = 0.119$. This gives

$$[H_2] = 0.214 - x = 0.214 - 0.119 = 0.095 \; M$$
$$[CO_2] = 0.214 - x = 0.214 - 0.119 = 0.095 \; M$$
$$[H_2O] = 0.214 + x = 0.214 + 0.119 = 0.333 \; M$$
$$[CO] = x = 0.119 \; M \quad ■$$

■ EXAMPLE 320.   At high temperatures phosgene, $COCl_2$, decomposes to give carbon monoxide, CO, and chlorine, Cl_2. In a typical experiment 0.631 g of $COCl_2$ is injected into a flask of volume 472.0 ml at 1000°K. When equilibrium has been established, it is found that the total pressure in the flask is 2.175 atm. Calculate K for the reaction $COCl_2(g) \rightleftarrows CO(g) + Cl_2(g)$ at 1000°K.

SOLUTION: (Use $PV = nRT$ to find out how many moles there are in the system at equilibrium. Then compare this with the number of moles injected into the system to find out how far the decomposition has proceeded.)

In the final state $P = 2.175$ atm, $V = 0.4720$ liter, and $T = 1000°K$:

$$n = \frac{PV}{RT} = \frac{(2.175)(0.4720)}{(0.08206)(1000)} = 0.01251 \text{ mole}$$

This represents the total number of moles in the final system and comprises $COCl_2$, CO, and Cl_2.

Initially, we put 0.631 g of $COCl_2$ into the system. One mole of $COCl_2$ weighs 98.917 g, so this amounts to

$$\frac{0.631 \text{ g}}{98.917 \text{ g/mole}} = 0.00638 \text{ mole of } COCl_2$$

If we let x = moles of $COCl_2$ that will decompose to establish equilibrium via $COCl_2 \rightarrow CO + Cl_2$, then we will have left $(0.00638 - x)$ moles of $COCl_2$ and will have formed x moles of CO and x moles of Cl_2. The total number of moles in the final system will be the sum of these: $(0.00638 - x)$ moles of $COCl_2$ plus x moles of CO plus x moles of $Cl_2 = 0.00638 + x$.

But, above, we calculated the total number of moles to be equal to 0.01251. We set the two equal to each other, $0.00638 + x = 0.01251$, and solve for $x = 0.00613$.

At equilibrium, we have then $x = 0.00613$ mole of CO, $x = 0.00613$ mole of Cl_2, and $(0.00638 - x) = (0.00638 - 0.00613) = 0.00025$ mole of $COCl_2$. Dividing each of these by the volume of the system, 0.472 liter, will give the concentrations at equilibrium:

$$[CO] = \frac{0.00613 \text{ mole}}{0.472 \text{ liter}} = 0.0130 \ M$$

$$[Cl_2] = \frac{0.00613 \text{ mole}}{0.472 \text{ liter}} = 0.0130 \ M$$

$$[COCl_2] = \frac{0.00025 \text{ mole}}{0.472 \text{ liter}} = 5.3 \times 10^{-4} \ M$$

Substitute these in the constant for the equilibrium for $COCl_2(g) \rightleftarrows CO(g) + Cl_2(g)$.

$$K = \frac{[CO][Cl_2]}{[COCl_2]} = \frac{(0.0130)(0.0130)}{5.3 \times 10^{-4}} = 0.32 \ \blacksquare$$

11.3   Different types of K: K_c, K_p, K_x

In all our preceding calculations we have gone along with the common convention in freshman chemistry that the equilibrium constant K is evaluated by using moles per liter as the units for expressing concentration. In subsequent chapters we shall continue to follow this convention; however, you should be aware that there is nothing magic in using C in moles per liter to describe the concentrations and that, in fact, two other conventions are in considerable use. One of these describes concentrations in terms of partial pressures, p (atm), in which case K is designated as K_p. The other describes concentrations in terms of mole fractions, x, in which case K is designated as K_x. Just plain K implies moles per liter but if you want to be dead-sure, you can mark it K_c. What you get as the numerical value of the equilibrium constant for a particular reaction at a given temperature depends on whether you use K_c, K_p, or K_x.

■ EXAMPLE 321.   (Compare Example 304.) You are given a box containing at equilibrium at 1000°K 0.102 mole/liter of NH_3, 1.03 moles/liter of N_2, and 1.62 moles/liter of H_2. For the reaction $N_2(g) + 3H_2(g) \rightleftarrows 2NH_3(g)$ calculate K_c, K_p, and K_x at 1000°K.

SOLUTION: To get K_c, we need moles per liter for each species:

$$[NH_3] = 0.102 \ M \qquad K_c = \frac{[NH_3]^2}{[N_2][H_2]^3} = \frac{(0.102)^2}{(1.03)(1.62)^3}$$

$$[N_2] = 1.03 \ M$$

$$[H_2] = 1.62 \ M \qquad\qquad = 2.37 \times 10^{-3}$$

To get K_p, we need the partial pressure of each species. We can calculate p by using $pV = nRT$, where $p = (n/V)RT$ and (n/V) is just the given concentration in moles per liter of each species:

$$p_{NH_3} = (0.102)(RT) = (0.102)(0.08206)(1000) = 8.37 \text{ atm}$$

$$p_{N_2} = (1.03)(RT) = (1.03)(0.08206)(1000) = 84.5 \text{ atm}$$

$$p_{H_2} = (1.62)(RT) = (1.62)(0.08206)(1000) = 133 \text{ atm}$$

For the reaction $N_2(g) + 3H_2(g) \rightleftarrows 2NH_3(g)$

$$K_p = \frac{p_{NH_3}^2}{p_{N_2} p_{H_2}^3} = \frac{(8.37)^2}{(84.5)(133)^3} = 3.52 \times 10^{-7}$$

To get K_x, we need the mole fraction of each species. If we take 1 liter for reference, it contains 0.102 mole of NH_3, 1.03 moles of N_2, and 1.62 moles of H_2, which comes to a total of 2.75 moles:

$$x_{NH_3} = \frac{0.102 \text{ mole of } NH_3}{2.75 \text{ moles total}} = 0.0371$$

$$x_{N_2} = \frac{1.03 \text{ moles of } N_2}{2.75 \text{ moles total}} = 0.375$$

$$x_{H_2} = \frac{1.62 \text{ moles of } H_2}{2.75 \text{ moles total}} = 0.589$$

For the reaction $N_2(g) + 3H_2(g) \rightleftarrows 2NH_3(g)$

$$K_x = \frac{x_{NH_3}^2}{x_{N_2} x_{H_2}^3} = \frac{(0.0371)^2}{(0.375)(0.589)^3} = 1.80 \times 10^{-2}$$

So, in summary, $K_c = 2.37 \times 10^{-3}$; $K_p = 3.52 \times 10^{-7}$; $K_x = 1.80 \times 10^{-2}$. They are all equally valid for describing the same equilibrium state and can be used for calculations where different kinds of concentration units are employed. ∎

∎ EXAMPLE 322.   Given that K_p for the equilibrium $2SO_2(g) + O_2(g) \rightleftarrows 2SO_3(g)$ equals 3.18 at 1000°K, figure out what must be the value of K_c for this same equilibrium at 1000°K. Assume ideal behavior.

SOLUTION: $2SO_2(g) + O_2(g) \rightleftarrows 2SO_3(g)$.

$$K_p = \frac{p_{SO_3}^2}{p_{SO_2}^2 p_{O_2}} = 3.18$$

For each gas $pV = nRT$, or $p = (n/V)RT$. But n/V, the number of moles per liter, is just the concentration C. So we can write

$$p_{SO_3} = \left(\frac{n_{SO_3}}{V}\right) RT = C_{SO_3} RT$$

$$p_{SO_2} = \left(\frac{n_{SO_2}}{V}\right) RT = C_{SO_2} RT$$

$$p_{O_2} = \left(\frac{n_{O_2}}{V}\right) RT = C_{O_2} RT$$

Substitute these in the foregoing expression for K_p:

$$K_p = \frac{(C_{SO_3} RT)^2}{(C_{SO_2} RT)^2 (C_{O_2} RT)} = 3.18$$

Solve for K_c:

$$K_c = \frac{C_{SO_3}^2}{C_{SO_2}^2 C_{O_2}} = 3.18(RT)$$

But $R = 0.08206$ liter atm/deg and $T = 1000°C$:

$$K_c = (3.18)(0.08206)(1000) = 261 \blacksquare$$

■ **EXAMPLE 323.** Suppose you have a 10-liter box containing only O_3 and O_2 at equilibrium at $2000°K$ with $K_p = 4.17 \times 10^{14}$ for $2O_3(g) \rightleftarrows 3O_2(g)$. If the total pressure in the box is 7.33 atm, what is the partial pressure of the ozone?

SOLUTION: $2O_3(g) \rightleftarrows 3O_2(g)$.

$$K_p = \frac{p_{O_2}^3}{p_{O_3}^2} = 4.17 \times 10^{14}$$

Total pressure $= 7.33$ atm $= p_{O_3} + p_{O_2}$.

K_p is such a large number that p_{O_2} must be much, much greater than p_{O_3}. So, practically all the pressure in the box must be due to O_2; that is, of the 7.33-atm pressure in the box, it is mostly p_{O_2}. Substitute $p_{O_2} \cong 7.33$ atm in K_p and solve for p_{O_3}:

$$K_p = \frac{p_{O_2}^3}{p_{O_3}^2} = \frac{(7.33)^3}{p_{O_3}^2} = 4.17 \times 10^{14}$$

$$p_{O_3} = 9.71 \times 10^{-7} \text{ atm.} \blacksquare$$

Exercises

*11.1. An equilibrium system at $480°K$ contains in a 10-liter volume 0.098 mole of PCl_3, 0.098 mole of Cl_2, and 0.047 mole of PCl_5. Calculate K at $480°K$ for the reaction

$$PCl_5(g) \rightleftarrows PCl_3(g) + Cl_2(g)$$

*11.2. A glass bulb of 1055-ml volume contains 0.0040 mole of NOBr, 0.0024 mole of Br_2, and 0.0063 mole of NO at equilibrium at $350°K$. Calculate K for each of the following:

(a) $2NO(g) + Br_2(g) \rightleftarrows 2NOBr(g)$
(b) $NOBr(g) \rightleftarrows NO(g) + \frac{1}{2}Br_2(g)$
(c) $2NOBr(g) \rightleftarrows 2NO(g) + Br_2(g)$
(d) $NO(g) + \frac{1}{2}Br_2(g) \rightleftarrows NOBr(g)$

11.3. At 3000°K and 1 atm total pressure CO_2 is 40.0% dissociated into CO and O_2. Calculate K for the reaction

$$CO_2(g) \rightleftarrows CO(g) + \tfrac{1}{2}O_2(g)$$

(*Answer*: 0.0174.)

11.4. At 2200°K and 1 atm total pressure, steam is 1.18% decomposed into H_2 and O_2. Calculate K for the reaction

$$2H_2O(g) \rightleftarrows 2H_2(g) + O_2(g)$$

11.5. Given that $K = 2.24 \times 10^{22}$ at 1000°K for $2CO(g) + O_2(g)$ $\rightleftarrows 2CO_2(g)$. What ratio of CO_2 to CO pressures would you need to set an equilibrium pressure for oxygen of 2.5×10^{-10} atm?

11.6. Given that $K = 3.76 \times 10^{-5}$ at 1000°K for the equilibrium $I_2(g)$ $\rightleftarrows 2I(g)$, calculate what per cent of the I_2 would be dissociated if 1 mole of I_2 were placed in a 1-liter flask at 1000°K and allowed to come to equilibrium. (*Answer*: 0.31%.)

11.7. Given that $K = 33.3$ at 760°K for the equilibrium $PCl_5(g) \rightleftarrows PCl_3(g)$ $+Cl_2(g)$, calculate what per cent of PCl_5 would be dissociated if 1 mole of PCl_5 were placed in a 1-liter flask at 760°K and allowed to come to equilibrium.

***11.8.** At 600°K the K for $NH_3(g) \rightleftarrows \tfrac{1}{2}N_2(g) + \tfrac{3}{2}H_2(g)$ is equal to 0.395. What pressure would be developed if 1.00 g of NH_3 were injected into a 1.00-liter box at 600°K and allowed to equilibrate? (*Answer*: 5.37 atm.)

11.9. When 6.28×10^{-3} mole of $N_2O_4(g)$ is placed in a 1.00-liter flask and allowed to come to equilibrium at 35°C, the final equilibrium pressure becomes 0.238 atm. Calculate K for the reaction $N_2O_4(g) \rightleftarrows 2NO_2(g)$.

***11.10.** Given that $K = 0.0269$ at 45°C for the equilibrium $N_2O_4(g)$ $\rightleftarrows 2NO_2(g)$, what concentrations of N_2O_4 and NO_2 would you find at equilibrium if you put 2.50×10^{-3} mole of NO_2 in a 350-ml flask? (*Answer*: $[N_2O_4] = 0.00099$ M; $[NO_2] = 0.00516$ M.)

*11.11.** Using $K = 1.37$ for the equilibrium $H_2(g) + CO_2(g) \rightleftarrows H_2O(g)$ $+ CO(g)$ at 1200°K, calculate the concentration of each species at equilibrium after 1.00 mole of H_2O and 1.00 mole of CO are simultaneously put into a 2.00-liter box at 1200°K.

11.12. Using $K = 1.37$ for the equilibrium $H_2(g) + CO_2(g) \rightleftarrows H_2O(g)$ $+ CO(g)$ at 1200°K, calculate the concentration of each species at equilibrium after 1.00 mole of H_2, 2.00 moles of CO_2, 3.00 moles of H_2O, and 4.00 moles of CO are simultaneously put into a 2.00-liter box at 1200°K. (*Answer*: $[H_2] = 0.905$ M; $[CO_2] = 1.405$ M; $[H_2O] = 1.10$ M; $[CO_2] = 1.60$ M.)

*11.13.** Given an equilibrium system at 464°C consisting of $I_2(g)$ at a pressure of 16.6 Torr, $Cl_2(g)$ at 9.5 Torr, and $ICl(g)$ at 316.2 Torr. Calculate K_c, K_p, and K_x for the reaction

$$I_2(g) + Cl_2(g) \rightleftarrows 2ICl(g)$$

**11.14. Given an equilibrium system at 440°C consisting of $Cl_2(g)$ at a pressure of 0.128 atm, $CO(g)$ at 0.116 atm, and $COCl_2(g)$ at 0.334 atm. Calculate K_c, K_p, and K_x for the reaction

$$COCl_2(g) \rightleftarrows CO(g) + Cl_2(g)$$

**11.15. Given for the reaction $2O_3(g) \rightleftarrows 3O_2(g)$ at 2000°K that $K_p = 4.17 \times 10^{14}$. In a 5.00-liter box containing 3.0×10^{20} oxygen atoms at equilibrium at 2000°K, how many ozone molecules would there be? (*Answer*: 6.7×10^{11}).

*11.16. For the reaction $2SO_2(g) + O_2(g) \rightleftarrows 2SO_3(g)$ at 1000°K, the value of $K_p = 3.18$. If you place 1.00 g of SO_2 into a 2.00-liter box at 1000°K, what pressure of oxygen would you need to maintain to assure that 90% of the SO_2 is converted to SO_3?

**11.17. At high temperatures $H_2S(g)$ is partly dissociated into $H_2(g)$ and $S_2(g)$. For a given experiment at 1132°C it was found that when the total pressure is 1.00 atm, 30.7% of the H_2S is dissociated. Calculate K_p for $2H_2S(g) \rightleftarrows 2H_2(g) + S_2(g)$.

**11.18. The equilibrium constant K_c for the decomposition of phosgene, $COCl_2(g) \rightleftarrows CO(g) + Cl_2(g)$, has a value of 0.0820 at 900°K. Suppose you put some $COCl_2$ at 900°K in a cylinder fitted with a moveable piston. To make half of the carbon atoms in the form of $COCl_2$ and half as CO, what external pressure must you place on the piston? (*Answer*: 18.2 atm.)

**11.19. The dissociation of $SbCl_5$ according to the equation $SbCl_5(g) \rightleftarrows SbCl_3(g) + Cl_2(g)$ increases from 8.2% at 128°C to 29.2% at 182°C. What change occurs in K_c from 128°C to 182°C?

***11.20. Suppose you are given a 5.25-liter box containing SO_2, O_2, and SO_3 at equilibrium at 1000°K. Initially the equilibrium pressures are $p_{O_2} = 1.00$ atm, $p_{SO_2} = 1.62$ atm, and $p_{SO_3} = 2.88$ atm. Being careful to keep the temperature constant at 1000°K, you squeeze the box until the volume gets reduced to 3.86 liters. Assuming nothing escapes, calculate the final pressures of the species in the box when equilibrium is reestablished. (*Answer*: $p_{SO_3} = 4.09$ atm; $p_{SO_2} = 2.03$ atm; $p_{O_2} = 1.28$ atm.)

12

STRONG AND WEAK
ELECTROLYTES

AN ELECTROLYTE is a substance that dissociates in water (or in some other suitable solvent) to produce an electrically conducting solution. Most electrolytes can be classified as *strong* or *weak*, the former being those that are extensively dissociated into ions and the latter, those dissociated to only a slight extent. This distinction breaks down in very dilute solutions, because as the concentration of a solute decreases, its per cent dissociation increases. In very dilute solutions, even weak electrolytes behave like strong electrolytes.

The interaction between an electrolyte AB and the solvent, generally water, is a chemical reaction that can be written

$$AB + \text{water} \longrightarrow A^+_{\text{hydrated}} + B^-_{\text{hydrated}}$$

Being reversible, this reaction comes to equilibrium, and in the equilibrium state there will be present undissociated AB as well as hydrated A^+ and hydrated B^-. For computation purposes, it is common to leave out the water and write the equilibrium as

$$AB \rightleftharpoons A^+ + B^-$$

with the understanding that all the species are hydrated. The equilibrium condition, as discussed in Chapter 11, is

$$K = \frac{C_{A^+} C_{B^-}}{C_{AB}} = \frac{[A^+][B^-]}{[AB]}$$

where we have used C as well as the square brackets to represent concentration in moles per liter.

236

12.1 Strong electrolytes

Most salts, a few acids, and a relatively few bases are strong electrolytes. For salts, representative compounds would be the sodium or potassium salts with almost any anion—for example, $NaCl$, KNO_3, Na_2SO_4, Na_2CO_3, $KAl(SO_4)_2$. For acids, the best example of a strong electrolyte would be perchloric acid, $HClO_4$; other possible examples are HNO_3, HCl, and the first proton to come off H_2SO_4 (via $H_2SO_4 + H_2O \rightarrow H_3O^+ + HSO_4^-$). For bases, $NaOH$ and KOH are good examples, as are $Ca(OH)_2$, $Sr(OH)_2$, and $Ba(OH)_2$. For these dihydroxides, both OH^- groups are normally assumed to be fully dissociated. In all these cases listed, dissociation is usually assumed to be 100% complete.

■ EXAMPLE 324. If you have a solution that is 0.010 M $NaCl$, what is the concentration of Na^+ and of Cl^- in this solution?

SOLUTION: Assume 100% dissociation since this is a typical salt:

$$NaCl \longrightarrow Na^+ + Cl^-$$

The 0.010 M $NaCl$ means 0.010 mole of $NaCl$ per liter, all of which is taken to be 100% dissociated into Na^+ and Cl^-. The dissociation equation shows that if 1 mole of $NaCl$ disappears on the left, we get 1 mole of Na^+ and 1 mole of Cl^- on the right. So, if we break up 0.010 mole of $NaCl$, we get 0.010 mole of Na^+ and 0.010 mole of Cl^-. Since the reference volume of solution is 1 liter, we can say the concentration of Na^+ is 0.010 mole/liter, or 0.010 M, and the concentration of Cl^- likewise equals 0.010 M. ■

■ EXAMPLE 325. In a solution that is 0.015 M K_2SO_4, what is the concentration of K^+ and of SO_4^{2-}?

SOLUTION: Assume 100% dissociation since this is a typical salt:

$$K_2SO_4 \longrightarrow 2K^+ + SO_4^{2-}$$

Each mole of K_2SO_4 produces 2 moles of K^+ and 1 mole of SO_4^{2-}. The 0.015 M K_2SO_4 means 0.015 mole of K_2SO_4 per liter of solution, which would give 0.030 mole of K^+ and 0.015 mole of SO_4^{2-}. So, $[K^+] = 0.030\ M$; $[SO_4^{2-}] = 0.015\ M$. ■

■ EXAMPLE 326. Suppose you add 3.63 g of $KAl(SO_4)_2$ to enough water to make 0.250 liter of solution. What is the concentration of each ion?

SOLUTION:

$$\frac{3.63 \text{ g of } KAl(SO_4)_2}{258.21 \text{ g/mole}} = 0.0141 \text{ mole}$$

Assume 100% dissociation:

$$KAl(SO_4)_2 \longrightarrow K^+ + Al^{3+} + 2SO_4^{2-}$$

If we dissociate 0.0141 mole of $KAl(SO_4)_2$ this way, the result will be 0.0141 mole of K^+, 0.0141 mole of Al^{3+}, and 0.0282 mole of SO_4^{2-}. To get concentrations, we divide the number of moles by the volume:

$$[K^+] = \frac{0.0141 \text{ mole}}{0.250 \text{ liter}} = 0.0564 \ M$$

$$[Al^{3+}] = \frac{0.0141 \text{ mole}}{0.250 \text{ liter}} = 0.0564 \ M$$

$$[SO_4^{2-}] = \frac{0.0282 \text{ mole}}{0.250 \text{ liter}} = 0.113 \ M \ \blacksquare$$

■ EXAMPLE 327.   Suppose you add 5.15 g of $HClO_4$ to 0.250 liter of 0.150 M $HClO_4$. Assuming the solution volume stays at 0.250 liter, calculate $[H_3O^+]$ and $[ClO_4^-]$.

SOLUTION: (Whenever you have a problem such as this, where more than one source contributes reagent, calculate how many moles of each species are contributed from each source. Then use the total to figure out the concentration.)
One mole of $HClO_4$ weighs 100.46 g.

$$\frac{5.15 \text{ g of } HClO_4}{100.46 \text{ g/mole}} = 0.0513 \text{ mole of } HClO_4$$

Since $HClO_4$ is a strong electrolyte, we assume 100% dissociation:

$$HClO_4 + H_2O \longrightarrow H_3O^+ + ClO_4^-$$

to give 0.0513 mole of H_3O^+ and 0.0513 mole of ClO_4^-. But the solution initially is 0.150 M $HClO_4$, which, corresponding to 100% dissociation via $HClO_4 + H_2O \rightarrow H_3O^+ + ClO_4^-$, would mean 0.150 M H_3O^+ and 0.150 M ClO_4^-. Since we start with 0.250 liter of 0.150 M $HClO_4$, we have at the start:

$$(0.250 \text{ liter})\left(0.150 \ \frac{\text{mole of } H_3O^+}{\text{liter}}\right) = 0.0375 \text{ mole of } H_3O^+$$

$$(0.250 \text{ liter})\left(0.150 \frac{\text{mole of ClO}_4^-}{\text{liter}}\right) = 0.0375 \text{ mole of ClO}_4^-$$

Total moles of $H_3O^+ = 0.0513 + 0.0375 = 0.0888$ mole of H_3O^+.
Total moles of $ClO_4^- = 0.0513 + 0.0375 = 0.0888$ mole of ClO_4^-.
Total volume of final solution equals 0.250 liter.

$$\text{Final } [H_3O^+] = \frac{0.0888 \text{ mole}}{0.250 \text{ liter}} = 0.355 \ M$$

$$[ClO_4^-] = \frac{0.0888 \text{ mole}}{0.250 \text{ liter}} = 0.355 \ M \ \blacksquare$$

■ EXAMPLE 328.   Suppose you add 1.65 g of $Ba(OH)_2$ to a mixture of 47.6 ml of 0.0562 M $Ba(OH)_2$ and 23.2 ml of 0.100 M $Ba(OH)_2$. Calculate the final concentrations of Ba^{2+} and OH^-, assuming complete dissociation and that the final volume of solution is just the sum of the two solutions mixed.

SOLUTION: Calculate the number of moles of $Ba(OH)_2$ from each source.

$$1.65 \text{ g of } Ba(OH)_2 = \frac{1.65 \text{ g}}{171.35 \text{ g/mole}} = 0.00963 \text{ mole}$$

47.6 ml of 0.0562 M $Ba(OH)_2$ furnishes

$$(0.0476 \text{ liter})\left(0.0562 \frac{\text{mole}}{\text{liter}}\right) = 0.00268 \text{ mole}$$

23.2 ml of 0.100 M $Ba(OH)_2$ furnishes

$$(0.0232 \text{ liter})\left(0.100 \frac{\text{mole}}{\text{liter}}\right) = 0.00232 \text{ mole}$$

Total $Ba(OH)_2 = 0.00963 + 0.00268 + 0.00232 = 0.0146$ mole.
Assume 100% dissociation via $Ba(OH)_2 \rightarrow Ba^{2+} + 2OH^-$. This will give 0.0146 mole of Ba^{2+} plus 0.0292 mole of OH^-.
Final volume = 47.6 ml + 23.2 ml = 70.8 ml.

$$\text{Final } [Ba^{2+}] = \frac{0.0146 \text{ mole}}{0.0708 \text{ liter}} = 0.206 \ M$$

$$[OH^-] = \frac{0.0292 \text{ mole}}{0.0708 \text{ liter}} = 0.412 \ M \ \blacksquare$$

12.2   Weak acids (monoprotic)

A "monoprotic" acid is one that can furnish but one H_3O^+ per molecule of acid. "Weak" means that at ordinary concentrations only a small fraction of the molecules are dissociated into H_3O^+ and another fragment. The per cent dissociation and the concentration of the various species can be figured out by applying the techniques for equilibrium calculations as outlined in Chapter 11. For the general example of a weak acid, HA, which can dissociate into H_3O^+ and A^-, the dissociation reaction is reversible and can be written

$$HA + H_2O \rightleftharpoons H_3O^+ + A^-$$

for which the condition at equilibrium is

$$K = \frac{C_{H_3O^+} \, C_{A^-}}{C_{HA}} = \frac{[H_3O^+]\,[A^-]}{[HA]}$$

Note that C_{H_2O} does not appear in this expression because any species whose concentration is not variable (e.g., H_2O in dilute aqueous solutions) is ignored.

The constant K is called the *dissociation constant*, or ionization constant, of the acid and is frequently designated K_{diss}. As before, we can use either C or square brackets to designate the concentration of each species in moles per liter.

Some typical weak acids are the following:

$$HNO_2 + H_2O \rightleftharpoons H_3O^+ + NO_2^- \qquad K_{diss} = 4.5 \times 10^{-4}$$
nitrous acid

$$HOCl + H_2O \rightleftharpoons H_3O^+ + OCl^- \qquad K_{diss} = 3.2 \times 10^{-8}$$
hypochlorous acid

$$HCN + H_2O \rightleftharpoons H_3O^+ + CN^- \qquad K_{diss} = 4 \times 10^{-10}$$
hydrocyanic acid

$$HOAc + H_2O \rightleftharpoons H_3O^+ + OAc^- \qquad K_{diss} = 1.8 \times 10^{-5}$$
acetic acid

In the last case we follow standard practice as done in the field of organic chemistry and use the designation Ac to represent the acetyl grouping, CH_3CO. The acetate ion, which contains an additional oxygen atom attached to the acetyl group, is CH_3COO^-, also written $C_2H_3O_2^-$, and is represented here as OAc^-.

■ EXAMPLE 329. Given a solution that is labeled $0.0200\ M$ HNO_2. What is the concentration of H_3O^+, NO_2^-, and HNO_2 in this solution? The dissociation constant of HNO_2 is 4.5×10^{-4}.

SOLUTION: The label $0.0200\ M$ HNO_2 means that the solution was made up by adding 0.0200 mole HNO_2 to enough water to make a liter of solution. Let $x =$ moles per liter of HNO_2 that will need to dissociate to establish equilibrium via the reaction $HNO_2 + H_2O \rightarrow H_3O^+ + NO_2^-$. The stoichiometry of this equation shows that every time 1 mole of HNO_2 breaks up, 1 mole of H_3O^+ and 1 mole of NO_2^- are produced. So, if x moles of HNO_2 break up, x moles of H_3O^+ and x moles of NO_2^- are produced. This means that, at equilibrium, there will be $(0.0200 - x)$ moles/liter of HNO_2 left undissociated along with x moles/liter of H_3O^+ and x moles/liter of NO_2^-. The equilibrium condition for

$$HNO_2 + H_2O \rightleftharpoons H_3O^+ + NO_2^-$$

is that

$$K = \frac{[H_3O^+][NO_2^-]}{[HNO_2]} = 4.5 \times 10^{-4}$$

We substitute $[H_3O^+] = x$; $[NO_2^-] = x$; $[HNO_2] = 0.0200 - x$:

$$K = \frac{(x)(x)}{0.0200 - x} = 4.5 \times 10^{-4}$$

This is a single equation with one unknown, so it can be solved for x. For exact solution, we rewrite the algebraic equation in the form $x^2 + 4.5 \times 10^{-4} x - 9.0 \times 10^{-6} = 0$ and apply the quadratic formula. It gives us

$$x = \frac{-4.5 \times 10^{-4} \pm \sqrt{(4.5 \times 10^{-4})^2 - (4)(1)(-9.0 \times 10^{-6})}}{2}$$

from which, discarding the negative root because it is impossible to have a negative concentration, we find that $x = 2.8 \times 10^{-3}$. Thus, at equilibrium, we have

$$[H_3O^+] = x = 2.8 \times 10^{-3}\ M$$

$$[NO_2^-] = x = 2.8 \times 10^{-3}\ M$$

$$[HNO_2] = 0.0200 - x = 0.0172\ M$$

[*Note:* We could also have gotten an approximate answer by noting that in the equation $K = (x)(x)/(0.0200 - x) = 4.5 \times 10^{-4}$ the x is probably small compared to 0.0200 in the denominator. If we neglect x with respect to 0.0200, we have the approximate relation $x^2/0.0200 \cong 4.5 \times 10^{-4}$, which quickly solves to $x = 3.0 \times 10^{-3}$. Checking our approximation $0.0200 - x \cong 0.0200$, we find that it did not stand up very well. With $x = 3.0 \times 10^{-3}$, the quantity $(0.0200 - x)$ is closer to 0.017 than it is to 0.0200. As a second approximation we could try that 0.017 value as a better guess for the denominator.

$$K = \frac{(x)(x)}{0.0200 - x} \cong \frac{(x)(x)}{0.017} \cong 4.5 \times 10^{-4}$$

It gives $x = 2.8 \times 10^{-3}$, which is not in bad agreement with the exact solution as found above]. ∎

∎ EXAMPLE 330.   If you dissolve 1.08 g of HBrO in enough water to make 42.7 ml of solution, what will be the concentration of H_3O^+, BrO^-, and HBrO in the final solution? $K_{diss} = 2.06 \times 10^{-9}$

SOLUTION:

$$\frac{1.08 \text{ g of HBrO}}{96.911 \text{ g/mole}} = 0.0111 \text{ mole of HBrO}$$

If none of the HBrO dissociated, then the 0.0111 mole of HBrO in 42.7 ml of solution would give a concentration of

$$\frac{0.0111 \text{ mole}}{0.0427 \text{ liter}} = 0.260 \, M$$

Let x mole/liter of HBrO dissociate. This produces x mole/liter of H_3O^+ and x mole/liter of BrO^-, leaving $0.260 - x$ mole/liter of HBrO undissociated. At equilibrium, we would have

$$[H_3O^+] = x; \ [BrO^-] = x; \ [HBrO] = 0.260 - x$$

For

$$HBrO + H_2O \rightleftharpoons H_3O^+ + BrO^-$$

the equilibrium condition is

$$K = \frac{[H_3O^+][BrO^-]}{[HBrO]} = \frac{(x)(x)}{0.260 - x} = 2.06 \times 10^{-9}$$

Since the value of K is so small, x is certainly negligible compared to 0.260 in the denominator. So, we solve

$$\frac{(x)(x)}{0.260} \cong 2.06 \times 10^{-9}$$

to get $x = 2.31 \times 10^{-5}$. At equilibrium, we will have

$$[H_3O^+] = 2.31 \times 10^{-5}\,M; \qquad [BrO^-] = 2.31 \times 10^{-5}\,M$$

$$[HBrO] = 0.260\,M \ \blacksquare$$

The big question in these calculations is "When can I neglect the x in the denominator?" Unfortunately, there is no simple answer; it depends on K_{diss} and the total concentration of solute. The bigger K_{diss} is, the less able you are to neglect x in the denominator; also, the more dilute the solution is, the less you can neglect x compared to total solute concentration. Generally, if K_{diss} is bigger than about 10^{-4} you cannot "neglect the x" over the whole range of concentration; if K_{diss} is less than about 10^{-5} you can usually "neglect the x" except in very dilute solutions (i.e., less concentrated than 0.01 M).

Instead of trying to remember when you can neglect x in the denominator, it is more reasonable to rely on the method of successive approximations in almost all cases: In the first step you neglect x in the denominator, then you calculate a value of x from the rest of the equation, put that value of x into the denominator, and recalculate. Keep doing that until the value of x settles down to a constant value. Only if K_{diss} is 10^{-2} or larger does it pay to go directly to the quadratic formula.

Another point to keep in mind is that dissociation constants are generally not known very precisely—rarely to three significant figures, occasionally to two figures, usually only to one. There is no point to beating your brains out for an exact answer beyond the number of significant figures allowed by the value of K_{diss}.

In the calculations above we have dealt only with the situation where all the H_3O^+ and all the A^- comes from the solute added to the solution. In practice, however, it is frequently the case that the solution already contains either H_3O^+ or A^- from some other source. This is sometimes referred to as the "common-ion effect" because any species already there needs to be taken into account in describing the equilibrium state.

■ EXAMPLE 331.   Suppose you add 1.08 g of $HClO_2$ to 427 ml of 0.0150 M $NaClO_2$ solution. Assuming no volume change, calculate the final concentrations of H_3O^+, ClO_2^-, and $HClO_2$. $K_{diss} = 1.1 \times 10^{-2}$.

SOLUTION: $NaClO_2$ is a strong electrolyte; it is 100% dissociated into Na^+ and ClO_2^-. This means that before you add any $HClO_2$, there is already 0.0150 M Na^+ and 0.0150 M ClO_2^- in the solution.

$$\frac{1.08 \text{ g of } HClO_2}{68.46 \text{ g/mole}} = 0.0158 \text{ mole of } HClO_2$$

In a volume of 0.427 liter, this comes to

$$\frac{0.0158 \text{ mole}}{0.427 \text{ liter}} = 0.0370 \text{ mole/liter of added } HClO_2.$$

If we let x moles/liter of this $HClO_2$ dissociate, we will get x moles/liter of H_3O^+ and x moles/liter of ClO_2^- from it, leaving $(0.0370 - x)$ moles/liter of $HClO_2$ undissociated. The x moles/liter of ClO_2^- will add to the 0.0150 M already there. At equilibrium, then, we will have

$$[H_3O^+] = x; [ClO_2^-] = x + 0.0150; [HClO_2] = 0.0370 - x$$

These values must satisfy the equilibrium condition, so we substitute in the expression for K_{diss}:

$$K = \frac{[H_3O^+][ClO_2^-]}{[HClO_2]} = \frac{(x)(x + 0.0150)}{0.0370 - x} = 1.1 \times 10^{-2}$$

The best way to solve this equation is through the quadratic formula, since x does not appear to be negligible compared to 0.0150 or 0.0370. The proper equation is $x^2 + 0.026x - 4.07 \times 10^{-4} = 0$, for which the quadratic formula gives $x = 0.0111$. Putting this x into the foregoing expressions for concentration, we get

$$[H_3O^+] = 0.011 \ M; [ClO_2^-] = 0.026 \ M; [HClO_2] = 0.026 \ M$$

It is just a coincidence that $[ClO_2^-]$ and $[HClO_2]$ come out to be equal; the thing to note is that $[H_3O^+]$ is *not* equal to $[ClO_2^-]$. ■

Commonly, the dissociation of a weak acid $HA + H_2O \rightarrow H_3O^+ + A^-$ is emphasized as a way of getting H_3O^+ and A^- in the solution at the expense of breakup of HA molecules. However, this reaction is a typical reversible reaction, where the same equilibrium involving H_3O^+, A^-, and HA can be "approached from the other side;" that is, by mixing H_3O^+ and A^- and allowing them to form as much HA as is consistent with the dissociation constant K_{diss} for HA. In other words, the equilibrium state is not dependent on how it is arrived at so long as equivalent amounts of reagent are involved. This business of approaching equilibrium from either side is particularly important to know when making computations for the final state expected on mixing some H_3O^+ and A^-.

■ EXAMPLE 332.   Hydrogen cyanide, HCN, is a weak acid with $K_{diss} =$ 4×10^{-10} for $HCN + H_2O \rightleftarrows H_3O^+ + CN^-$. If you mix 5.01 g of HCl and 6.74 g of NaCN in enough water to make 0.275 liter of solution, what will be the final concentrations of H_3O^+, CN^-, and HCN?

SOLUTION: HCl and NaCN are strong electrolytes that are 100% dissociated into $H_3O^+ + Cl^-$ and $Na^+ + CN^-$, respectively. K_{diss} for HCN is so small that HCN is but very little dissociated. This means that if H_3O^+ and CN^- are mixed together, they combine almost 100% to form HCN. The best way to solve the present problem is to take the available H_3O^+ and CN^-, convert them (hypothetically) 100% to HCN, then allow this HCN to "back-dissociate" as much as it needs to establish equilibrium. The advantage of working this way is we will be able to introduce an unknown x that is small compared to the number from which it will be subtracted.

$$5.01 \text{ g of HCl} = \frac{5.01 \text{ g}}{36.46 \text{ g/mole}} = 0.137 \text{ mole of HCl}$$

$$6.74 \text{ g of NaCN} = \frac{6.74}{49.01 \text{ g/mole}} = 0.137 \text{ mole of NaCN}$$

Assuming 100% dissociation:

$$0.137 \text{ mole of HCl} \longrightarrow 0.137 \text{ mole of } H_3O^+ + 0.137 \text{ mole of } Cl^-$$

$$0.137 \text{ mole of NaCN} \longrightarrow 0.137 \text{ mole of } Na^+ + 0.137 \text{ mole of } CN^-$$

We now let the 0.137 mole H_3O^+ and the 0.137 mole CN^- combine completely to give us 0.137 mole HCN. Since the volume of the solution is 0.275 liter, our hypothetical HCN would have a concentration of

$$\frac{0.137 \text{ mole}}{0.275 \text{ liter}} = 0.498 \ M$$

Let x moles per liter of this HCN dissociate via

$$HCN + H_2O \longrightarrow H_3O^+ + CN^-.$$

This will give at equilibrium:

$$[H_3O^+] = x; \qquad [CN^-] = x; \qquad [HCN] = 0.498 - x.$$

The equilibrium to be satisfied is

$$HCN + H_2O \; \rightleftharpoons \; H_3O^+ + CN^-$$

for which

$$K_{diss} = \frac{[H_3O^+][CN^-]}{[HCN]} = \frac{(x)(x)}{0.498 - x} = 4 \times 10^{-10}$$

Because K_{diss} is so small, x is small and can be neglected in the denominator where it is subtracted from 0.498. Solving

$$\frac{(x)(x)}{0.498} \cong 4 \times 10^{-10}$$

we get $x = 1.41 \times 10^{-5}$. However, since we are given K to only one significant figure, we should quote $x = 1 \times 10^{-5}$. Final concentrations, then, are

$$[H_3O^+] = 1 \times 10^{-5} \, M; \qquad [CN^-] = 1 \times 10^{-5} \, M; \qquad [HCN] = 0.498 \, M \; \blacksquare$$

12.3   Weak bases

A common way to define a "base" is to say it is a compound of the type MOH that can dissociate to give M^+ and OH^-. A "weak" base would be one for which the per cent dissociation of MOH into M^+ and OH^- is small. There are relatively few bases of the type MOH, but for computation purposes it is frequently advantageous to postulate their existence to explain observed concentrations.

■ EXAMPLE 333.   When ammonia is dissolved in water, the solution has basic properties and contains a small but appreciable concentration of NH_4^+ and OH^-. To account for this, it is frequently assumed that NH_3 reacts with H_2O to form NH_4OH, which acts as a weak base in being slightly dissociated into NH_4^+ and OH^-. If 0.106 mole of NH_3 in enough water to produce a liter of solution shows NH_4^+ and OH^- concentrations each equal to $1.38 \times 10^{-3} \, M$, what would be the dissociation constant for "NH_4OH," assuming all the dissolved NH_3 gets converted to NH_4OH?

SOLUTION: Assuming 100% conversion, 0.106 mole of NH_3 would react with water via $NH_3 + H_2O \rightarrow NH_4OH$ to produce 0.106 mole of

NH_4OH. Enough of this has to dissociate to produce 1.38×10^{-3} mole/liter of NH_4^+ and OH^-. From the equation $NH_4OH \rightarrow NH_4^+ + OH^-$ we see that 1 mole of NH_4OH disappears for every mole of NH_4^+ and OH^- produced. At equilibrium, then, the concentration of undissociated NH_4OH would be $(0.106 - 1.38 \times 10^{-3}) = 0.105\,M$; the concentration of NH_4^+ is $1.38 \times 10^{-3}M$; the concentration of OH^- is $1.38 \times 10^{-3}\,M$.

$$NH_4OH \quad \rightleftharpoons \quad NH_4^+ + OH^-$$

$$K = \frac{[NH_4^+]\,[OH^-]}{[NH_4OH]} = \frac{(1.38 \times 10^{-3})(1.38 \times 10^{-3})}{0.105}$$

$$= 1.81 \times 10^{-5} \blacksquare$$

Actually, there is considerable evidence against the existence of NH_4OH in solution and there is a growing trend to discard it as a description of a species in solution. Probably the best way to describe aqueous ammonia solutions at the present time is to write the chemical reaction as

$$NH_3 + H_2O \quad \rightleftharpoons \quad NH_4^+ + OH^-$$

and the condition for equilibrium as

$$K = \frac{[NH_4^+]\,[OH^-]}{[NH_3]} = 1.81 \times 10^{-5}$$

This expression for K, which is still referred to as the dissociation constant of ammonia, is the same as that given in Example 333 except that the denominator shows the molar concentration of NH_3 instead of the molar concentration of a hypothetical NH_4OH. We do *not* show the concentration of H_2O in the denominator because it is essentially constant (as is true in all dilute aqueous solutions) and its activity stays at unity. In dealing with aqueous solutions, we shall always omit the H_2O from the equilibrium constant expression whether it shows up on the right or left side of the chemical equation.

■ EXAMPLE 334.   Given that $K = 1.81 \times 10^{-5}$ for the reaction

$$NH_3 + H_2O \quad \rightleftharpoons \quad NH_4^+ + OH^-$$

calculate the concentrations of NH_3, NH_4^+, and OH^- in a solution made by dissolving 2.63 g of NH_3 in enough water to give 768 ml of solution.

SOLUTION:

$$\frac{2.63 \text{ g of NH}_3}{17.03 \text{ g/mole}} = 0.154 \text{ mole of NH}_3$$

$$\frac{0.154 \text{ mole of NH}_3}{0.768 \text{ liter of solution}} = 0.201 \text{ mole/liter of NH}_3$$

Let x = moles per liter of NH_3 that react via

$$NH_3 + H_2O \longrightarrow NH_4^+ + OH^-$$

This will give at equilibrium:

$$[NH_4^+] = x; \quad [OH^-] = x; \quad [NH_3] = 0.201 - x$$

Substitute in the expression for K:

$$K = \frac{[NH_4^+][OH^-]}{[NH_3]} = \frac{(x)(x)}{0.201 - x} = 1.81 \times 10^{-5}$$

Solve to get $x = 1.90 \times 10^{-3}$. It is not a bad approximation to neglect x in the denominator. We get finally

$$[NH_4^+] = 1.90 \times 10^{-3} \ M; \quad [OH^-] = 1.90 \times 10^{-3} \ M$$

$$[NH_3] = 0.199 \ M \ \blacksquare$$

In the preceding section, it was stressed that the weak-acid dissociation equilibrium $HA + H_2O \rightleftarrows H_3O^+ + A^-$ can be approached from either side; that is, by putting HA in water or by mixing H_3O^+ and A^-. In either case, the missing species will be produced in sufficient concentration to assure equilibrium. The same reasoning applies to bases. For the general base BOH, which sets up the equilibrium $BOH \rightleftarrows B^+ + OH^-$, the same dissociation equilibrium can be set up by dissolving BOH in water or by mixing B^+ and OH^-.

■ EXAMPLE 335. Given that the weak base BOH has a dissociation constant of 1.50×10^{-8}, what will be the concentrations of B^+, OH^-, and BOH in a solution made by dissolving 1.25×10^{-3} mole of BCl (assumed to be a strong electrolyte) and 1.25×10^{-3} mole of NaOH in enough water to make 0.100 liter of solution?

SOLUTION: Since BCl and NaOH are strong electrolytes, they can be assumed to be 100% dissociated.

1.25×10^{-3} mole of BCl gives 1.25×10^{-3} mole of B^+ and 1.25×10^{-3} mole of Cl^-.

1.25×10^{-3} mole of NaOH gives 1.25×10^{-3} mole of Na^+ and 1.25×10^{-3} mole of OH^-.

The B^+ and OH^-, being the pieces of a weak electrolyte, are likely to combine together to form BOH. For the moment, let us assume this reaction goes 100% via $B^+ + OH^- \rightarrow BOH$ to produce 1.25×10^{-3} mole of BOH. In 0.100 liter of solution, this would give

$$\frac{1.25 \times 10^{-3} \text{ mole of BOH}}{0.100 \text{ liter}} = 1.25 \times 10^{-2} \, M \text{ BOH}.$$

Let x = moles/liter of this BOH that "back-dissociates" to set up the equilibrium. This will give x moles/liter of B^+ and x moles/liter of OH^- and leave $(1.25 \times 10^{-2} - x)$ mole/liter of BOH. For the equilibrium

$$BOH \quad \rightleftharpoons \quad B^+ + OH^-$$

the equilibrium condition is

$$K = \frac{[B^+]\,[OH^-]}{[BOH]} = 1.50 \times 10^{-8}$$

Substitute the concentration of each species in terms of x and get

$$\frac{(x)(x)}{0.0125 - x} = 1.50 \times 10^{-8}$$

from which we get $x = 1.37 \times 10^{-5}$. At equilibrium we have

$$[B^+] = x = 1.37 \times 19^{-5} \, M$$

$$[OH^-] = x = 1.37 \times 10^{-5} \, M$$

$$[BOH] = 1.25 \times 10^{-2} - x = 0.0125 \, M \quad \blacksquare$$

12.4 *Weak salts*

In general, in the absence of information to the contrary, we assume that salts are strong electrolytes and are 100% dissociated in all but the most concentrated solutions. This assumption is a good one and holds no matter how insoluble the salt is. A common error is to confuse low solubility with low per cent dissociation. They are different things. Barium sulfate, for example, is very low in solubility but the small amount that goes into solution is 100% dissociated into Ba^{2+} and SO_4^{2-}. Consequently, barium sulfate is a *strong* electrolyte.

However, not all salts are 100% dissociated into their component ions. Prime examples are the salts of zinc, cadmium, and mercury, particularly with the halides, and some of the salts of lead, particularly lead acetate. Most of these cases lend themselves to computation only with great difficulty because they usually involve more than one equilibrium at a time. Thus, for example, in a solution of mercuric chloride, $HgCl_2$, there is not only the equilibrium represented by

$$HgCl_2 \;\rightleftharpoons\; HgCl^+ + Cl^-$$

corresponding to the split-off of one chloride ion, but also

$$HgCl^+ \;\rightleftharpoons\; Hg^{2+} + Cl^-$$

corresponding to the split-off of the second chloride ion. Not only that— $HgCl_2$ can also pick up a Cl^- to form $HgCl_3^-$, which gives a third possible equilibrium:

$$HgCl_3^- \;\rightleftharpoons\; HgCl_2 + Cl^-$$

Similarly, $HgCl_3^-$ can pick up a chloride ion to form $HgCl_4^{2-}$. To work with all these equilibria simultaneously is obviously quite a job, since all have to be satisfied at the same time by the same set of concentrations. Such complex problems can be treated by the methods to be discussed in Chapter 16. At this point we can only scratch the surface of the problem so as to get a little know-how about weak salts. We shall have to make some large simplifications.

■ EXAMPLE 336.   A saturated solution of the weak salt mercuric chloride, $HgCl_2$, is found to have a concentration of 0.26 mole/liter of undissociated $HgCl_2$, 2.7×10^{-4} M $HgCl^+$, and 3.2×10^{-4} M chloride ion. Calculate the dissociation K for the reaction $HgCl_2 \rightleftharpoons HgCl^+ + Cl^-$.

SOLUTION: The equilibrium constant for this reaction is

$$K = \frac{[HgCl^+][Cl^-]}{[HgCl_2]}$$

All we have to do is to substitute the equilibrium concentrations that are given to us: $[HgCl_2] = 0.26\ M$; $[HgCl^+] = 2.7 \times 10^{-4}\ M$; $[Cl^-] = 3.2 \times 10^{-4}\ M$.

$$K = \frac{(2.7 \times 10^{-4})(3.2 \times 10^{-4})}{0.26} = 3.3 \times 10^{-7}\ \blacksquare$$

■ **EXAMPLE 337.** Suppose you are given a solution that is labeled $0.10\ M\ HgCl_2$. Assuming that only the first step of the dissociation is important, calculate the concentration of chloride ion in this solution. $K = 3.3 \times 10^{-7}$ for $HgCl_2 \rightleftarrows HgCl^+ + Cl^-$.

SOLUTION: Let $x =$ moles per liter of $HgCl_2$ that dissociate to give $HgCl^+$ and Cl^-. This will give x moles/liter of $HgCl^+$, x moles/liter of Cl^-, and $(0.10 - x)$ mole/liter of undissociated $HgCl_2$:

$$[HgCl^+] = x; \qquad [Cl^-] = x; \qquad [HgCl_2] = 0.10 - x$$

For the dissociation

$$HgCl_2 \quad \rightleftharpoons \quad HgCl^+ + Cl^-$$

the equilibrium condition is

$$K = \frac{[HgCl^+][Cl^-]}{[HgCl_2]} = 3.3 \times 10^{-7} = \frac{(x)(x)}{0.10 - x}$$

This solves to $x = 1.8 \times 10^{-4}$, which gives us at equilibrium:

$$[HgCl^+] = 1.8 \times 10^{-4}\ M; \qquad [Cl^-] = 1.8 \times 10^{-4}\ M$$

$$[HgCl_2] = 0.10\ M$$

[*Note:* The second dissociation step will not change these numbers very much because (a) there is not very much $HgCl^+$ from which to start the second dissociation reaction and (b) the K constant for the second dissociation is not very big.] ■

■ **EXAMPLE 338.** Suppose you mix 0.100 mole of $Hg(ClO_4)_2$ and 0.080 mole of NaCl in enough water to make 1.25 liters of solution. Given that these are strong electrolytes, but that $K = 1.8 \times 10^{-7}$ for $HgCl^+ \rightleftarrows Hg^{2+} + Cl^-$, calculate the final chloride-ion concentration in the final solution.

SOLUTION: 0.100 mole of $Hg(ClO_4)_2$ gives 0.100 mole of Hg^{2+} 0.080 mole of NaCl gives 0.080 mole of Cl^-.

Since the K_{diss} for $HgCl^+$ is so small, Hg^{2+} and Cl^- will have to combine to an appreciable extent. As a first guess, let us call for 100% association between the Hg^{2+} and the Cl^-. The 0.100 mole of Hg^{2+} + 0.080 mole of Cl^- yields 0.080 mole of $HgCl^+$ and 0.020 mole of Hg^{2+} left over. In 1.25 liters this would correspond to these concentrations:

$$\frac{0.080 \text{ mole of } HgCl^+}{1.25 \text{ liters}} = 0.064 \ M \ HgCl^+$$

$$\frac{0.020 \text{ mole of } Hg^{2+}}{1.25 \text{ liters}} = 0.016 \ M \ Hg^{2+}$$

But some of this $HgCl^+$ needs to dissociate to furnish Cl^-. Let $x =$ moles/liter of this $HgCl^+$ that dissociates. This gives x moles/liter of Cl^- and x moles/liter of Hg^{2+} to add to the 0.016 M Hg^{2+} we already have. The $HgCl^+$ concentration will be reduced from 0.064 to $0.064 - x$.

$$[HgCl^+] = 0.064 - x; \qquad [Hg^{2+}] = 0.016 + x; \qquad [Cl^-] = x$$

For the equilibrium $HgCl^+ \rightleftarrows Hg^{2+} + Cl^-$ we have

$$K = \frac{[Hg^{2+}][Cl^-]}{[HgCl^+]} = 1.8 \times 10^{-7} = \frac{(0.016 + x)(x)}{0.064 - x}$$

Solving this for x is simple, since the x is small compared to 0.016 or 0.064. We get $x = 7.2 \times 10^{-7}$.

Therefore, the final chloride-ion concentration is 7.2×10^{-7} M. ∎

12.5 Diprotic acids

A diprotic acid is one that can furnish two protons in neutralization reactions. In general, release of the first proton is much easier than that of the second. If we write H_2X as the general formula of a diprotic acid, then the dissociation of the first proton can be written

$$H_2X + H_2O \ \rightleftharpoons \ H_3O^+ + HX^-$$

and the subsequent dissociation of the second proton can be written

$$HX^- + H_2O \ \rightleftharpoons \ H_3O^+ + X^{2-}$$

The double arrows again emphasize the reversibility of both reactions. The equilibrium constant for the first reaction is frequently designated K_I and has the form

$$K_I = \frac{[H_3O^+][HX^-]}{[H_2X]}$$

The equilibrium constant for the second reaction is frequently designated K_{II}; it has the form

$$K_{II} = \frac{[H_3O^+][X^{2-}]}{[HX^-]}$$

As a very rough rule, the value of K_I is usually about 10^4 or 10^5 times as great as the value of K_{II}. This circumstance is what makes computations with diprotic acid systems so easy, since it makes one of the foregoing reactions dominant over the other.

■ EXAMPLE 339. Hydrogen sulfide, H_2S, is a diprotic acid with $K_I = 1.1 \times 10^{-7}$ and $K_{II} = 1 \times 10^{-14}$. Calculate the concentrations of H_2S, HS^-, S^{2-}, and H_3O^+ in a solution labeled 0.10 M H_2S.

SOLUTION: When H_2S is placed in solution, the first thing that happens is that H_2S dissociates to give H_3O^+ and HS^- in accord with the reaction

$$H_2S + H_2O \rightleftharpoons H_3O^+ + HS^-$$

Because K_I for this reaction is much less than 1 only a small per cent of H_2S will so dissociate. Then, some of the HS^- produced can also dissociate to give more H_3O^+ and some S^{2-} via the reaction

$$HS^- + H_2O \rightleftharpoons H_3O^+ + S^{2-}$$

However, the amount of this second dissociation cannot be very great, because (a) there is only a small amount of HS^- available for the dissociation and (b) the K_{II} for this second reaction is even smaller than K_I. So, in the final solution, there should be mostly undissociated H_2S, considerably less HS^-, and only a trace of S^{2-}.

The easiest way to solve the problem numerically is to neglect first the effect of the second dissociation as being trivial compared to the first. Given 0.10 M H_2S. Let $x =$ moles per liter of H_2S that dissociate via $H_2S + H_2O \rightarrow H_3O^+ + HS^-$. This will give us x moles/liter of H_3O^+ and x moles/liter of HS^- and leave $(0.10 - x)$ moles/liter of undissociated H_2S.

$$[H_3O^+] = x; \quad [HS^-] = x; \quad [H_2S] = 0.10 - x$$

The condition for equilibrium for the reaction

$$H_2S + H_2O \rightleftharpoons H_3O^+ + HS^-$$

is given by

$$K_I = \frac{[H_3O^+][HS^-]}{[H_2S]} = 1.1 \times 10^{-7}$$

Substituting the concentrations above, we get

$$\frac{(x)(x)}{0.10 - x} = 1.1 \times 10^{-7}$$

which easily solves to $x = 1.05 \times 10^{-4}$. Because K_I is given only to two significant figures, we trim the answer to $x = 1.0 \times 10^{-4}$.

Substituting in the foregoing concentration expressions, we get

$$[H_3O^+] = 1.0 \times 10^{-4} M; \qquad [HS^-] = 1.0 \times 10^{-4} M$$
$$[H_2S] = 0.10 M$$

Now we consider the second dissociation,

$$HS^- + H_2O \rightleftharpoons H_3O^+ + S^{2-}$$

for which the equilibrium condition is

$$K_{II} = \frac{[H_3O^+][S^{2-}]}{[HS^-]} = 1 \times 10^{-14}$$

Suppose we let $y =$ moles per liter of the HS^- that will dissociate to set up this second equilibrium. This will reduce $[HS^-]$ from the $1.0 \times 10^{-4} M$ value we just calculated to $(1.0 \times 10^{-4} - y)$ and will increase the $[H_3O^+]$ from the $1.0 \times 10^{-4} M$ we just calculated to $(1.0 \times 10^{-4} + y)$. At the same time, it will produce y moles of S^{2-}. So, at equilibrium, we will have

$$[H_3O^+] = 1.0 \times 10^{-4} + y; \qquad [HS^-] = 1.0 \times 10^{-4} - y; \qquad [S^{2-}] = y$$

Substitute these values in the expression for K_{II}:

$$K_{II} = \frac{[H_3O^+][S^{2-}]}{[HS^-]} = \frac{(1.0 \times 10^{-4} + y)(y)}{1.0 \times 10^{-4} - y} = 1 \times 10^{-14}$$

This looks quite ominous as an equation to solve, but note the very small value of $K_{II} = 1 \times 10^{-14}$. It tells you that y is indeed a very small number, so small that it can easily be neglected when added to or subtracted from 1.0×10^{-4}. If we scratch the y where it can be neglected, we get the equation

$$\frac{(1.0 \times 10^{-4})(y)}{1.0 \times 10^{-4}} \cong 1 \times 10^{-14}$$

which quickly solves to $y = 1 \times 10^{-14}$, truly a small number. In summary, then, we find in 0.10 M H_2S solution:

$$[H_2S] = 0.10\ M; \quad [HS^-] = 1.0 \times 10^{-4}\ M; \quad [S^{2-}] = 1 \times 10^{-14}\ M$$

$$[H_3O^+] = 1.0 \times 10^{-4}\ M$$

Some points to note especially are these: The second dissociation is important only as a source of S^{2-}. It contributes negligibly to raising the H_3O^+ or reducing the HS^- concentrations. Second, if you substitute the calculated concentrations in the expression for K_I

$$\frac{[H_3O^+][HS^-]}{[H_2S]} = \frac{(1.0 \times 10^{-4})(1.0 \times 10^{-4})}{0.10} = 1.0 \times 10^{-7}$$

you do not get $K_I = 1.1 \times 10^{-7}$, as given in the original problem. Do not fret about it. It is a straightforward consequence of our limitation here to two significant figures, the second of which is a doubtful digit. If you can reproduce the K to ± 1 in the last digit, you have done as much as you are entitled to. In the foregoing problem, we would need to use 1.05×10^{-4} for $[H_3O^+]$ and $[HS^-]$ to reproduce $K_I = 1.1 \times 10^{-7}$. This would go beyond the precision allowed by the given data. If you insist on doing that, you should at least write $1.0_5 \times 10^{-4}$, following the convention that the digit dropped below the line is dubious. ∎

One of the most important diprotic acids is sulfuric acid, H_2SO_4. It differs from the examples above in that the first dissociation is complete; that is, 100% of the H_2SO_4 is broken down into H_3O^+ and HSO_4^-. The other point of difference is that the second dissociation—that is, the breakdown of HSO_4^- into more H_3O^+ and some SO_4^{2-}—is not at all negligible. We can write the chemical equations for the stepwise dissociation as follows;

$$H_2SO_4 + H_2O \longrightarrow H_3O^+ + HSO_4^-$$

$$HSO_4^- + H_2O \rightleftharpoons H_3O^+ + SO_4^{2-}$$

The single arrow to the right in the first equation emphasizes that H_2SO_4 is a strong acid for this dissociation. The first dissociation is assumed to go 100% to the right, although strictly speaking it also comes to equilibrium. However, K_I is of the order of 10^3, which for our purposes corresponds to 100% conversion to the right. K_{II}, for the second reaction, is equal to 1.26 $\times 10^{-2}$, which is considerably larger than for most "weak" acids. For this reason, HSO_4^- is frequently referred to as a "moderately strong" acid.

■ EXAMPLE 340.   Given a solution that is 0.150 M $NaHSO_4$. What are the concentrations of HSO_4^-, SO_4^{2-}, and H_3O^+ in this solution?

SOLUTION: $NaHSO_4$ is a typical sodium salt and is a strong electrolyte in the sense that it is completely dissociated into Na^+ and HSO_4^-. So, our solution of 0.150 M $NaHSO_4$ can be regarded initially as 0.150 M Na^+ and 0.150 M HSO_4^-. However, some of the HSO^- will be dissociated into H_3O^+ and SO_4^{2-}. Let $x =$ moles per liter of HSO_4^- that are dissociated. This will leave $(0.150 - x)$ moles/liter of undissociated HSO_4^- and will produce x moles/liter of H_3O^+ and x moles/liter of SO_4^{2-}. At equilibrium, then, we will have the following:

$$[H_3O^+] = x; \quad [SO_4^{2-}] = x; \quad [HSO_4^-] = 0.150 - x$$

These concentrations must satisfy the equilibrium condition, which for the reaction

$$HSO_4^- + H_2O \rightleftharpoons H_3O^+ + SO_4^{2-}$$

is given by

$$K_{II} = \frac{[H_3O^+][SO_4^{2-}]}{[HSO_4^-]} = 1.26 \times 10^{-2} = \frac{(x)(x)}{0.150 - x}$$

With a K this big, it pays to go directly to the quadratic formula, since x will certainly not be negligible compared to 0.150. The equation converts to $x^2 = 0.189 \times 10^{-2} - 1.26 \times 10^{-2} x$, which solves to $x = 3.76 \times 10^{-2}$. Putting this into the foregoing concentration expressions, we get for the equilibrium concentrations:

$$[H_3O^-] = 3.76 \times 10^{-2} M; \quad [SO_4^{2-}] = 3.76 \times 10^{-2} M$$

$$[HSO_4^-] = 0.112 M ■$$

■ EXAMPLE 341.   Given a solution that is 0.150 M H_2SO_4. What are the concentrations of H_3O^+, SO_4^{2-}, and HSO_4^- in this solution?

SOLUTION: Compared to Example 340, this one differs in that H_2SO_4 is completely dissociated into H_3O^+ and HSO_4^-, and the H_3O^+ from this dissociation tends to repress the further dissociation of HSO_4^-. Assume 100% dissociation of $H_2SO_4 + H_2O \rightarrow H_3O^+ + HSO_4^-$. This means that the 0.150 M H_2SO_4 can be regarded initially as 0.150 M H_3O^+ and 0.150 M HSO_4^-. However, the HSO_4^- will dissociate slightly to drive the concentration of H_3O^+ somewhat higher than 0.150 M. Let x = moles per liter of HSO_4^- that dissociate. This leaves $(0.150 - x)$ moles/liter of HSO_4^- undissociated and produces x moles/liter of SO_4^{2-} while adding x moles/liter of H_3O^+ to the 0.150 M already there from the first dissociation. The equilibrium concentrations will be

$$[H_3O^+] = 0.150 + x; \quad [SO_4^{2-}] = x; \quad [HSO_4^-] = 0.150 - x$$

Substituting in the equilibrium condition, we get

$$K = \frac{[H_3O^+][SO_4^{2-}]}{[HSO_4^-]} = 1.26 \times 10^{-2} = \frac{(0.150 + x)(x)}{0.150 - x}$$

The resulting algebraic equation $x^2 + 0.163x - 0.00189 = 0$ solves by the quadratic formula to $x = 0.0109$. Substituting above, we get for the final equilibrium concentrations:

$$[H_3O^+] = 0.150 + x = 0.150 + 0.0109 = 0.161 \ M$$

$$[SO_4^{2-}] = x = 0.0109 \ M$$

$$[HSO_4^-] = 0.150 - x = 0.150 - 0.0109 = 0.139 \ M \quad \blacksquare$$

Some of the stiffest problems involving diprotic acids arise when we try to approach some of the foregoing equilibria from the other direction—for example, by mixing 0.1 mole of H_3O^+ and 0.1 mole of S^{2-}. Such a problem *cannot* be solved by just computing the dissociation of an equivalent amount of HS^-. Why it cannot be done this easily we shall have to leave until Chapter 16. It has to do with the fact that HS^- can not only dissociate to give H_3O^+ and S^{2-} but can also react with H_3O^+ to form H_2S. Both equilibria need to be considered simultaneously. At this stage, we shall simply consider a few cases of "back reaction" where we do not run into this complication.

■ EXAMPLE 342. Suppose you mix 1.00 liter of 0.100 M HCl with 1.00 liter of 0.100 M Na_2SO_4. What will be the concentrations of H_3O^+, SO_4^{2-}, and HSO_4^- in the final solution? Assume volumes are additive.

SOLUTION: One liter of 0.100 M HCl gives

$$(1.00 \text{ liter})\left(0.100 \frac{\text{mole}}{\text{liter}}\right) = 0.100 \text{ mole of HCl}$$

One liter of 0.100 M Na_2SO_4 gives

$$(1.00 \text{ liter})\left(0.100 \frac{\text{mole}}{\text{liter}}\right) = 0.100 \text{ mole of } Na_2SO_4$$

Both HCl and Na_2SO_4 are strong electrolytes and are 100% dissociated. From 0.100 mole of HCl we get 0.100 mole of H_3O^+ and 0.100 mole of Cl^-; 0.100 mole of Na_2SO_4 gives 0.200 mole of Na^+ and 0.100 mole of SO_4^{2-}. Initially, we assume that all the H_3O^+ and all the SO_4^{2-} convert to HSO_4^- via the reaction $H_3O^+ + SO_4^{2-} \rightarrow HSO_4^- + H_2O$. This would make the 0.100 mole of H_3O^+ plus 0.100 mole SO_4^{2-} equivalent to 0.100 mole of HSO_4^-. The total volume of the solution is $(1.00 + 1.00) = 2.00$ liters, so we have the equivalent of

$$\frac{0.100 \text{ mole of } HSO_4^-}{2.00 \text{ liters}} = 0.0500 \text{ } M \text{ } HSO_4^-$$

But this would make no allowance for dissociation. [*Note:* We do not have to worry about reaction of H_3O^+ with HSO_4^-, since H_2SO_4 would be 100% dissociated.] Let x = moles per liter of this hypothetical 0.0500 M HSO_4^- that dissociate. This gives

$$[H_3O^+] = x; \qquad [SO_4^{2-}] = x; \qquad [HSO_4^-] = 0.0500 - x$$

$$HSO_4^- + H_2O \rightleftharpoons H_3O^+ + SO_4^{2-}$$

$$K = \frac{[H_3O^+][SO_4^{2-}]}{[HSO_4^-]} = \frac{(x)(x)}{0.0500 - x} = 1.26 \times 10^{-2}$$

$$x = 1.96 \times 10^{-2}$$

Final equilibrium concentrations are

$$[H_3O^+] = 1.96 \times 10^{-2} \text{ } M; \qquad [SO_4^{2-}] = 1.96 \times 10^{-2} \text{ } M;$$

$$[HSO_4^-] = 3.04 \times 10^{-2} \text{ } M \blacksquare$$

■ EXAMPLE 343.   Oxalic acid, $H_2C_2O_4$, is a diprotic acid for which K_I is equal to 6.5×10^{-2} and K_{II} is equal to 6.1×10^{-5}. What will be the concentrations of H_3O^+, $C_2O_4^{2-}$, and $HC_2O_4^-$ in a solution made by dissolving 1.0×10^{-2} mole of HCl and 3.0×10^{-2} mole of $Na_2C_2O_4$ in enough water to make 0.250 liter of solution?

SOLUTION: Both HCl and $Na_2C_2O_4$ are strong electrolytes. From 1.0×10^{-2} mole of HCl would be formed 1.0×10^{-2} mole of H_3O^+ and 1.0×10^{-2} mole of Cl^-.

From 3.0×10^{-2} mole of $Na_2C_2O_4$ would be formed 6.0×10^{-2} mole of Na^+ and 3.0×10^{-2} mole of $C_2O_4^{2-}$.

H_3O^+ and $C_2O_4^{2-}$ will extensively combine.

Assume 100% reaction by $H_3O^+ + C_2O_4^{2-} \rightarrow HC_2O_4^- + H_2O$. From 1.0×10^{-2} mole of H_3O^+ and 3.0×10^{-2} mole of $C_2O_4^{2-}$ we would get formed 1.0×10^{-2} mole of $HC_2O_4^-$ and 2.0×10^{-2} mole of $C_2O_4^{2-}$ left over. In 0.250 liter this would give

$$\frac{1.0 \times 10^{-2} \text{ mole of } HC_2O_4^-}{0.250 \text{ liter}} = 0.040 \ M \ HC_2O_4^-$$

$$\frac{2.0 \times 10^{-2} \text{ mole of } C_2O_4^{2-}}{0.250 \text{ liter}} = 0.080 \ M \ C_2O_4^{2-}$$

But this allows for no free H_3O^+, which must be there if we are to have equilibrium. So, we need to let some of the $HC_2O_4^-$ dissociate to produce some H_3O^+ and, incidentally, some additional $C_2O_4^{2-}$. Let $x = $ moles per liter of the $HC_2O_4^-$ that dissociate. At equilibrium we would have

$$[H_3O^+] = x; \qquad [C_2O_4^{2-}] = 0.080 + x; \qquad [HC_2O_4^-] = 0.040 - x$$

The reaction is

$$HC_2O_4^- + H_2O \ \rightleftharpoons \ H_3O^+ + C_2O_4^{2-}$$

for which the equilibrium condition is

$$K = \frac{[H_3O^+][C_2O_4^{2-}]}{[HC_2O_4^-]} = 6.1 \times 10^{-5} = \frac{(x)(0.080 + x)}{0.040 - x}$$

No need to use the quadratic formula, since the rather small K value suggests an x small compared to 0.080 or 0.040. Approximating $0.080 + x \cong 0.080$ and $0.040 - x \cong 0.040$, we get $x = 3.0 \times 10^{-5}$. The final concentrations will be

$$[H_3O^+] = 3.0 \times 10^{-5} \ M; \qquad [C_2O_4^{2-}] = 0.080 \ M$$

$$[HC_2O_4^-] = 0.040 \ M \ \blacksquare$$

An important but very complicated example of diprotic acid behavior is given by solutions containing dissolved carbon dioxide. These solutions contain H_3O^+, HCO_3^-, and CO_3^{2-}, and for many years it was assumed that the principal equilibria in the solution were

$$H_2CO_3 + H_2O \rightleftharpoons H_3O^+ + HCO_3^- \qquad K_I = 4.16 \times 10^{-7}$$

$$HCO_3^- + H_2O \rightleftharpoons H_3O^+ + CO_3^{2-} \qquad K_{II} = 4.84 \times 10^{-11}$$

The reasoning was that CO_2 reacts with H_2O to form H_2CO_3, and H_2CO_3 then dissociates stepwise to give HCO_3^- and then CO_3^{2-}. At present it is known that less than 1% of the CO_2 actually gets converted to H_2CO_3, so that the foregoing K_I is not really for H_2CO_3 but for dissolved CO_2. Consequently, we are more accurate if we write

$$CO_2 + 2H_2O \rightleftharpoons H_3O^+ + HCO_3^- \qquad K_I = 4.16 \times 10^{-7}$$

$$HCO_3^- + H_2O \rightleftharpoons H_3O^+ + CO_3^{2-} \qquad K_{II} = 4.84 \times 10^{-11}$$

We refer to the first reaction and its constant as applying to the first step in the " dissociation " of aqueous CO_2. Leaving out the water, as we did in the case of aqueous ammonia equilibria, we would have

$$K_I = \frac{[H_3O^+][HCO_3^-]}{[CO_2]} = 4.16 \times 10^{-7}$$

$$K_{II} = \frac{[H_3O^+][CO_3^{2-}]}{[HCO_3^-]} = 4.84 \times 10^{-11}$$

■ EXAMPLE 344.   Given the constants above, what would be the concentrations of H_3O^+, CO_3^{2-}, and HCO_3^- in a solution labeled 0.034 M CO_2?

SOLUTION: Let x = moles per liter of CO_2 converted to HCO_3^-. Ignoring for the moment the second dissociation, we would have

$$[H_3O^+] = x; \qquad [HCO_3^-] = x; \qquad [CO_2] = 0.034 - x$$

For the reaction

$$CO_2 + 2H_2O \rightleftharpoons H_3O^+ + HCO_3^-$$

the equilibrium condition is

$$K = \frac{[H_3O^+][HCO_3^-]}{[CO_2]} = \frac{(x)(x)}{0.034 - x} = 4.16 \times 10^{-7}$$

which solves to $x = 1.2 \times 10^{-4}$. Substituting gives us

$$[H_3O^+] = 1.2 \times 10^{-4}\ M; \qquad [HCO_3^-] = 1.2 \times 10^{-4}\ M$$

$$[CO_2] = 0.034\ M$$

Now worry about the second dissociation. Let $y =$ moles per liter of the HCO_3^- that will be dissociated to CO_3^{2-} and more H_3O^+ to add to that from the first step

$$[H_3O^+] = 1.2 \times 10^{-4} + y; \qquad [CO_3^{2-}] = y; \qquad [HCO_3^-] = 1.2 \times 10^{-4} - y$$

$$HCO_3^- + H_2O \;\rightleftharpoons\; H_3O^+ + CO_3^{2-}$$

$$K_{II} = \frac{[H_3O^+][CO_3^{2-}]}{[HCO_3^-]} = 4.84 \times 10^{-11} = \frac{(1.2 \times 10^{-4} + y)(y)}{1.2 \times 10^{-4} - y}$$

With such a very small K, y is bound to be very small. Neglecting y where it adds or subtracts, we get $y = 4.84 \times 10^{-11}$:

$$[H^+] = 1.2 \times 10^{-4}\ M; \qquad [CO_3^-] = 4.84 \times 10^{-11}\ M$$

$$[HCO_3^-] = 1.2 \times 10^{-4}\ M \;\blacksquare$$

12.6 Triprotic acids

A triprotic acid is one that can furnish three protons in neutralization reactions. Again the reactions occur stepwise and the release of the first proton is easier than that of the second, which is easier than release of the third. The general situation can be summarized in terms of a hypothetical triprotic acid H_3X, which can dissociate stepwise as follows:

$$H_3X + H_2O \;\rightleftharpoons\; H_3O^+ + H_2X^-$$

$$H_2X^- + H_2O \;\rightleftharpoons\; H_3O^+ + HX^{2-}$$

$$HX^{2-} + H_2O \;\rightleftharpoons\; H_3O^+ + X^{3-}$$

A solution of H_3X will contain, besides undissociated H_3X, some H_2X^-, some HX^{2-}, some X^{3-}, and some H_3O^+. The H_3O^+ looks as if it is produced by three separate reactions, but, of course, there is only one hydrogen-ion concentration in the solution and it must be in equilibrium with all the other species present. The equilibrium conditions for the reactions above are

$$K_I = \frac{[H_3O^+][H_2X^-]}{[H_3X]}; \qquad K_{II} = \frac{[H_3O^+][HX^{2-}]}{[H_2X^-]}; \qquad K_{III} = \frac{[H_3O^+][X^{3-}]}{[HX^{2-}]}$$

In general, K_I is about 10^5 times as big as K_{II}, and K_{II} is about 10^5 times as big as K_{III}.

Typical triprotic acids are H_3PO_4 and H_3AsO_4, but you must not jump to the obvious conclusion that all acids showing three hydrogen atoms in the formula are triprotic. For example, H_3PO_3 is only diprotic, the third hydrogen being stuck directly on the phosphorus and not neutralizable by regular bases.

■ EXAMPLE 345. A typical solution of H_3PO_4 contains the following species at equilibrium: $0.076\ M\ H_3PO_4$; $0.0239\ M\ H_2PO_4^-$; $6.2 \times 10^{-8}\ M$ HPO_4^{2-}; $3 \times 10^{-18}\ M\ PO_4^{3-}$; $0.0239\ M\ H_3O^+$. Calculate K_I, K_{II}, and K_{III} for this acid.

SOLUTION: The main equilibrium involves

$$H_3PO_4 + H_2O \rightleftharpoons H_3O^+ + H_2PO_4^-$$

The equilibrium condition for it is given by K_I:

$$K_I = \frac{[H_3O^+][H_2PO_4^-]}{[H_3PO_4]} = \frac{(0.0239)(0.0239)}{0.076} = 7.5 \times 10^{-3}$$

where we have simply substituted the given concentrations.

The second step of the dissociation is written

$$H_2PO_4^- + H_2O \rightleftharpoons H_3O^+ + HPO_4^{2-}$$

and has

$$K_{II} = \frac{[H_3O^+][HPO_4^{2-}]}{[H_2PO_4^-]} = \frac{(0.0239)(6.2 \times 10^{-8})}{0.0239} = 6.2 \times 10^{-8}$$

The third step is

$$HPO_4^{2-} + H_2O \rightleftharpoons H_3O^+ + PO_4^{3-}$$

and has

$$K_{III} = \frac{[H_3O^+][PO_4^{3-}]}{[HPO_4^{2-}]} = \frac{(0.0239)(3 \times 10^{-18})}{6.2 \times 10^{-8}} = 1 \times 10^{-12}\ ■$$

In dealing with the dissociation of triprotic acids, the procedure is much like that used for diprotic acids. Since the first dissociation is usually dominant, the second and third can be ignored in the first stage of calculation and then those results can be fed into the calculation of the second and third stages.

■ EXAMPLE 346.   Given a solution that is labeled 0.100 M H_3PO_4; calculate the concentration of H_3O^+, $H_2PO_4^-$, HPO_4^{2-} and PO_4^{3-} in this solution, using $K_I = 7.5 \times 10^{-3}$; $K_{II} = 6.2 \times 10^{-8}$; $K_{III} = 1 \times 10^{-12}$.

SOLUTION: First consider the dissociation of the first proton:

$$H_3PO_4 + H_2O \rightleftharpoons H_3O^+ + H_2PO_4^-$$

Let x = moles per liter of H_3PO_4 that dissociate to establish the equilibrium. This will give us x moles/liter of H_3O^+, x moles/liter of $H_2PO_4^-$, and will leave $(0.100 - x)$ mole/liter of H_3PO_4 in the undissociated state. If we assume that the subsequent steps of ionization will be negligible compared to the first step, then we can write for the equilibrium concentrations:

$$[H_3PO_4] = 0.100 - x; \qquad [H_3O^+] = x; \qquad [H_2PO_4^-] = x$$

The equilibrium condition

$$K_I = \frac{[H_3O^+][H_2PO_4^-]}{[H_3PO_4]} = 7.5 \times 10^{-3} = \frac{(x)(x)}{0.100 - x}$$

Solve this by the quadratic formula or by successive approximation. (First: Neglect the x in the denominator. Solve the rest of the equation to give $x \cong 0.027$. Second: Put $x \cong 0.027$ for x in the denominator. Solve the rest to give $x \cong 0.023$. Third: Put $x \cong 0.023$ for x in the denominator. Solve the rest to get $x \cong 0.024$. Fourth: Put $x \cong 0.024$ for x in the denominator. Solve the rest to give $x = 0.024$. That must be it!)

Substitute the solution $x = 0.024$ in the concentration expressions as set up above:

$$[H_3PO_4] = 0.100 - x = 0.100 - 0.024 = 0.076 \ M$$

$$[H_3O^+] = x = 0.024 \ M$$

$$[H_2PO_4^-] = x = 0.024 \ M$$

Now tackle the second dissociation step:

$$H_2PO_4^- + H_2O \rightleftharpoons H_3O^+ + HPO_4^{2-}$$

It can not go very far because (a) there is not much $H_2PO_4^-$ to start with and (b) K_{II} is considerably smaller than K_I. Let y = moles per liter of $H_2PO_4^-$ that dissociate. This will give y more moles per liter of H_3O^+ to add to the 0.024 M we just calculated. At the same time, we will get y moles/liter of HPO_4^{2-} and reduce the $H_2PO_4^-$ concentration from 0.024 to $0.024 - y$. At equilibrium, then, we would have

$$[H_2PO_4^-] = 0.024 - y; \quad [H_3O^+] = 0.024 + y; \quad [HPO_4^{2-}] = y$$

$$K_{II} = \frac{[H_3O^+]\,[HPO_4^{2-}]}{[H_2PO_4^-]} = 6.2 \times 10^{-8} = \frac{0.024 - y}{(0.024 + y)(y)}$$

Because of the smallness of K_{II}, y must be small and probably negligible when added to or subtracted from 0.024. The equation simplifies to

$$6.2 \times 10^{-8} \cong \frac{(0.024)(y)}{0.024}$$

which solves to $y = 6.2 \times 10^{-8}$. Substituting as above, we get

$$[H_2PO_4^-] = 0.024\ M; \quad [H_3O^+] = 0.024\ M; \quad [HPO_4^{2-}] = 6.2 \times 10^{-8}\ M$$

Finally we consider the third step of the dissociation:

$$HPO_4^{2-} + H_2O \rightleftharpoons H_3O^+ + PO_4^{3-}$$

which must be trivial, for the same reasons outlined above, except it is the only source of PO_4^{3-} concentration for our solution. Let z = moles per liter of HPO_4^{2-} that dissociate.

$$[HPO_4^{2-}] = 6.2 \times 10^{-8} - z; \quad [H_3O^+] = 0.024; \quad [PO_4^{3-}] = z$$

$$K_{III} = \frac{[H_3O^+]\,[PO_4^{3-}]}{[HPO_4^{2-}]} = 1 \times 10^{-12} = \frac{(0.024)(z)}{6.2 \times 10^{-8} - z}$$

Neglecting z in the denominator, we solve for $z = 3 \times 10^{-18}$. Summarizing, we have as the final equilibrium concentrations in the solution

$$[H_3O^+] = 0.024\ M; \quad [H_2PO_4^-] = 0.024\ M$$

$$[HPO_4^{2-}] = 6.2 \times 10^{-8}\ M; \quad [PO_4^{3-}] = 3 \times 10^{-18}\ M \ \blacksquare$$

The solutions above are fairly complicated in that there are three equilibria that need to be satisfied simultaneously. The only reason we could handle the problems at all is that the first step in the dissociation is the dominant equilibrium and can be treated alone. The arithmetic gets considerably more complicated when this is not the case, and we shall take up these more complex situations in Chapter 14, which deals with buffer solutions, and Chapter 16, which deals with complex equilibria. Also, as we shall see in Chapter 13, it becomes necessary to allow for the dissociation of water via the reaction $H_2O + H_2O \rightleftharpoons H_3O^+ + OH^-$ in order to make a full quantitative treatment of these solutions.

The following problems illustrate some of the more complex calculations involving triprotic acid equilibria but still based on dominance of the first step in the dissociation.

■ EXAMPLE 347.   Suppose you make a solution by dissolving 0.100 mole of H_3PO_4 and 0.200 mole of NaH_2PO_4 in enough water to make 1.00 liter of solution. Given $K_I = 7.5 \times 10^{-3}$, $K_{II} = 6.2 \times 10^{-8}$, and $K_{III} = 1 \times 10^{-12}$, calculate the concentrations of H_3PO_4, $H_2PO_4^-$, HPO_4^{2-}, PO_4^{3-}, H_3O^+, and Na^+ in this solution.

SOLUTION: NaH_2PO_4 is a strong electrolyte and is 100% dissociated in solution to give 0.200 mole of Na^+ and 0.200 mole of $H_2PO_4^-$. Let us ignore for the moment the slight amount of this $H_2PO_4^-$ that might further dissociate to H_3O^+ and HPO_4^{2-}. Regarding our solution initially as containing 0.100 mole of H_3PO_4 and 0.200 mole of $H_2PO_4^-$, their concentrations would be

$$\frac{0.100 \text{ mole of } H_3PO_4}{1.00 \text{ liter}} = 0.100 \ M \ H_3PO_4$$

$$\frac{0.200 \text{ mole of } H_2PO_4^-}{1.00 \text{ liter}} = 0.200 \ M \ H_2PO_4^-$$

Some of the H_3PO_4 must be dissociated to give H_3O^+ and $H_2PO_4^-$. Let $x =$ moles per liter of H_3PO_4 that must be dissociated to establish equilibrium. This will reduce the concentration of H_3PO_4 from 0.100 to $0.100 - x$, increase the concentration of $H_2PO_4^-$ from 0.200 to $0.200 + x$, and produce x moles per liter of H_3O^+. When equilibrium is established we shall have

$$[H_3PO_4] = 0.100 - x; \quad [H_2PO_4^-] = 0.200 + x; \quad [H_3O^+] = x$$

The equilibrium is

$$H_3PO_4 + H_2O \quad \rightleftharpoons \quad H_3O^+ + H_2PO_4^-$$

and the condition for equilibrium is that

$$K_I = \frac{[H_3O^+][H_2PO_4^-]}{[H_3PO_4]} = 7.5 \times 10^{-3} = \frac{(x)(0.200 + x)}{0.100 - x}$$

This can be solved by the quadratic formula or by successive approximations. The latter is quicker in this case because only two simple steps are needed. First, assume x is negligible when added to 0.200 or subtracted from 0.100. This gives the approximate relation

$$7.5 \times 10^{-3} \cong \frac{(x)(0.200)}{0.100}$$

which solves to $x = 3.8 \times 10^{-3}$. Then feed this x into the terms $(0.200 + x)$ and $(0.100 - x)$ to get an improved approximate relation

$$7.5 \times 10^{-3} \cong \frac{(x)(0.204)}{0.096}$$

which solves to $x = 3.5 \times 10^{-3}$. This is close enough to the fed-in value to warrant confidence. (If in doubt, try another step in the approximation, feeding $x = 3.5 \times 10^{-3}$ into the $0.200 + x$ and $0.100 - x$. You get no change.) Substituting $x = 3.5 \times 10^{-3}$ into the concentration expressions above, we get

$$[H_3PO_4] = 0.100 - x = 0.100 - 0.0035 = 0.096 \; M$$

$$[H_2PO_4^-] = 0.200 + x = 0.200 + 0.0035 = 0.204 \; M$$

$$[H_3O^+] = x = 3.5 \times 10^{-3} \; M$$

Once you have calculated the principal equilibrium, all the others need to be consistent. If we let $y =$ moles per liter of the $H_2PO_4^-$ that will be dissociated, we have

$$[H_2PO_4^-] = 0.204 - y; \quad [H_3O^+] = 3.5 \times 10^{-3} + y; \quad [HPO_4^{2-}] = y$$

$$H_2PO_4^- + H_2O \; \rightleftharpoons \; H_3O^+ + HPO_4^{2-}$$

$$K_{II} = \frac{[H_3O^+][HPO_4^{2-}]}{[H_2PO_4^-]} = \frac{(3.5 \times 10^{-3} + y)(y)}{0.204 - y} = 6.2 \times 10^{-8}$$

Assuming that y is very small and can be neglected when added to 3.5×10^{-3} or subtracted from 0.304, we get $y = 3.6 \times 10^{-6}$. Substituting above, we get

$$[H_2PO_4^-] = 0.204 \ M; \qquad [H_3O^+] = 3.5 \times 10^{-3} \ M$$

$$[HPO_4^{2-}] = 3.6 \times 10^{-6} \ M$$

Finally, for the third dissociation step, let z = moles per liter of HPO_4^{2-} that break up. This will give

$$[HPO_4^-] = 3.6 \times 10^{-6} - z; \qquad [H_3O^+] = 3.5 \times 10^{-3} + z; \qquad [PO_4^{3-}] = z$$

$$HPO_4^{2-} + H_2O \ \rightleftharpoons \ H_3O^+ + PO_4^{3-}$$

$$K_{III} = \frac{[H_3O^+][PO_4^{3-}]}{[HPO_4^{2-}]} = \frac{(3.5 \times 10^{-3} + z)(z)}{3.6 \times 10^{-6} - z} = 1 \times 10^{-12}$$

$$z = 1 \times 10^{-15}$$

Summarizing our results, we would have in the final solution the following equilibrium concentrations:

$$[H_3PO_4] = 0.096 \ M; \qquad [H_2PO_4^-] = 0.204 \ M$$

$$[HPO_4^{2-}] = 3.6 \times 10^{-6} \ M; \qquad [PO_4^{3-}] = 1 \times 10^{-15} \ M$$

$$[H_3O^+] = 3.5 \times 10^{-3} \ M; \qquad [Na^+] = 0.200 \ M \ \blacksquare$$

Exercises

*12.1. What is the concentration of Ba^{2+} and of NO_3^- in a solution that is made by dissolving 0.0280 mole of $Ba(NO_3)_2$ in enough water to make a liter of solution?

*12.2. What is the concentration of Ba^{2+} and of Cl^- in a solution that is made by dissolving 58.7 g of $BaCl_2 \cdot 2H_2O$ in enough water to make 0.350 liter of solution?

**12.3. What would be the concentration of Na^+ and of SO_4^{2-} in a solution made by adding 1.86 g of Na_2SO_4 to 36.3 ml of 0.133 M Na_2SO_4? Assume no change in volume of solution. (*Answer:* 0.987 M Na^+ and 0.494 M SO_4^{2-}).

*12.4. Assuming complete dissociation, what would be the final concentrations of Ba^{2+} and of OH^- on mixing 25.0 ml of 0.030 M $Ba(OH)_2$ plus 35.0 ml of 0.056 M $Ba(OH)_2$ with enough water to make a total volume of 75.0 ml?

**12.5. How many grams of $Al_2(SO_4)_3$ would you need to add to 0.660 liter of 0.100 M $KAl(SO_4)_2$ to make the SO_4^{2-} concentration equal to 0.300 M? Assume no change in volume of solution and complete dissociation.

12.6. Given 25 ml of 0.10 M K_2SO_4, 50 ml of 0.20 M $Al_2(SO_4)_3$, and 25 ml of H_2O. You wish to make a solution that is simultaneously 0.10 M K^+, 0.10 M Al^{3+}, and 0.20 M SO_4^{2-}. In what ratio should you mix the given reagents? Assume additive volumes and complete dissociation. (*Answer:* 2:1:1.)

12.7. If the dissociation constant of HF is 6.71×10^{-4}, what will be the concentration of H_3O^+ and of F^- in a solution made by dissolving 0.015 mole of HF in enough water to make 0.100 liter of solution?

12.8. If the dissociation constant of HF is 6.71×10^{-4}, what will be the concentration of H_3O^+ and of F^- in a solution made by dissolving 0.015 mole of HF in 0.100 liter of 0.50 M NaF solution?

12.9. If the dissociation constant of HF is 6.71×10^{-4}, how many grams of NaF would you need to dissolve in 0.100 liter of 0.500 M HF to make the H_3O^+ concentration equal to 2.00×10^{-3} M? Assume no change in solution volume. (*Answer:* 0.697 g.)

12.10. Given that the dissociation constant of HF is 6.71×10^{-4}, what will be the concentration of H_3O^+ in a solution made by mixing 25.0 ml of 0.100 M NaF with 25.0 ml of 0.100 M HF? Assume additive volumes.

12.11. Given that the dissociation constant of HF is 6.71×10^{-4}, how much water would you need to add to 50.0 ml of 0.270 M HF to cut the H_3O^+ concentration by a factor of 2? Assume additive volumes.

12.12. Given that $K_{diss} = 6.71 \times 10^{-4}$ for HF, what H_3O^+ concentration would you expect to get on mixing 0.10 mole of HCl and 0.10 mole of KF in enough water to make 0.235 liter of solution? (*Answer:* 0.0166 M.)

12.13. If $K = 1.81 \times 10^{-5}$ for $NH_3 + H_2O \rightleftarrows NH_4^+ + OH^-$, how many grams of NH_3 would you need to add to 10.0 liters of H_2O to make the OH^- concentration be equal to 1.50×10^{-3} M? Assume no change in volume.

12.14. If $K = 1.8 \times 10^{-5}$ for $NH_3 + H_2O \rightleftarrows NH_4^+ + OH^-$, what fraction of the NH_3 is actually present as NH_4^+ in a solution that is 6.0 M NH_3?

***12.15.** Given that $K = 1.81 \times 10^{-5}$ for $NH_3 + H_2O \rightleftarrows NH_4^+ + OH^-$, how much H_2O would you need to add to 50.0 ml of 6.0 M NH_3 to double the per cent conversion of NH_3 into NH_4^+? Assume additive volumes. (*Answer:* 151 ml.)

12.16. Suppose you mix 0.285 mole of NH_4NO_3 and 0.285 mole of KOH in enough water to make 5.25 liters of solution. Given that $K = 1.81 \times 10^{-5}$ for $NH_3 + H_2O \rightleftarrows NH_4^+ + OH^-$, calculate the final concentration of NH_3, NH_4^+, and OH^-.

12.17. Considering only the first dissociation step, $HgCl_2 \rightleftarrows HgCl^+ + Cl^-$, for which $K = 3.3 \times 10^{-7}$, calculate what concentration of $BaCl_2$ (a strong electrolyte for both steps of dissociation) you would need to get the same Cl^- concentration as there is in 0.10 M $HgCl_2$.

12.18. For its dissociation as a weak salt $HgCl_2$ has $K_1 = 3.3 \times 10^{-7}$ and

$K_{II} = 1.8 \times 10^{-7}$. Calculate the concentration of Hg^{2+} in 0.10 M $HgCl_2$. (*Answer:* 1.8×10^{-7} M.)

12.19. Given the diprotic acid H_2X for which $K_I = 4.0 \times 10^{-6}$ and $K_{II} = 6.0 \times 10^{-10}$. What will be the ratio of HX^- to X^{2-} concentrations in 0.10 M H_2X solution?

12.20. Given the diprotic acid H_2Te for which $K_I = 2.3 \times 10^{-3}$ and $K_{II} = 1 \times 10^{-11}$. What concentration of H_2Te would you need to take to ensure that 10.0% of the tellurium atoms are as the species HTe^-?

12.21. How many grams of $KHSO_4$ would you need to add to 440 ml of water to produce a H_3O^+ concentration of 2.20×10^{-3} M? Assume no change in volume. (*Answer:* 0.155 g.)

12.22. If you mix 20 drops of water and 1 drop of concentrated sulfuric acid (18 M H_2SO_4), what will be the H_3O^+ concentration in the resulting solution? Assume additive volumes.

12.23. A solution is made by mixing 10.0 ml of 1.00 M H_2SO_4 and 10.0 ml of 1.00 M $KHSO_4$. Assuming additive volumes, what will be the final concentration of H_3O^+?

***12.24.** A solution is made by mixing 10.0 ml of 1.00 M H_2SO_4 and 30.0 ml of 1.00 M K_2SO_4. Assuming additive volumes, what will be the final concentration of H_3O^+? (*Answer:* 0.0120 M.)

12.25. For the triprotic acid H_3AsO_4 the successive dissociation constants are $K_I = 2.5 \times 10^{-4}$, $K_{II} = 5.6 \times 10^{-8}$, and $K_{III} = 3 \times 10^{-13}$. What concentration of H_3AsO_4 should you take to make $[AsO_4^{3-}] = 4 \times 10^{-18}$ M?

12.26. For the triprotic acid H_3PO_4 the successive dissociation constants are $K_I = 7.5 \times 10^{-3}$, $K_{II} = 6.2 \times 10^{-8}$, $K_{III} = 1 \times 10^{-12}$. What would be the concentration of H_3O^+ and of $H_2PO_4^-$ in a solution made by mixing 30.0 ml of 1.00 M HCl with 10.0 ml of 0.50 M Na_3PO_4? Assume additive volumes.

***12.27.** The Henry's law constant (see Section 10.8) for $CO_2(g)$ in water is 1.08×10^6 (Torr) at 20°C. Suppose you want to make an aqueous CO_2 solution in which 99.0% of the dissolved carbon dioxide is as $[CO_2]$ and 1.0% is as $[HCO_3^-]$. Given that the first and second dissociation constants of aqueous CO_2 are 4.16×10^{-7} and 4.84×10^{-11}, respectively, calculate what pressure of CO_2 gas above the solution you would need to achieve the desired condition. (*Answer:* 79 Torr.)

***12.28.** Generally there is a 10^4 or 10^5 factor between the successive dissociation constants of an acid, but this rule breaks down when successive protons come off different places in a molecular chain. This happens, for example, with pyrophosphoric acid, $H_4P_2O_7$, a tetraprotic acid having $K_I = 1.4 \times 10^{-1}$, $K_{II} = 1.1 \times 10^{-2}$, $K_{III} = 2.9 \times 10^{-7}$, and $K_{IV} = 3.6 \times 10^{-9}$. Given a solution labeled 0.250 M $H_4P_2O_7$, calculate the equilibrium concentrations of H_3O^+, $H_4P_2O_7$, $H_3P_2O_7^-$, $H_2P_2O_7^{2-}$, $HP_2O_7^{3-}$, and $P_2O_7^{4-}$. (*Answer:* $[H_3O^+] = 0.140$ M.)

13

pH

WHEN A BASE, such as 0.1 M NaOH, is added gradually to an acid, such as 0.1 M HCl, the hydrogen-ion (H^+) concentration, or, more precisely, the hydronium-ion (H_3O^+) concentration, goes through a change involving 12 powers of 10, or 12 orders of magnitude. In order to describe such a change, especially graphically, we take refuge, as one normally does in such cases, in representation by logarithms. In fact, the pH function is nothing more than a logarithmic way of representing the hydronium-ion concentration. For this purpose, it would be highly sensible to review briefly the discussion on logarithms given in Section 1.3, since there is the heart of the whole pH problem.

13.1 Definition of pH

It is customary to define the pH as the "negative of the logarithm of the hydronium-ion concentration." Another way of saying the same thing is the "logarithm of the reciprocal of the hydronium-ion concentration." Symbolically, these can be represented as follows:

$$pH = -\log C_{H_3O^+} = -\log [H_3O^+]$$

or

$$pH = \log \frac{1}{C_{H_3O^+}} = \log \frac{1}{[H_3O^+]}$$

where we have shown both $C_{H_3O^+}$ and $[H_3O^+]$ as ways of representing concentration of hydronium ion. The usual units for $C_{H_3O^+}$ are moles per liter, or molarity. (This is one place where some trouble creeps in. Logarithms of $C_{H_3O^+}$ do not have units, and, in fact, pH does not have units. There is no way at the freshman level that we can completely resolve this difficulty except to refer to the discussion on "activity," given in Section 11.1. In the strict definition, pH is the negative logarithm of the hydronium-ion *activity*. Activity is a unitless quantity that we can regard as the actual concentration divided by unit concentration of a reference state. If it strains your mind, do not worry too much about this elegance. If you simply define pH = −log [H_3O^+] and do not worry about the units, you will be doing what most chemists do in practice.)

■ EXAMPLE 348. Given that the hydronium-ion concentration of a particular solution is 3.0×10^{-2} *M*, what is its pH?

SOLUTION: pH = −log [H_3O^+] = −log (3.0×10^{-2}).
Recall that the log of a product is the sum of the logs.

log (3.0×10^{-2}) = log (3.0) + log (10^{-2}).

Recall also that the log of (10 raised to a power) is just equal to that power. For the log of 3.0 we look either in the table (Appendix A) or get it off the slide rule, Section 1.4j. We find log 3.0 = 0.48.

log (3.0) + log (10^{-2}) = 0.48 + (−2) = 0.48 − 2 = −1.52
pH = −(−1.52) = +1.52 ■

■ EXAMPLE 349. Given a solution with a hydronium-ion concentration of 4.8×10^{-6} mole per liter, what is its pH?

SOLUTION: pH = −log (4.8×10^{-6}) = −[0.68 + (−6)] = −[− 5.32] = +5.32. ■

There is always some confusion as to what is the proper number of significant figures when working with logarithms. The simple rule to follow is that the logarithm gets looked up to as many significant digits as are in the *coefficient* of the exponential number. For example, in Example 349, there are two significant figures in the number 4.8×10^{-6}, the two decided by the coefficient 4.8. The power 10^{-6} has nothing to do with significant figures; it just indicates where the decimal point goes. So, we look up the log of 4.8 and we find 0.6812. We take only two digits 0.68 and add them to −6. The −6 is an exact number, so we end up with −5.32. We end up with three digits, but only the two to the right of the decimal count as significant figures. This kind of business is special for the case of logarithms and should not be confused with the ordinary counting of significant figures.

■ EXAMPLE 350. You have a solution for which the hydronium-ion concentration is 4.8×10^{-13} mole per liter. What is its pH?

SOLUTION:

$$pH = -\log [H_3O^+] = -\log (4.8 \times 10^{-13})$$
$$= -(\log 4.8 + \log 10^{-13})$$
$$= -(0.68 - 13) = 12.32 \ \blacksquare$$

Many students have trouble in making the reverse calculation of going from a given pH to the hydronium-ion concentration. For this computation, take a look again at Section 1.3, especially Examples 37–39.

■ EXAMPLE 351.   You are told that a solution has a pH of 4.00. What is its hydronium-ion concentration?

SOLUTION:

$$pH = -\log [H_3O^+] = 4.00$$
$$\log [H_3O^+] = -4.00$$
$$[H_3O^+] = 10^{-4.00} = (10^{-4})(10^{0.00})$$

The number whose log is 0.00 is 1.0:

$$[H_3O^+] = 1.0 \times 10^{-4} \ M \ \blacksquare$$

■ EXAMPLE 352.   A typical blood sample shows a pH of 7.4. What is its hydronium-ion concentration?

SOLUTION:

$$pH = -\log [H_3O^+] = 7.4$$
$$\log [H_3O^+] = -7.4 = -8 + 0.6$$

Note how you take the pH number before the decimal (the -7 of -7.4) and carry it to the next higher digit (-8) so the fractional part after the decimal (the -0.4 of -7.4) can be converted to a positive number ($+0.6$):

$$[H_3O^+] = 10^{-8} \times 10^{0.6}$$

The antilog of 0.6 is 4:

$$[H_3O^+] = 4 \times 10^{-8} \ M \ \blacksquare$$

There is no theoretical limit as to what pH values are allowed, although some people have the notion that negative pH values are automatically wrong. As the following problem shows, there is no a priori reason for rejecting a negative pH value.

■ EXAMPLE 353.   A given solution has a hydronium-ion concentration of 2.89 M. What is its pH?

SOLUTION: pH $= -\log [H_3O^+] = -\log (2.89) = -0.461.$ ■

13.2 The water equilibrium

Water is a weak electrolyte that is dissociated into hydronium ion, H_3O^+, and hydroxide ion, OH^-. The equilibrium reaction can be written

$$H_2O + H_2O \rightleftharpoons H_3O^+ + OH^-$$

and the condition for equilibrium is given by

$$K_w = [H_3O^+][OH^-] = 1.00 \times 10^{-14}$$

K_w is called the dissociation constant, or ion product, of water. Since in all dilute aqueous solutions the H_2O stays at unit activity, its concentration does not appear in the condition for equilibrium.

The importance of the ion product of water comes from the fact that once the H_3O^+ concentration of a solution is specified, the OH^- can be calculated. Conversely, once you know the OH^- concentration, you can calculate the H_3O^+.

■ EXAMPLE 354.   What is the pH of pure water?

SOLUTION: Let $x =$ moles per liter of H_2O that transfers protons to H_2O. This will give x moles/liter of H_3O^+ and x moles/liter of OH^- via the reaction $H_2O + H_2O \rightarrow H_3O^+ + OH^-$. The condition for equilibrium is the following:

$$K_w = [H_3O^+][OH^-] = 1.00 \times 10^{-14}$$
$$(x)(x) = 1.00 \times 10^{-14}$$
$$x = 1.00 \times 10^{-7}$$
$$pH = -\log [H_3O^+] = -\log (1.00 \times 10^{-7})$$
$$= -(\log 1.00 + \log 10^{-7})$$
$$= -(0.000 - 7) = 7.000$$ ■

■ EXAMPLE 355. Given a solution in which the concentration of OH⁻ is $2.50 \times 10^{-3}\ M$, what is the pH of the solution?

SOLUTION: $K_w = [H_3O^+][OH^-] = 1.00 \times 10^{-14}$

Substitute $[OH^-] = 2.50 \times 10^{-3}$.

Get

$$[H_3O^+] = \frac{1.00 \times 10^{-14}}{2.50 \times 10^{-3}} = 0.400 \times 10^{-11} = 4.00 \times 10^{-12}$$

$$pH = -\log[H_3O^+] = -\log(4.00 \times 10^{-12}) = -(0.602 - 12)$$
$$= 11.398 \ ■$$

■ EXAMPLE 356. A solution has a pH of 12.68. What is its hydroxide-ion concentration?

SOLUTION:

$$pH = -\log[H_3O^+] = 12.68$$
$$\log[H_3O^+] = -12.68 = -13 + 0.32$$
$$[H_3O^+] = (10^{-13})(10^{0.32}) = (10^{-13})(2.1)$$

Recall that the number whose log is -13 is just 10^{-13} and that the number whose log is 0.32 can be written as $10^{0.32}$. From the tables, the antilog of 0.32 is found to be 2.1. The condition for equilibrium is

$$K_w = [H_3O^+][OH^-] = 1.00 \times 10^{-14}$$

If $[H_3O^+] = 2.1 \times 10^{-13}$ as found above, then

$$[OH^-] = \frac{K_w}{[H_3O^+]} = \frac{1.00 \times 10^{-14}}{2.1 \times 10^{-13}} = 4.8 \times 10^{-2}\ M \ ■$$

13.3 Solutions of acids

One way of defining an acid is to say it is a substance that raises the hydronium-ion concentration beyond that of pure water. This means $[H_3O^+]$ has to go greater than $10^{-7}\ M$; this means the pH has to fall below 7. With strong acids, we have no problem. We simply count the concentration of H_3O^+ added. With weak acids, there may be trouble because the $H_2O + H_2O \rightleftharpoons H_3O^+ + OH^-$ equilibrium needs to be solved simultaneously with the $HX + H_2O \rightleftarrows H_3O^+ + X^-$ equilibrium. The exact solution to this problem we shall postpone until Chapter 16. For the present, we just point out that if K_{diss} of the weak acid HX is appreciably greater than K_w and if the concentration of HX is not too small, then we shall be able to get away with the approximation that the HX equilibrium dominates the situation and effectively controls the $H_2O + H_2O \rightleftarrows H_3O^+ + OH^-$ equilibrium.

■ EXAMPLE 357.   Given a solution that is 0.235 M HCl. What is its pH?

SOLUTION: HCl is a strong electrolyte. 0.235 M HCl is completely dissociated to give 0.235 M H_3O^+.

$$pH = -\log [H_3O^+] = -\log (0.235) = -\log (2.35 \times 10^{-1})$$
$$= -(0.371 - 1) = 0.629 \ ■$$

■ EXAMPLE 358.   You have a solution that is 0.150 M HNO_2. Given that K_{diss} is 4.5×10^{-4}, calculate the pH of the solution.

SOLUTION: (This is a weak-acid problem. Refer to Section 12.2 for the general discussion.)

$$HNO_2 + H_2O \ \rightleftarrows \ H_3O^+ + NO_2^-$$

Let $x =$ moles per liter of HNO_2 that dissociate. At equilibrium: $[HNO_2] = 0.150 - x$; $[H_3O^+] = x$; $[NO_2^-] = x$. Note that we implicitly neglect any H_3O^+ contribution from the water dissociation.

$$K_{diss} = \frac{[H_3O^+][NO_2^-]}{[HNO_2]} = 4.5 \times 10^{-4} = \frac{(x)(x)}{0.150 - x}$$
$$x = 8.0 \times 10^{-3} = [H_3O^+]$$
$$pH = -\log [H_3O^+] = -\log (8.0 \times 10^{-3}) = -(0.90 - 3) = 2.10 \ ■$$

■ EXAMPLE 359.   You are told that the K_{diss} of HNO_2 is 4.5×10^{-4}. You want a solution of HNO_2 that has a pH of 2.50. What concentration of HNO_2 would you need to take?

SOLUTION:

$$pH = 2.50 = -\log [H_3O^+]$$
$$\log (H_3O^+) = -2.50 = -3 + 0.50$$
$$[H_3O^+] = 10^{-3} \times 10^{0.50} = 3.2 \times 10^{-3} \ M$$

You want this $[H_3O^+]$ to come from the dissociation of HNO_2.

$$HNO_2 + H_2O \rightleftharpoons H_3O^+ + NO_2^-$$

But each H_3O^+ formed must be accompanied by formation of NO_2^-. Hence, we know that the solution must also contain $[NO_2^-] = 3.2 \times 10^{-3}\ M$. We substitute these known values of the concentration in the expression for K_{diss}

$$K_{diss} = \frac{[H_3O^+][NO_2^-]}{[HNO_2]} = 4.5 \times 10^{-4} = \frac{(3.2 \times 10^{-3})(3.2 \times 10^{-3})}{[HNO_2]}$$

$$[HNO_2] = \frac{(3.2 \times 10^{-3})(3.2 \times 10^{-3})}{4.5 \times 10^{-4}} = 2.3 \times 10^{-2}\ M$$

We need to furnish 2.3×10^{-2} mole/liter of HNO_2 to build up this concentration of HNO_2 plus 3.2×10^{-3} mole/liter of HNO_2, which will dissociate to give $3.2 \times 10^{-3}\ M\ H_3O^+$ and $3.2 \times 10^{-3}\ M\ NO_2^-$. Total concentration of HNO_2 needs to be $2.3 \times 10^{-2} + 3.2 \times 10^{-3} = 2.6 \times 10^{-2}\ M$. So, the solution needs to be $0.026\ M\ HNO_2$. ∎

■ EXAMPLE 360. What will be the pH of $0.216\ M\ H_3PO_4$ solution? For this acid $K_I = 7.5 \times 10^{-3}$, $K_{II} = 6.2 \times 10^{-8}$, and $K_{III} = 1 \times 10^{-12}$.

SOLUTION: First calculate the hydrogen-ion concentration of the solution.

$$H_3PO_4 + H_2O \rightleftharpoons H_3O^+ + H_2PO_4^-$$

Let x = moles per liter of H_3PO_4 that dissociate.

$$[H_3PO_4] = 0.216 - x; \quad [H_3O^+] = x; \quad [H_2PO_4^-] = x$$

$$K_I = \frac{[H_3O^+][H_2PO_4^-]}{[H_3PO_4]} = \frac{(x)(x)}{0.216 - x} = 7.5 \times 10^{-3}$$

By successive approximations, $x = 3.7 \times 10^{-2}$.

As discussed previously (see Section 12.6), the hydronium-ion contribution from the second and third steps of the dissociation is negligible. So, we take $[H_3O^+] = x = 3.7 \times 10^{-2}\ M$.

$$pH = -\log [H_3O^+] = -\log (3.7 \times 10^{-2}) = -(0.57 - 2) = 1.43 \quad ∎$$

■ EXAMPLE 361. What would be the pH of $0.360\ M\ H_2SO_4$? The K_{II} for sulfuric acid is 1.26×10^{-2}.

SOLUTION: (Compare Example 341): Recall that H_2SO_4 is a strong electrolyte for the first dissociation. So, 0.360 M H_2SO_4 is 100% dissociated to 0.360 M H_3O^+ and 0.360 M HSO_4^-. Now let x moles/liter of HSO_4^- dissociate to produce x moles/liter more of H_3O^+. At equilibrium:

$$[HSO_4^-] = 0.360 - x; \quad [H_3O^+] = 0.360 + x; \quad [SO_4^{2-}] = x$$
$$HSO_4^- + H_2O \rightleftharpoons H_3O^+ + SO_4^{2-}$$
$$K_{II} = \frac{[H_3O^+][SO_4^{2-}]}{[HSO_4^-]} = 1.26 \times 10^{-2} = \frac{(0.360 + x)(x)}{0.360 - x}$$

By the quadratic formula, $x = 0.0118$. Substitute this in the above expression to get $[H_3O^+] = 0.360 + x = 0.372\ M$.

$$pH = -\log [H_3O^+] = -\log (0.372) = -\log (3.72 \times 10^{-1})$$
$$= -(0.571 - 1) = 0.429 \blacksquare$$

■ EXAMPLE 362. Suppose you wanted a sulfuric acid solution that had a pH of 1.000. Given that K_{II} of H_2SO_4 is 1.26×10^{-2}, what molarity H_2SO_4 solution would you take?

SOLUTION:

$$pH = 1.000 = -\log [H_3O^+]$$
$$\log [H_3O^+] = -1.000 = -1 + 0.000$$
$$[H_3O^+] = (10^{-1})(1.00) = 1.00 \times 10^{-1}$$

The principal equilibrium in the solution is

$$HSO_4^- + H_2O \rightleftharpoons H_3O^+ + SO_4^{2-}$$

for which

$$K_{II} = \frac{[H_3O^+][SO_4^{2-}]}{[HSO_4^-]} = 1.26 \times 10^{-2}$$

But we know that $[H_3O^+]$ must be $1.00 \times 10^{-1}\ M$, so we substitute

$$[H_3O^+] = 1.00 \times 10^{-1}$$
$$\frac{(1.00 \times 10^{-1})[SO_4^{2-}]}{[HSO_4^-]} = 1.26 \times 10^{-2}$$

which tells us that in this solution

$$\frac{[SO_4^{2-}]}{[HSO_4^-]} = \frac{1.26 \times 10^{-2}}{1.00 \times 10^{-1}} = 1.26 \times 10^{-1} = 0.126$$

But we end up with two unknowns, $[SO_4^{2-}]$ and $[HSO_4^-]$, and only one equation. Can we find another equation? The solution must be electrically neutral. This means that the concentration of positive charge (i.e., H_3O^+) must equal the total concentration of negative charge (i.e., from SO_4^{2-} and HSO_4^-). The only point to be careful about is to make sure we allow for the fact that SO_4^{2-} is doubly charged whereas HSO_4^- is only singly charged. We get finally as the electrical neutrality condition

$$[H_3O^+] = 2[SO_4^{2-}] + [HSO_4^-]$$

where the concentration of sulfate is counted twice as heavily as the concentration of HSO_4^-. But we know that in this solution $[H_3O^+] = 1.00 \times 10^{-1}\ M$, or $0.100\ M$, so we can write

$$0.100 = 2[SO_4^{2-}] + [HSO_4^-] \quad \text{and} \quad [SO_4^{2-}] = 0.126[HSO_4^-]$$

which gives us two simultaneous equations to solve. By substituting the second equation into the first, we get

$$0.100 = (2)(0.126[HSO_4^-]) + [HSO_4^-] = 1.252[HSO_4^-]$$

This solves to $[HSO_4^-] = 0.0799\ M$ and gives $[SO_4^{2-}] = 0.0101\ M$.
The solute H_2SO_4 must show up either as HSO_4^- or SO_4^{2-}, so we can calculate the total moles of solute needed simply as the sum of the HSO_4^- and SO_4^{2-}. This gives $0.0799 + 0.0101 = 0.0900$ mole/liter. Thus, to get pH $= 1.000$, we need $0.900\ M$ H_2SO_4. ∎

13.4 Solutions of bases

The same remarks prefacing Section 13.3 on solutions of acids can equally well be made for solutions of bases; that is, strong bases can be taken care of by counting up the concentration of OH^- added to the solution; weak bases can be considered as dominating over the $H_2O + H_2O \rightleftarrows H_3O^+ + OH^-$ equilibrium, so long as K_{diss} of the base is bigger than K_w of the water and so long as the concentration of base is not too small. The only extra step we shall need to get the pH of basic solutions is to go from $[OH^-]$ to $[H_3O^+]$ via $K_w = [H_3O^+][OH^-]$.

■ EXAMPLE 363. What is the pH of $0.150\ M$ NaOH solution?

SOLUTION: NaOH is a strong electrolyte and is assumed to be 100% dissociated. This means that 0.150 M NaOH is taken to be 100% dissociated into the form 0.150 M Na$^+$ and 0.150 M OH$^-$. Once we know the concentration of OH$^-$, we can calculate the hydronium-ion concentration from $K_w = [H_3O^+][OH^-] = 1.00 \times 10^{-14}$:

$$[H_3O^+] = \frac{1.00 \times 10^{-14}}{[OH^-]} = \frac{1.00 \times 10^{-14}}{0.150} = 6.67 \times 10^{-14} \, M$$

$$pH = -\log [H_3O^+] = -\log (6.67 \times 10^{-14}) = -(0.824 - 14)$$
$$= 13.176 \ \blacksquare$$

■ EXAMPLE 364.   What would be the pH of 3.86×10^{-3} M Ba(OH)$_2$ solution? Assume complete dissociation.

SOLUTION: Ba(OH)$_2$ is a strong electrolyte and is 100% dissociated into Ba^{2+} and 2OH$^-$. From 3.86×10^{-3} M Ba(OH)$_2$, complete dissociation would give 3.86×10^{-3} M Ba^{2+} and $(2)(3.86 \times 10^{-3})$, or 7.72×10^{-3}, M OH$^-$.

$$[H_3O^+] = \frac{K_w}{[OH^-]} = \frac{1.00 \times 10^{-14}}{7.22 \times 10^{-3}} = 1.38 \times 10^{-12} \, M$$

$$pH = -\log [H_3O^+] = -\log (1.38 \times 10^{-12})$$
$$= -(0.140 - 12) = 11.860 \ \blacksquare$$

■ EXAMPLE 365.   Aqueous solutions of NH$_3$ are slightly basic because of the equilibrium NH$_3$ + H$_2$O $\rightleftarrows$ NH$_4^+$ + OH$^-$, for which $K = 1.81 \times 10^{-5}$. What will be the pH of 0.55 M NH$_3$ solution?

SOLUTION: (First calculate the [OH$^-$] as a problem in weak base dissociation. Then use the result to compute the [H$_3$O$^+$].)
Let x = moles per liter of NH$_3$ that convert to NH$_4^+$. At equilibrium, this will leave $(0.55 - x)$ mole/liter of NH$_3$ unchanged and will have produced x moles/liter of NH$_4^+$ and x moles/liter of OH$^-$.

$$[NH_3] = 0.55 - x; \quad [NH_4^+] = x; \quad [OH^-] = x$$

$$NH_3 + H_2O \quad \rightleftarrows \quad NH_4^+ + OH^-$$

$$K = \frac{[NH_4^+][OH^-]}{[NH_3]} = 1.81 \times 10^{-5} = \frac{(x)(x)}{0.55 - x}$$

$$x = 3.2 \times 10^{-3} = [OH^-]$$

$$[H_3O^+] = \frac{K_w}{[OH^-]} = \frac{1.00 \times 10^{-14}}{3.2 \times 10^{-3}} = 3.1 \times 10^{-12}$$

$$pH = -\log(H_3O^+) = -\log(3.1 \times 10^{-12}) = -(0.49 - 12) = 11.51 \blacksquare$$

■ **EXAMPLE 366.** You would like to have an aqueous ammonia solution with a pH of 11.111. Given that $K = 1.81 \times 10^{-5}$ for $NH_3 + H_2O \rightleftarrows NH_4^+ + OH^-$, what molarity NH_3 solution would you take?

SOLUTION: You want the pH to be 11.111.

$$pH = 11.111 = -\log[H_3O^+]$$
$$\log[H_3O^+] = -11.111 = -12 + 0.889$$
$$[H_3O^+] = 7.74 \times 10^{-12}$$

Use $K_w = [H_3O^+][OH^-]$ to find out what $[OH^-]$ this corresponds to.

$$[OH^-] = \frac{K_w}{[H_3O^+]} = \frac{1.00 \times 10^{-14}}{7.74 \times 10^{-12}} = 1.29 \times 10^{-3}\ M$$

According to the reaction $NH_3 + H_2O \rightarrow NH_4^+ + OH^-$, there is produced one NH_4^+ for each OH^-. Therefore, the concentration of NH_4^+ in the solution must also be $1.29 \times 10^{-3}\ M$. We can use K for the NH_3 "dissociation" to find the actual concentration of NH_3 needed for equilibrium with the foregoing values of NH_4^+ and OH^-.

$$NH_3 + H_2O \rightleftharpoons NH_4^+ + OH^-$$

$$K = \frac{[NH_4^+][OH^-]}{[NH_3]} = 1.81 \times 10^{-5} = \frac{(1.29 \times 10^{-3})(1.29 \times 10^{-3})}{[NH_3]}$$

$$[NH_3] = \frac{(1.29 \times 10^{-3})^2}{1.81 \times 10^{-5}} = 0.0919\ M$$

The molarity of the NH_3 solution to be prepared needs to be great enough to supply the 0.0919 M of free NH_3 plus $1.29 \times 10^{-3}\ M$ to take care of that part that converts to NH_4^+. So, total NH_3 needed $= 0.0919 + 1.29 \times 10^{-3} = 0.0932\ M. \blacksquare$

■ **EXAMPLE 367.** Suppose you dissolve 1.00 g of NH_3 and 1.00 g of NH_4NO_3 in enough water to make 0.250 liter of solution. Given that $K = 1.81 \times 10^{-5}$ for $NH_3 + H_2O \rightleftarrows NH_4^+ + OH^-$, what pH would you get for this solution?

SOLUTION: (The only new point here is to note that the solution already has NH_4^+ from the NH_4NO_3 and this must be allowed for when calculating the NH_3 dissociation.)

NH_4NO_3 is a strong electrolyte and is 100% dissociated into NH_4^+ and NO_3^-.

$$1.00 \text{ g of } NH_4NO_3 = \frac{1.00 \text{ g}}{80.04 \text{ g/mole}} = 0.0125 \text{ mole of } NH_4NO_3$$

which gives 0.0125 mole of NH_4^+. In a volume of 0.250 liter, this gives an ammonium-ion concentration of 0.0125 mole/0.250 liter = 0.0500 M.

$$1.00 \text{ g of } NH_3 = \frac{1.00 \text{ g}}{17.03 \text{ g/mole}} = 0.0587 \text{ mole of } NH_3$$

In a volume of 0.250 liter, this gives 0.0587 mole/0.250 liter = 0.235 M NH_3.

Let x = moles per liter of the NH_3 that react $NH_3 + H_2O \rightarrow NH_4^+ + OH^-$. This will give at equilibrium

$$[NH_3] = 0.235 - x; \quad [NH_4^+] = 0.0500 + x; \quad [OH^-] = x$$
$$NH_3 + H_2O \rightleftharpoons NH_4^+ + OH^-$$

$$K = \frac{[NH_4^+][OH^-]}{[NH_3]} = 1.81 \times 10^{-5} = \frac{(0.0500 + x)(x)}{0.235 - x}$$

Solve, using the very good approximation that x is small enough to be neglected when added to 0.0500 or subtracted from 0.235. The result is $x = 8.51 \times 10^{-5} = [OH^-]$.

$$[H_3O^+] = \frac{K_w}{[OH^-]} = \frac{1.00 \times 10^{-14}}{8.51 \times 10^{-5}} = 1.18 \times 10^{-10} \text{ M}$$

$$pH = -\log [H_3O^+] = -\log (1.18 \times 10^{-10}) = -(0.072 - 10) = 9.928 \blacksquare$$

13.5 Solutions of salts

The influence of an added salt on the pH of water depends on what effect the added salt has on the water dissociation equilibrium, $H_2O + H_2O \rightleftharpoons H_3O^+ + OH^-$. In general, cations that come from strong bases (e.g., Na^+, K^+, Ba^{2+}) will not affect the water equilibrium, nor will anions that come from strong acids (e.g., ClO_4^-, Cl^-, NO_3^-). The reason for this is that such cations have no tendency to associate with OH^- and such anions have no tendency to associate with H_3O^+. Hence, the water equilibrium in such salt solution is considered to stay unchanged from what it is in pure water.

■ EXAMPLE 368.   What is the pH of 0.10 M KNO$_3$ solution?

SOLUTION: The equilibrium is $H_2O + H_2O \rightleftarrows H_3O^+ + OH^-$.
Let $x =$ moles per liter of H_2O that dissociate. Then $x = [H_3O^+] = [OH^-]$.

$K_w = [H_3O^+][OH^-] = 1.00 \times 10^{-14} = (x)(x)$

$x = 1.00 \times 10^{-7}$

$pH = -\log[H_3O^+] = -\log(1.00 \times 10^{-7}) = -(0.000 - 7) = 7.000$ ■

Cations of weak bases (e.g., NH_4^+) *do* influence the $H_2O + H_2O \rightleftarrows H_3O^+ + OH^-$ by tying up some of the OH^- ion. Similarly, anions of weak acids (e.g., CO_3^{2-}) perturb the equilibrium $H_2O + H_2O \rightleftarrows H_3O^+ + OH^-$ by tying up some H_3O^+. In the first case, the pH of the solution drops below 7; in the second case, the pH rises above 7. Salts that produce such action are said to undergo hydrolysis. The computation of pH in hydrolysis is taken up in detail in Chapter 16.

A special kind of situation occurs with certain salts in which the anion can act as an acid. This occurs, for example, with $NaHSO_4$, where HSO_4^- is a moderately strong acid ($K_{diss} = 1.26 \times 10^{-2}$). We are carefully avoiding at this time such cases as $NaHCO_3$, where HCO_3^- can also act as an acid but where the situation is extensively complicated by its simultaneous ability to act as a base. $NaHSO_4$ does not have this complication, since HSO_4^- has no tendency to react with H_3O^+ to form H_2SO_4.

■ EXAMPLE 369.   What is the pH of 0.168 M NaHSO$_4$ solution, given that K_{diss} is 1.26×10^{-2}?

SOLUTION: $NaHSO_4$ is a strong electrolyte and is taken to be 100% dissociated to Na^+ and HSO_4^-. Therefore, 0.168 M NaHSO$_4$ can be considered 0.168 M Na$^+$ and 0.168 M HSO$_4^-$.
Suppose we let $x =$ moles per liter of HSO_4^- that are dissociated.

$[HSO_4^-] = 0.168 - x;$ $[H_3O^+] = x;$ $[SO_4^{2-}] = x$

$HSO_4^- + H_2O \rightleftarrows H_3O^+ + SO_4^{2-}$

$K = \dfrac{[H_3O^+][SO_4^{2-}]}{[HSO_4^-]} = 1.26 \times 10^{-2} = \dfrac{(x)(x)}{0.168 - x}$

This solves by the quadratic formula to give $x = 0.0401 = [H_3O^+]$.

$pH = -\log[H_3O^+] = -\log(0.0401) = -\log(0.0401)$
$= -\log(4.01 \times 10^{-2}) = -(0.603 - 2) = +1.397$ ■

Exercises

*13.1. Given a solution with a hydronium-ion concentration of 1.35×10^{-3} M, what is its pH?

*13.2. Given a solution with a hydronium-ion concentration of 1.35 M, what is its pH?

*13.3. Given a solution of pH = 12.89, what is its hydronium-ion concentration? (*Answer*: 1.3×10^{-13} M).

*13.4. What would be the hydronium-ion concentration corresponding to a pH of -1.09?

*13.5. If you are told a certain solution has a pH of zero, what must be its concentration of hydronium ion?

*13.6. The dissociation constant of water K_w changes from 1.00×10^{-14} at 25°C to a value of 5.47×10^{-14} at 50°C. What would be the pH of pure water at 50°C? (*Answer*: 6.631.)

*13.7. What is the pH of a solution having a hydroxide-ion concentration of 5.47×10^{-7} M?

*13.8. The pH of gooseberries is 2.9. To what hydroxide-ion concentration does this correspond?

*13.9. What pH would you get by dissolving 3.58 g of HCl in enough water to make 0.85 liter of solution? (*Answer*: 0.94.)

*13.10. What pH would you get by dissolving 3.58 g of NaOH in enough water to make 0.85 liter of solution?

*13.11. Assuming complete dissociation, how many grams of $Ba(OH)_2$ would you need to dissolve in 0.850 liter of solution to get a pH of 12.00?

**13.12. Suppose you mix equal volumes of HCl solutions that have pH's of 1.00 and 2.00, respectively. What will be the pH of the resulting solution? (*Answer*: 1.26.)

**13.13. Given that the dissociation constant of HF is 6.71×10^{-4}, what would be the pH of a solution made by dissolving 3.58 g of HF in enough water to make 0.850 liter of solution?

**13.14. Given that the dissociation constant of HF is 6.71×10^{-4}, how many grams of HF would you need to dissolve in 0.850 liter of solution to get a pH of 2.22?

**13.15. Given that the dissociation constant of HF is 6.71×10^{-4} what would be the pH of a solution made by mixing equal volumes of HF solutions that have pH's of 2.00 and 3.00 respectively? (*Answer*: 2.155.)

**13.16. Given that $K = 1.81 \times 10^{-5}$ for $NH_3 + H_2O \rightleftarrows NH_4^+ + OH^-$, what would be the pH of 0.50 M NH_3 solution?

**13.17. How many grams of NH_3 would you need to add to a liter of water to raise the pH from 7.00 to 10.00? Assume no change in volume.

**13.18. Given that K_{II} of H_2SO_4 is 1.26×10^{-2}, what is the pH of 0.100 N H_2SO_4 solution? (*Answer*: 1.23.)

***13.19. How many milliliters of concentrated sulfuric acid (18 M) would you need to add to a liter of water to make the pH = 3.00? K_{II} of H_2SO_4 is 1.26×10^{-2}.

***13.20. Given a liter of sulfuric acid solution that has a pH of 1.00, how much water do you need to add to bring the pH to 2.00? K_{II} of H_2SO_4 is 1.26×10^{-2}.

**13.21. Given that $K_{II} = 1.26 \times 10^{-2}$ for H_2SO_4, what molarity $NaHSO_4$ solution would you need to take to get a pH of 1.26? (*Answer*: 0.294 M.)

***13.22. Given that $K_{II} = 1.26 \times 10^{-2}$ for H_2SO_4, what would be the pH of the final solution made by mixing 10.0 ml of H_2SO_4 solution with pH = 1.500 and 20.0 ml of $NaHSO_4$ solution with pH = 1.500? Assume additive volumes.

*** 13.23. Given that $K_{II} = 1.26 \times 10^{-2}$ for H_2SO_4, what would be the pH of the final solution made by mixing 10.0 ml of 0.10 M HCl with 20.0 ml of 0.20 M Na_2SO_4? Assume additive volumes.

***13.24. How many milliliters of 0.10 M Na_2SO_4 do you need to add to 10.0 ml of 0.10 M HCl to change the pH from 1.00 to 1.50? Assume additive volumes. $K_{II} = 1.26 \times 10^{-2}$ for H_2SO_4. (*Answer*: 6.6 ml.)

14

BUFFER SOLUTIONS

A BUFFER SOLUTION is one to which either acid or base can be added without drastically changing the pH. In general, a buffer solution consists of a weak acid plus one of its salts, a weak base plus one of its salts, or a combination of weak acids and bases. The efficiency of the buffering action against added acid or base depends on the concentrations of the ingredients and is most effective when they are present in roughly equal amounts. Most of the discussion and computations that follow apply to buffers consisting of an acid HX and a salt NaX, since this is the type commonly encountered in freshman chemistry and qualitative analysis. However, we shall also do some computations for the NH_3–NH_4^+ system, an important and practical buffer in the basic region. Some of the more complicated aspects of buffer calculations are postponed until Chapter 16.

14.1 The acetic acid-sodium acetate buffer

Acetic acid is one of the common, easy-to-work-with weak acids with a dissociation constant of 1.8×10^{-5} for the reaction

$$HOAc + H_2O \rightleftharpoons H_3O^+ + OAc^-$$

(where the symbol Ac stands for the acetyl grouping CH_3CO). Any solution containing an appreciable amount of HOAc and of OAc^- provides buffering action against added H_3O^+ (which would be neutralized by free OAc^-) or against added OH^- (which would be neutralized by HOAc). Neither of these actions is 100% perfect, so the question is: How "constant" can the pH be held?

■ EXAMPLE 370.   What would be the pH of a buffer solution made by adding 0.350 mole of HOAc and 0.350 mole of NaOAc to enough water to make 0.600 liter of solution?

285

SOLUTION:

$$\frac{0.350 \text{ mole of HOAc}}{0.600 \text{ liter}} = 0.583 \ M \text{ HOAc}$$

$$\frac{0.350 \text{ mole of NaOAc}}{0.600 \text{ liter}} = 0.583 \ M \text{ NaOAc}$$

But sodium acetate, NaOAc, is a strong electrolyte, as most sodium salts are. So, we can assume that the NaOAc is completely dissociated into Na^+ and OAc^-. From 0.583 M NaOAc, we would then have 0.583 M Na^+ and 0.583 M OAc^-.

Let x = moles per liter of the HOAc that is dissociated. This will mean that the equilibrium concentration of HOAc has been reduced from 0.583 to $0.583 - x$. Likewise, the OAc^- concentration has been increased from 0.583 to $0.583 + x$.

$$[\text{HOAc}] = 0.583 - x; \quad [H_3O^+] = x; \quad [OAc^-] = 0.583 + x$$

$$\text{HOAc} + H_2O \ \rightleftharpoons \ H_3O^+ + OAc^-$$

$$K = \frac{[H_3O^+][OAc^-]}{[\text{HOAc}]} = 1.8 \times 10^{-5} = \frac{(x)(0.583 + x)}{0.583 - x}$$

Solve this equation for x, noting that x is small enough to be neglected when added to or subtracted from 0.583.

$$x \cong \frac{0.583}{0.583}(1.8 \times 10^{-5}) = 1.8 \times 10^{-5} = [H_3O^+]$$

$$\text{pH} = -\log[H_3O^+] = -\log(1.8 \times 10^{-5}) = -(0.26 - 5) = 4.74 \ \blacksquare$$

The preceding problem illustrates one important point: The pH of a buffer solution is fixed by the K_{diss} of the weak acid involved and, in fact, is equal to $-\log K_{\text{diss}}$ for all buffer solutions in which acid and salt are present in *equal* concentrations. What happens if they are not present in equal concentrations?

■ EXAMPLE 371. What would be the pH of a buffer solution made by adding 0.350 mole of HOAc and 0.225 mole of NaOAc to enough water to make 0.600 liter solution?

SOLUTION:

$$\frac{0.350 \text{ mole of HOAc}}{0.600 \text{ liter}} = 0.583 \ M \text{ HOAc}$$

$$\frac{0.225 \text{ mole of NaOAc}}{0.600 \text{ liter}} = 0.375 \ M \text{ NaOAc}$$

The 0.375 M NaOAc will be dissociated 100% into 0.375 M Na^+ and 0.375 M OAc^-. Let x = moles per liter of HOAc that are dissociated. This will give at equilibrium:

$$[HOAc] = 0.583 - x; \quad [H_3O^+] = x; \quad [OAc^-] = 0.375 + x$$

$$HOAc + H_2O \rightleftharpoons H_3O^+ + OAc^-$$

$$K = \frac{[H_3O^+][OAc^-]}{[HOAc]} = 1.8 \times 10^{-5} = \frac{(x)(0.375 + x)}{0.583 - x}$$

$$x \cong \frac{0.583}{0.375}(1.8 \times 10^{-5}) = 2.8 \times 10^{-5} = [H_3O^+]$$

$$pH = -\log [H_3O^+] = -\log (2.8 \times 10^{-5}) = -(0.45 - 5) = 4.55 \ \blacksquare$$

It should be evident at this stage that a buffer problem is nothing more than a weak electrolyte problem with the added complication that one of the dissociation products is already present in the solution. For rapid calculation of buffer hydronium-ion concentration, it might be noted that the expression for a dissociation constant of weak acid HX

$$K_{diss} = \frac{[H_3O^+][X^-]}{[HX]}$$

can be rewritten in the more convenient form

$$[H_3O^+] = \frac{[HX]}{[X^-]} K_{diss}$$

This states that the hydronium-ion concentration equals the ratio of weak acid-to-salt concentration times the dissociation constant. This form of the equation is particularly useful when other acid or base is added to the buffer solution.

■ EXAMPLE 372.   Suppose you have 0.250 liter of a buffer solution that contains acetic acid at 0.350 M concentration and sodium acetate at 0.350 M concentration. What would be the pH change if 30.0 ml of 0.100 M HCl is added to this buffer? Assume volumes are additive. $K_{diss} = 1.8 \times 10^{-5}$ for HOAc.

SOLUTION: (First calculate the initial solution. Then add the acid.) In the initial solution, let $x =$ moles per liter of HOAc that dissociate.

$$[HOAc] = 0.350 - x; \qquad [H_3O^+] = x; \qquad [OAc^-] = 0.350 + x$$

$$x = [H_3O^+] = \frac{[HOAc]}{[OAc^-]} K_{diss} = \frac{0.350 - x}{0.350 + x}(1.8 \times 10^{-5}) \cong 1.8 \times 10^{-5}$$

where we have neglected the x as being small compared to 0.350.

$$pH = -\log [H_3O^+] = -\log (1.8 \times 10^{-5}) = -(0.26 - 5) = 4.74$$

Now add the 30.0 ml of 0.100 M HCl. Since HCl is a strong electrolyte, we are effectively adding 30.0 ml of 0.100 M H_3O^+, or

$$(0.0300 \text{ liter})\left(0.100 \text{ } \frac{\text{mole of } H_3O^+}{\text{liter}}\right) = 0.00300 \text{ mole of } H_3O^+$$

Let us assume that all of this H_3O^+ reacts with OAc^- to form HOAc. What will this do to the OAc^- and HOAc in the buffer? The buffer initially contains

$$(0.250 \text{ liter})\left(0.350 \text{ } \frac{\text{mole of HOAc}}{\text{liter}}\right) = 0.0875 \text{ mole of HOAc}$$

and

$$(0.250 \text{ liter})\left(0.350 \text{ } \frac{\text{mole of } OAc^-}{\text{liter}}\right) = 0.0875 \text{ mole of } OAc^-$$

The effect of the added 0.00300 mole of H_3O^+ would be to decrease the moles of OAc^- by 0.00300 and increase the moles of HOAc by 0.00300 by the reaction

$$H_3O^+ + OAc^- \longrightarrow HOAc + H_2O$$

Assuming 100% conversion, this will give us $0.0875 - 0.00300 = 0.0845$ mole OAc^- and $0.0875 + 0.00300 = 0.0905$ mole HOAc. The total volume of the solution is now 0.280 liter made up of the 0.250 liter original buffer plus the 30.0 ml added solution. So far as concentration is concerned we now have

$$\frac{0.0845 \text{ mole of OAc}^-}{0.280 \text{ liter}} = 0.302 \ M \text{ OAc}^-$$

and

$$\frac{0.0905 \text{ mole of HOAc}}{0.280 \text{ liter}} = 0.323 \ M \text{ HOAc}$$

But this does not allow for any dissociation to give H_3O^+.
Let y = moles per liter of HOAc that dissociate.

$$[\text{HOAc}] = 0.323 - y; \qquad [H_3O^+] = y; \qquad [\text{OAc}^-] = 0.302 + y$$

$$y = [H_3O^+] = \frac{[\text{HOAc}]}{[\text{OAc}^-]} K_{\text{diss}} = \frac{0.323 - y}{0.302 + y}(1.8 \times 10^{-5}) \cong 1.9 \times 10^{-5}.$$

where, again, we take advantage of the fact that y is small enough to neglect compared to 0.323 or 0.302.

$$\text{pH} = -\log [H_3O^+] = -\log (1.9 \times 10^{-5}) = -(0.28 - 5) = 4.72$$

So, the pH changes from 4.74 to 4.72. ∎

■ EXAMPLE 373.   Suppose you have 0.250 liter of a buffer solution that contains acetic acid at 0.225 M concentration and sodium acetate at 0.225 M concentration. What would be the pH change if 30.0 ml of 0.100 M HCl is added to this buffer? Assume volumes are additive. $K_{\text{diss}} = 1.8 \times 10^{-5}$ for HOAc.

ANSWER: pH changes from 4.74 to 4.70. ∎

Comparison of Examples 382 and 383 shows they are identical except that the former has higher concentrations of the buffer ingredients and suffers a smaller pH change when acid is added. This is generally true. The more concentrated a buffer is, the more efficient it is at keeping pH constant.

■ EXAMPLE 374.   Suppose you have 0.125 liter of a buffer solution that contains acetic acid at 0.225 M concentration and sodium acetate at 0.225 M concentration. What would be the pH change if 30.0 ml of 0.100 M HCl is added to this buffer? Assume volumes are additive. $K_{\text{diss}} = 1.8 \times 10^{-5}$ for HOAc.

ANSWER: pH changes from 4.74 to 4.65. ∎

Comparison of Examples 373 and 374 shows they are identical except that the former has a higher volume of buffer solution and suffers a smaller pH change when acid is added. This is an example of the general rule that the more of a buffer solution you take, the more efficient it is at keeping the pH constant.

■ EXAMPLE 375.   Just for comparison with Example 374, if you had 0.125 liter of pure water to which you added 30.0 ml of 0.100 M HCl, what would the pH change be? This is an example of an unbuffered solution. Assume additive volumes.

ANSWER: pH changes from 7.000 to 1.713. ■

The preceding examples involve addition of acid to a buffer. True buffering action requires that the same system be equally good at resisting change of pH on addition of base.

■ EXAMPLE 376.   (Compare to Example 372.) Suppose you have 0.250 liter of a buffer solution that contains acetic acid at 0.350 M and sodium acetate at 0.350 M. What would be the pH change if 30.0 ml of 0.100 M NaOH is added? Assume volumes are additive. $K_{diss} = 1.8 \times 10^{-5}$ for HOAc.

SOLUTION: The initial buffer has a pH of 4.74 as calculated in Example 382. Now we add 30.0 ml of 0.100 M NaOH. Since NaOH is a strong electrolyte, it is taken to be 100% dissociated into Na^+ and OH^-. From the 30.0 ml of 0.100 M OH^- we would get

$$(0.0300 \text{ liter})\left(0.100 \frac{\text{mole}}{\text{liter}}\right) = 0.00300 \text{ mole of } OH^- \text{ added}$$

The effect of the added OH^- is to convert HOAc to OAc^- via the net reaction $OH^- + HOAc \rightarrow H_2O + OAc^-$.
The initial buffer has

$$(0.250 \text{ liter})\left(0.350 \frac{\text{mole of HOAc}}{\text{liter}}\right) = 0.0875 \text{ mole of HOAc}$$

$$(0.250 \text{ liter})\left(0.350 \frac{\text{mole of } OAc^-}{\text{liter}}\right) = 0.0875 \text{ mole of } OAc^-$$

Assuming all 0.00300 mole of the OH^- reacts with HOAc to form more OAc^-, we would get $0.0875 - 0.00300 = 0.0845$ mole of HOAc left, along with $0.0875 + 0.00300 = 0.0905$ mole of OAc^-. In a total solution volume of $0.250 + 0.030 = 0.280$ liter, we would have

$$\frac{0.0845 \text{ mole of HOAc}}{0.280 \text{ liter}} = 0.302 \, M \text{ HOAc}$$

$$\frac{0.0905 \text{ mole of OAc}^-}{0.280 \text{ liter}} = 0.323 \, M \text{ OAc}^-$$

Let y = moles per liter of this HOAc that is dissociated.

$$[\text{HOAc}] = 0.302 - y; \quad [\text{H}_3\text{O}^+] = y; \quad [\text{OAc}^-] = 0.323 + y$$

$$y = [\text{H}_3\text{O}^+] = \frac{[\text{HOAc}]}{[\text{OAc}^-]} K_{\text{diss}} = \frac{(0.302 - y)}{(0.323 + y)}(1.8 \times 10^{-5}) \cong 1.7 \times 10^{-5}$$

$$\text{pH} = -\log [\text{H}_3\text{O}^+] = -\log (1.7 \times 10^{-5}) = -(0.23 - 5) = 4.77$$

So, pH changes from 4.74 to 4.77. ■

■ EXAMPLE 377. (Compare to Example 373.) Given 0.250 liter of a solution that is 0.225 M HOAc and 0.225 M NaOAc. What will be the pH change on addition of 30.0 ml of 0.100 M NaOH? Assume additive volumes. $K_{\text{diss}} = 1.8 \times 10^{-5}$ for HOAc.

ANSWER: pH goes from 4.74 to 4.79. ■

■ EXAMPLE 378. (Compare to Example 374.) Given 0.125 liter of a solution that is 0.225 M HOAc and 0.225 M NaOAc. What will be the pH change on addition of 30.0 ml of 0.100 M NaOH? Assume additive volumes. $K_{\text{diss}} = 1.8 \times 10^{-5}$ for HOAc.

ANSWER: pH goes from 4.74 to 4.84. ■

■ EXAMPLE 379. For comparison with the preceding problem, what will be the pH change if 30.0 ml of 0.100 M NaOH is added to 0.125 liter of pure water? Assume additive volumes.

ANSWER: pH goes from 7.00 to 12.287. ■

14.2   The ammonia-ammonium salt buffer

A buffer solution is most effective at keeping the pH constant at about the value that is given by $(-\log K_{\text{diss}})$ for the appropriate weak electrolyte. Thus, if we want to buffer around pH = 5, we select an acid with $K_{\text{diss}} = 10^{-5}$ and make up a solution of this acid and its sodium salt. If we want to buffer around pH = 7, we look for an acid with $K_{\text{diss}} = 10^{-7}$. What about the basic side? Suppose we want to buffer around pH = 9. We could do the same thing. Look for a weak acid with $K_{diss} = 10^{-9}$ and make up a solution containing equal concentrations of this weak acid and its sodium salt. There is an alternative method, however. We could select a weak base with $K_{diss} = 10^{-5}$ and make up a solution containing equal concentrations of this weak base and one of its salts. The OH^- concentration would be buffered at 10^{-5} M, which, of course, would be the same as keeping the pH constant at 9.

The ammonia "dissociation," which can be written

$$NH_3 + H_2O \rightleftharpoons NH_4^+ + OH^-$$

and which has

$$K_{diss} = \frac{[NH_4^+][OH^-]}{[NH_3]} = 1.81 \times 10^{-5}$$

is frequently used to buffer in this basic range.

- EXAMPLE 380.   What will be the pH of a buffer solution consisting of 0.150 mole of NH_3 plus 0.250 mole of NH_4Cl in enough water to make 0.750 liter of solution? $K_{diss} = 1.81 \times 10^{-5}$ for NH_3 in water.

SOLUTION:

$$\frac{0.150 \text{ mole of } NH_3}{0.750 \text{ liter}} = 0.200 \ M \ NH_3$$

NH_4Cl is a strong electrolyte and is 100% dissociated into NH_4^+ and Cl^-.

$$\frac{0.250 \text{ mole of } NH_4^+}{0.750 \text{ liter}} = 0.333 \ M \ NH_4^+$$

[*Note:* There is no absolute requirement that the components of a buffer be present in equal concentrations. Equal concentrations are usually chosen so as to maximize the efficiency of the buffer to resisting pH change on addition of acid *or* base.]

Let x = moles per liter of NH_3 that react via $NH_3 + H_2O \rightarrow NH_4^+ + OH^-$. This will give us x moles of OH^- and will raise NH_4^+ from 0.333 to $0.333 + x$, at the same time decreasing NH_3 from 0.200 to $0.200 - x$.

$$[NH_3] = 0.200 - x; \qquad [NH_4^+] = 0.333 + x; \qquad [OH^-] = x$$

$$NH_3 + H_2O \rightleftharpoons NH_4^+ + OH^-$$

$$K = \frac{[NH_4^+][OH^-]}{[NH_3]} = 1.81 \times 10^{-5} = \frac{(0.333 + x)(x)}{0.200 - x}$$

Neglecting x where it adds or subtracts, we get $x = 1.09 \times 10^{-5}$. But this is the OH^- concentration. To get H_3O^+, we need to use K_w.

$$[H_3O^+] = \frac{K_w}{[OH^-]} = \frac{1.00 \times 10^{-14}}{1.09 \times 10^{-5}} = 9.17 \times 10^{-10}$$

$$pH = -\log[H_3O^+] = -\log(9.17 \times 10^{-10}) = -(0.962 - 10) = 9.038 \blacksquare$$

■ EXAMPLE 381. Suppose you have 80.0 ml of a buffer solution consisting of 0.169 M NH_3 and 0.183 M NH_4Cl. If you add 10.0 ml of 0.100 M HCl, what will be the pH change? $K_{diss} = 1.81 \times 10^{-5}$. Assume additive volumes.

SOLUTION: Initial solution has 0.169 M NH_3 and 0.183 M NH_4^+. Let $x =$ moles per liter of NH_3 that react $NH_3 + H_2O \rightarrow NH_4^+ + OH^-$. This will give at equilibrium:

$$[NH_3] = 0.169 - x; \quad [NH_4^+] = 0.183 + x; \quad [OH^-] = x$$

$$NH_3 + H_2O \rightleftharpoons NH_4^+ + OH^-$$

$$K = \frac{[NH_4^+][OH^-]}{[NH_3]} = 1.81 \times 10^{-5} = \frac{(0.183 + x)(x)}{0.169 - x}$$

$$x \cong \frac{0.169}{0.183}(1.81 \times 10^{-5}) = 1.67 \times 10^{-5} = [OH^-]$$

$$[H_3O^+] = \frac{K_w}{[OH^-]} = \frac{1.00 \times 10^{-14}}{1.67 \times 10^{-5}} = 5.99 \times 10^{-10} M$$

$$pH = -\log[H_3O^+] = -\log(5.99 \times 10^{-10})$$
$$= -(0.777 - 10) = 9.223$$

Now add the 10.0 ml of 0.100 M HCl, which, being a strong electrolyte gives

$$(0.0100 \text{ liter})\left(0.100 \frac{\text{mole of } H_3O^+}{\text{liter}}\right) = 0.00100 \text{ mole } H_3O^+$$

This H_3O^+ will effectively be neutralized by the base NH_3. We assume the reaction $NH_3 + H_3O^+ \rightarrow NH_4^+ + H_2O$ goes 100%. This converts the 0.00100 mole of H_3O^+ into 0.00100 mole of NH_4^+ and uses up 0.00100 mole of NH_3. The initial solution contained

$$(0.0800 \text{ liter})\left(0.169 \frac{\text{mole of NH}_3}{\text{liter}}\right) = 0.0135 \text{ mole of NH}_3$$

and

$$(0.0800 \text{ liter})\left(0.183 \frac{\text{mole of NH}_4^+}{\text{liter}}\right) = 0.0146 \text{ mole of NH}_4^+$$

These will now be decreased and increased, respectively, to give $0.0135 - 0.00100 = 0.0125$ mole of NH_3 and $0.0146 + 0.00100 = 0.0156$ mole of NH_4^+. The total volume of the solution has gone from 80.0 ml to $80.0 + 10.0 = 90.0$ ml, so the concentrations are

$$\frac{0.0125 \text{ mole of NH}_3}{0.0900 \text{ liter}} = 0.139 \ M \ NH_3$$

and

$$\frac{0.0156 \text{ mole of NH}_4^+}{0.0900 \text{ liter}} = 0.173 \ M \ NH_4^+$$

Let $y =$ moles per liter of NH_3 that "dissociate" via the reaction $NH_3 + H_2O \rightarrow NH_4^+ + OH^-$ to give

$$[NH_3] = 0.139 - y; \qquad [NH_4^+] = 0.173 + y; \qquad [OH^-] = y$$

$$NH_3 + H_2O \rightleftharpoons NH_4^+ + OH^-$$

$$K = \frac{[NH_4^+][OH^-]}{[NH_3]} = 1.81 \times 10^{-5} = \frac{(0.173 + y)(y)}{0.139 - y}$$

$$y \cong \frac{0.139}{0.173}(1.81 \times 10^{-5}) = 1.45 \times 10^{-5} \ M = [OH^-]$$

$$[H_3O^+] = \frac{K_w}{[OH^-]} = \frac{1.00 \times 10^{-14}}{1.45 \times 10^{-5}} = 6.09 \times 10^{-10} \ M$$

$$pH = -\log[H_3O^+] = -\log(6.90 \times 10^{-10})$$
$$= -(0.839 - 10) = 9.161$$

So, the pH goes from 9.223 to 9.161. ∎

■ EXAMPLE 382.   What will be the pH change that occurs when 20.0 ml of 0.100 M NaOH is added to 80.0 ml of a buffer solution consisting of 0.169 M NH_3 and 0.183 M NH_4Cl? Assume additive volumes. $K_{diss} = 1.81 \times 10^{-5}$ for $NH_3 + H_2O \rightleftarrows NH_4^+ + OH^-$.

SOLUTION: The pH of the initial buffer is 9.223, as calculated in Example 391. When 20.0 ml of 0.100 M NaOH is added, the following happens: NaOH is a strong electrolyte and is taken to be 100% dissociated. Thus, the 20.0 ml of 0.100 M NaOH furnishes

$$(0.0200 \text{ liter})\left(0.100 \ \frac{\text{mole of OH}^-}{\text{liter}}\right) = 0.00200 \text{ mole of OH}^-$$

This OH^- will be largely sponged up by the NH_4^+ present. We assume that 100% of the OH^- is so used to convert some NH_4^+ into NH_3 via the reaction $NH_4^+ + OH^- \rightarrow NH_3 + H_2O$. The initial solution was 80.0 ml of 0.169 M NH_3 and 0.183 M NH_4Cl. This would give

$$(0.0800 \text{ liter})\left(0.169 \ \frac{\text{mole of NH}_3}{\text{liter}}\right) = 0.0135 \text{ mole of NH}_3$$

and

$$(0.0800 \text{ liter})\left(0.183 \ \frac{\text{mole of NH}_4^+}{\text{liter}}\right) = 0.0146 \text{ mole of NH}_4^+$$

The reaction $NH_4^+ + OH^- \rightarrow NH_3 + H_2O$ to consume the OH^- reduces the moles of NH_4^+ from 0.0146 to $0.0146 - 0.00200 = 0.0126$ mole NH_4^+ and increases the moles of NH_3 from 0.0135 to $0.0135 + 0.00200 = 0.0155$ mole NH_3. The total volume of the solution is 80.0 ml + 20.0 ml = 0.100 liter. Therefore, the concentrations will be

$$\frac{0.0126 \text{ mole of NH}_4^+}{0.100 \text{ liter}} = 0.126 \ M \ NH_4^+$$

$$\frac{0.0155 \text{ mole of NH}_3}{0.100 \text{ liter}} = 0.155 \ M \ NH_3$$

Let x = moles per liter of NH_3 that "dissociate" to form NH_4^+ and OH^-. This will give at equilibrium:

$$[NH_3] = 0.155 - x; \qquad [NH_4^+] = 0.126 + x; \qquad [OH^-] = x$$

$$NH_3 + H_2O \; \rightleftharpoons \; NH_4^+ + OH^-$$

$$K = \frac{[NH_4^+][OH^-]}{[NH_3]} = 1.81 \times 10^{-5} = \frac{(0.126 + x)(x)}{0.155 - x}$$

$$x \cong \frac{0.155}{0.126}(1.81 \times 10^{-5}) = 2.23 \times 10^{-5} \, M = [OH^-]$$

$$[H_3O^+] = \frac{K_w}{[OH^-]} = \frac{1.00 \times 10^{-14}}{2.23 \times 10^{-5}} = 4.48 \times 10^{-10} \, M$$

$$pH = -\log[H_3O^+] = -\log(4.48 \times 10^{-10})$$
$$= -(0.651 - 10) = 9.349$$

So, the pH changes from 9.223 to 9.349. ∎

14.3   The sulfate-hydrogen sulfate buffer

The buffer solution containing SO_4^{2-} and HSO_4^- is an important one for several schemes of identifying cations by precipitation procedures. The buffer is a little out of ordinary because its effect is primarily in the fairly acid region around $pH = 1.9$. Calculations concerning it are somewhat more difficult than in the two preceding sections because the quadratic formula needs to be used here, whereas obvious approximations could be made in the HOAc–NaOAc and NH_3–NH_4^+ cases.

■ EXAMPLE 383.   Given a buffer solution made by dissolving 0.150 mole of $NaHSO_4$ and 0.150 mole of Na_2SO_4 in enough water to make 0.250 liter of solution. What is the pH of this buffer? $K_{diss} = 1.26 \times 10^{-2}$ for $HSO_4^- + H_2O \rightleftharpoons H_3O^+ + SO_4^{2-}$.

SOLUTION: $NaHSO_4$ is a strong electrolyte; 0.150 mole of $NaHSO_4$ gives 0.150 mole of Na^+ and 0.150 mole of HSO_4^-. Na_2SO_4 is also a strong electrolyte; 0.150 mole of Na_2SO_4 gives 0.300 mole of Na^+ and 0.150 mole of SO_4^{2-}. Not allowing for any dissociation of HSO_4^-, we would have in the solution

$$\frac{0.150 \text{ mole of } HSO_4^-}{0.250 \text{ liter}} = 0.600 \, M \; HSO_4^-$$

and

$$\frac{0.150 \text{ mole of } SO_4^{2-}}{0.250 \text{ liter}} = 0.600 \, M \; SO_4^{2-}$$

Now, let x = moles per liter of HSO_4^- that are dissociated. This will cut down the HSO_4^- from 0.600 to $0.600 - x$ and will increase the SO_4^{2-} from 0.600 to $0.600 + x$, at the same time releasing x moles/liter of H_3O^+ into the solution. At equilibrium, we would have

$$[HSO_4^-] = 0.600 - x; \quad [SO_4^{2-}] = 0.600 + x; \quad [H_3O^+] = x$$

$$HSO_4^- + H_2O \;\rightleftharpoons\; H_3O^+ + SO_4^{2-}$$

$$K = \frac{[H_3O^+][SO_4^{2-}]}{[HSO_4^-]} = 1.26 \times 10^{-2} = \frac{(x)(0.600 + x)}{0.600 - x}$$

$$x = 1.21 \times 10^{-2} = [H_3O^+]$$

$$pH = -\log[H_3O^+] = -\log(1.21 \times 10^{-2})$$

$$= -(0.083 - 2) = 1.917 \;\blacksquare$$

∎ **EXAMPLE 384.**   Given 0.120 liter of a buffer solution consisting of 0.150 M $NaHSO_4$ and 0.150 M Na_2SO_4. What pH change will be produced on addition of 10.0 ml of 0.100 M NaOH to this solution? Assume volumes are additive; $K_{diss} = 1.26 \times 10^{-2}$.

SOLUTION: In the initial solution, we have 0.150 M HSO_4^- and 0.150 M SO_4^{2-}. Let x = moles per liter of HSO_4^- that are dissociated. This gives at equilibrium:

$$[HSO_4^-] = 0.150 - x; \quad [SO_4^{2-}] = 0.150 + x; \quad [H_3O^+] = x$$

$$HSO_4^- + H_2O \;\rightleftharpoons\; H_3O^+ + SO_4^{2-}$$

$$K = \frac{[H_3O^+][SO_4^{2-}]}{[HSO_4^-]} = 1.26 \times 10^{-2} = \frac{(x)(0.150 + x)}{0.150 - x}$$

$$x = 1.09 \times 10^{-2} = [H_3O^+]$$

$$pH = -\log[H_3O^+] = -\log(1.09 \times 10^{-2})$$

$$= -(0.037 - 2) = 1.963$$

Now we add the 10.0 ml of 0.100 M NaOH. Inasmuch as NaOH is a strong electrolyte and is 100% dissociated into Na^+ and OH^-, this will give us

$$(0.0100 \text{ liter})\left(0.100 \; \frac{\text{mole of } OH^-}{\text{liter}}\right) = 0.00100 \text{ mole of } OH^-$$

which will convert HSO_4^- to SO_4^{2-} by neutralization. As a first approximation, we assume 100% conversion via the reaction

$$HSO_4^- + OH^- \longrightarrow H_2O + SO_4^{2-}$$

From 0.00100 mole of OH^-, we will get 0.00100 mole of SO_4^{2-} and use up 0.00100 mole of HSO_4^-.

The initial solution is 0.120 liter of 0.150 M $NaHSO_4$ and 0.150 M Na_2SO_4. Therefore we have initially

$$(0.120 \text{ liter})\left(0.150 \frac{\text{mole of } HSO_4^-}{\text{liter}}\right) = 0.0180 \text{ mole } HSO_4^-$$

and

$$(0.120 \text{ liter})\left(0.150 \frac{\text{mole of } SO_4^{2-}}{\text{liter}}\right) = 0.0180 \text{ mole } SO_4^{2-}$$

By adding 0.00100 mole OH^-, we raise SO_4^{2-} from 0.0180 to 0.0180 + 0.0010 = 0.0190 mole and decrease HSO_4^- from 0.0180 to 0.0180 − 0.0010 = 0.0170 mole. The total volume of solution is 0.120 liter + 0.010 liter = 0.130 liter. Concentrations are

$$\frac{0.0190 \text{ mole of } SO_4^{2-}}{0.130 \text{ liter}} = 0.146 \text{ } M \text{ } SO_4^{2-}$$

$$\frac{0.0170 \text{ mole of } HSO_4^-}{0.130 \text{ liter}} = 0.131 \text{ } M \text{ } HSO_4^-$$

Let y = moles per liter of HSO_4^- that are dissociated.

$[HSO_4^-] = 0.131 - y;$ $[SO_4^{2-}] = 0.146 + y;$ $[H_3O^+] = y$

$$HSO_4^- + H_2O \rightleftharpoons H_3O^+ + SO_4^{2-}$$

$$K = \frac{[H_3O^+][SO_4^{2-}]}{[HSO_4^-]} = 1.26 \times 10^{-2} = \frac{(y)(0.146 + y)}{0.131 - y}$$

$$y = 9.80 \times 10^{-3} = [H_3O^+]$$

$$pH = -\log[H_3O^+] = -\log(9.80 \times 10^{-3}) = -(0.991 - 3) = 2.009$$

So, pH changes from 1.963 to 2.009. ∎

Exercises

14.1. Given that K_{diss} of HNO_2 is 4.5×10^{-4}, calculate the pH of a buffer solution made by mixing 0.225 mole of HNO_2 and 0.225 mole of $NaNO_2$ in enough water to make 0.400 liter of solution.

14.2. Given that K_{diss} of HNO_2 is 4.5×10^{-4}, calculate the pH of a buffer solution made by mixing 0.450 mole of HNO_2 and 0.450 mole of $NaNO_2$ in enough water to make 0.400 liter of solution.

*14.3. Given that K_{diss} of HNO_2 is 4.5×10^{-4}, calculate the pH of a buffer solution made by mixing 0.225 mole of HNO_2 and 0.450 mole of $NaNO_2$ in enough water to make 0.400 liter of solution. (*Answer*: 3.65.)

**14.4. Given that K_{diss} of $HClO$ is 3.2×10^{-8}, calculate the pH of a buffer solution made by mixing 10.0 ml of 0.10 M $HClO$ and 20.0 ml of 0.10 M $NaClO$.

**14.5. Given that K_{diss} of $HClO$ is 3.2×10^{-8}, what pH change occurs on addition of 0.001 mole of $NaOH$ to a buffer solution consisting of 0.100 liter of 0.25 M $HClO$ plus 0.100 liter of 0.25 M $NaClO$?

**14.6. Suppose you are given 0.5 liter of 0.500 M HOAc and 0.5 liter of 0.250 M NaOAc. What is the maximum volume you can make of a buffer solution having a pH of 4.58? (*Answer*: 0.86 liter.)

**14.7. Given that K_{diss} of HF is 6.71×10^{-4}, calculate what pH change occurs when (a) 10.0 ml of 0.10 M HCl or (b) 10.0 ml of 0.10 M NaOH is added to 100.0 ml of a mixture that is simultaneously 0.10 M HF and 0.10 M NaF.

**14.8. You are given two separate HOAc–NaOAc buffer solutions, both of which have the same HOAc concentration but which differ in NaOAc concentration. If the respective pH's are 4.94 and 4.84, what must be the ratio of NaOAc concentrations?

**14.9. Suppose you mix equal volumes of the following two buffers: Buffer I, pH = 4.58, containing 1.00 M NaOAc and 1.46 M HOAc; Buffer II, pH = 4.78, containing 0.500 M NaOAc and 0.46 M HOAc. What will be the pH of the resulting solution? (*Answer*: 4.64.)

**14.10. You are given two buffer solutions: Buffer I, pH = 4.58, containing 1.00 M NaOAc and 1.46 M HOAc; Buffer II, pH = 4.78, containing 0.500 M NaOAc and 0.46 M HOAc. In what ratio should you mix Buffer I and Buffer II to come out with pH = 4.68?

**14.11. Given 0.50 liter of a buffer containing 0.80 M HOAc and 0.60 M NaOAc, how many milliliters of 0.10 M HCl would you have to add to change the pH by 0.10 unit? By 0.20 unit?

**14.12. Given 1.00 liter of 6.0 M NH_3 solution. How many milliliters of 1.0 M NH_4Cl do you have to add to make a pH = 9.00 buffer? (*Answer*: 11 liters.)

**14.13. In order to see how the capacity of a buffer to absorb added H_3O^+ or OH^- depends on its composition, calculate the pH for each of the following cases: (a) the original solution: (b) after addition of 1.0 ml of 0.10 M HCl to the original; (c) after addition of 1.0 ml of 0.10 M NaOH to the original. The original solutions are

Case I: 10.0 ml of 1.8×10^{-5} M NaOH

Case II: 10.0 ml of (0.10 M NH$_3$ and 0.10 M NH$_4$Cl)
Case III: 10.0 ml of (0.20 M NH$_3$ and 0.20 M NH$_4$Cl)
Case IV: 20.0 ml of (0.20 M NH$_3$ and 0.20 M NH$_4$Cl)

Summarize your results in a table.

**14.14. Given that $K = 4.16 \times 10^{-7}$ for $CO + 2H_2O \rightleftarrows H_3O^+ + HCO_3^-$, what fraction of the carbon in a CO_2/HCO_3^- buffer would need to be as CO_2 in order to produce a neutral solution?

**14.15. One of the principal equilibria in blood plasma, where the pH stays remarkably constant in the range 7.35–7.45, is $CO_2 + 2H_2O \rightleftarrows H_3O^+ + HCO_3^-$ for which $K_1 = 4.16 \times 10^{-7}$. Given that the concentration of HCO_3^- is 0.025 M, which range of CO_2 concentration is possible? (*Answer:* $0.0027 - 0.0021$ M.)

***14.16. Suppose you are given 0.250 liter of 1.00 M HOAc and 0.250 liter of 1.00 M NaOAc. You are told to make up a buffer that has a pH of 4.800 and that will not change more than 0.010 pH unit when 10.0 ml of 0.100 M NaOH or 10.0 ml of 0.100 M HCl is added. What is the minimal recipe?

**14.17. Given 0.100 liter of a buffer solution consisting of 0.150 M NaHSO$_4$ and 0.150 M Na$_2$SO$_4$; what pH change will occur on addition of 10.0 ml of 0.100 M NaOH to this solution? The dissociation constant of HSO_4^- is 1.26×10^{-2}.

**14.18. How many ml of 1.00 M NaOH would you need to add to 100.0 ml of 1.00 M H$_2$SO$_4$ to produce a buffer having pH = 1.900? (*Answer:* 149 ml.)

**14.19. How many milliliters of 1.00 M H$_2$SO$_4$ would you need to add to 100.0 ml of 1.00 M NaOH to produce a buffer having pH = 2.000?

**14.20. What pH change would be produced on adding 10.0 ml of 0.100 M H$_2$SO$_4$ to 0.100 liter of a buffer containing 0.150 M NaHSO$_4$ and 0.150 M Na$_2$SO$_4$? (*Answer:* 1.963 to 1.895.)

15

SOLUBILITY PRODUCTS

IN THIS CHAPTER we turn our attention to the problem of making calculations for a system consisting of excess solid salt in contact with its saturated solution—for example, solid $BaSO_4$ in contact with saturated $BaSO_4$ solution. We shall limit ourselves to cases of *slightly soluble* salts, because highly soluble salts would involve high concentrations of ions, which produce significant departures from ideal behavior. Furthermore, we shall limit ourselves to strong electrolytes (which covers most salts) so that we shall not have to worry about other equilibria, as would be involved, for example, with weak salts (Section 12.4).

With many insoluble salts—for example, the carbonates—there is a complication in that the anions interact with the water equilibrium $(H_2O + H_2O \rightleftharpoons H_3O^+ + OH^-)$ in the process of hydrolysis. This means that we again would be faced with the problem of simultaneous equilibria, which is discussed in Chapter 16. To avoid too many complications at this time, we restrict ourselves to cases where hydrolysis can be neglected, leaving the more elegant calculations for Chapter 16. Thus, in this chapter we confine ourselves to equilibria of the following types:

$$AB(s) \quad \rightleftharpoons \quad A^+ + B^-$$

$$AB_2(s) \quad \rightleftharpoons \quad A^{2+} + 2B^-$$

$$A_2B(s) \quad \rightleftharpoons \quad 2A^+ + B^{2-}$$

15.1   Evaluation of the solubility product

The equilibrium

$$AB(s) \quad \rightleftharpoons \quad A^+ + B^-$$

can be described by the equilibrium condition $K_{sp} = [A^+][B^-]$, where K_{sp} (usually read "kay-ess-pee" and sometimes called the "ion-product") is a conventional equilibrium constant except that there is no concentration term in the denominator corresponding to the left side of the chemical equation. The rationale behind leaving this term out is that the concentration of AB(s) in the solid state is a constant, no matter how much of the solid AB we take; and if we set a reference state as AB(s) we can regard solid AB as permanently staying at unit activity.

What $K_{sp} = [A^+][B^-]$ means is that the concentration of A^+ in the solution times the concentration of B^- in that same solution is equal numerically to the characteristic number K_{sp}, and this must be true for any equilibrium system involving solid AB, A^+, and B^-. Usually K_{sp} can be evaluated by noting how much AB dissolves in giving a saturated solution. (All of our following computations are for 25°C, which needs to be mentioned, since K_{sp} generally changes with temperature.)

■ EXAMPLE 385.   When you try to dissolve barium sulfate in water, you find that only 0.0017 g of $BaSO_4$ will dissolve in 187 ml of H_2O. What is the K_{sp} for $BaSO_4$?

SOLUTION: The equilibrium for the saturated solution is

$$BaSO_4(s) \rightleftharpoons Ba^{2+} + SO_4^{2-}$$

and the K_{sp} has the form

$$K_{sp} = [Ba^{2+}][SO_4^{2-}]$$

so we need to know the concentration of Ba^{2+} and of SO_4^{2-} in the saturated solution.

One mole of $BaSO_4$ weighs 233.4 g.

$$\frac{0.0017 \text{ g of } BaSO_4}{233.4 \text{ g/mole}} = 7.3 \times 10^{-6} \text{ mole}$$

The volume of the solution is 0.187 liter. (Strictly speaking, we should worry about the volume after mixing 0.187 liter of water plus 7.3×10^{-6} mole of $BaSO_4$. But the latter is such a trifling amount that it contributes a negligible change of volume.)

$$\text{Concentration of dissolved } BaSO_4 = \frac{7.3 \times 10^{-6} \text{ mole}}{0.187 \text{ liter}} = 3.9 \times 10^{-5} \text{ } M$$

$BaSO_4$, like most salts, is a strong electrolyte, so we can assume that all the $BaSO_4$ that goes into solution is dissociated into Ba^{2+} and SO_4^{2-} ions. From 3.9×10^{-5} M $BaSO_4$ completely dissociated we would have 3.9×10^{-5} M Ba^{2+} and 3.9×10^{-5} M SO_4^{2-}. Substitute these values in the expression for K_{sp}

$$K_{sp} = [Ba^{2+}][SO_4^{2-}] = (3.9 \times 10^{-5})(3.9 \times 10^{-5}) = 1.5 \times 10^{-10} \blacksquare$$

■ EXAMPLE 386.   To produce a saturated solution of calcium fluoride, CaF_2, you need to dissolve 0.0068 g of CaF_2 per 0.250 liter. What is the K_{sp} for CaF_2?

SOLUTION: The equilibrium for the saturated solution is

$$CaF_2(s) \rightleftharpoons Ca^{2+} + 2F^-$$

and the equilibrium condition is given by

$$K_{sp} = [Ca^{2+}][F^-]^2$$

Note that the concentration of fluoride ion is taken to the second power in conformance with the coefficient 2 in the foregoing chemical equation. All we need to know now is what is the concentration of calcium ion and what is the concentration of fluoride ion in the saturated solution.
One mole of CaF_2 weighs 78.08 g.

$$\frac{0.0068 \text{ g of } CaF_2}{78.08 \text{ g/mole}} = 8.7 \times 10^{-5} \text{ mole}$$

The volume of the solution is 0.250 liter.

Concentration of dissolved CaF_2 is $\dfrac{8.7 \times 10^{-5} \text{ mole}}{0.250 \text{ liter}} = 3.5 \times 10^{-4} M$

We assume that CaF_2, being a strong electrolyte, is 100% dissociated into Ca^{2+} and F^-. This means that, conforming to the chemical equation $CaF_2(s) \rightarrow Ca^{2+} + 2F^-$, the dissolving of 3.5×10^{-4} mole/liter of CaF_2 will produce 3.5×10^{-4} mole/liter of Ca^{2+} and $2 \times (3.5 \times 10^{-4})$, or 7.0×10^{-4}, mole/liter of F^-.

$$[Ca^{2+}] = 3.5 \times 10^{-4} M; [F^-] = 7.0 \times 10^{-4} M$$
$$K_{sp} = [Ca^{2+}][F^-]^2 = (3.5 \times 10^{-4})(7.0 \times 10^{-4})^2 = 1.7 \times 10^{-10} \blacksquare$$

■ EXAMPLE 387. Suppose you are told that a given solution in equilibrium with solid Ag_2CrO_4 contains, besides other things, Ag^+ at a concentration of 4.4×10^{-6} mole/liter and CrO_4^{2-} at a concentration of 0.100 M. What would be the corresponding K_{sp} of Ag_2CrO_4?

SOLUTION: We are told there is an equilibrium between solid Ag_2CrO_4 and Ag^+ and CrO_4^{2-} in solution. We can write this as follows:

$$Ag_2CrO_4(s) \rightleftharpoons 2Ag^+ + CrO_4^{2-}$$

The requirement for equilibrium is given by

$$K_{sp} = [Ag^+]^2[CrO_4^{2-}]$$

We are told that equilibrium exists when

$$[Ag^+] = 4.4 \times 10^{-6} \, M; \quad [CrO_4^{2-}] = 0.100 \, M$$

so we simply substitute these values in the expression for K_{sp}:

$$K_{sp} = [Ag^+]^2[CrO_4^{2-}] = (4.4 \times 10^{-6})^2(0.100) = 1.9 \times 10^{-12} \blacksquare$$

■ EXAMPLE 388. You are given an equilibrium system consisting of solid $Mg(OH)_2$ and solid MgF_2 in contact with a saturated solution containing Mg^{2+}, OH^-, and F^-. Analysis of the solution shows that it contains Mg^{2+} at a concentration of 0.0027 M, OH^- at 5.75×10^{-5} M, and F^- at 0.0054 M. Calculate the K_{sp} for $Mg(OH)_2$ and for MgF_2.

SOLUTION: Both equilibria exist simultaneously.

$$Mg(OH)_2(s) \rightleftharpoons Mg^{2+} + 2OH^-$$
$$MgF_2(s) \rightleftharpoons Mg^{2+} + 2F^-$$

The equilibrium requirements are

$$K_{sp} \text{ of } Mg(OH)_2 = [Mg^{2+}][OH^-]^2$$
$$K_{sp} \text{ of } MgF_2 = [Mg^{2+}][F^-]^2$$

We are told what some typical equilibrium concentrations are:

$$[Mg^{2+}] = 0.0027 \, M; \quad [OH^-] = 5.75 \times 10^{-5} \, M; \quad [F^-] = 0.0054 \, M$$

All we need to do is substitute these values in the expressions above:

K_{sp} of $Mg(OH)_2 = [Mg^{2+}][OH^-]^2 = (0.0027)(5.75 \times 10^{-5})^2 =$
8.9×10^{-12}

K_{sp} of $MgF_2 = [Mg^{2+}][F^-]^2 = (0.0027)(0.0054)^2 = 7.9 \times 10^{-8}$

The important point to be learned from the problems above is that the ion concentration that feeds into the K_{sp} expression is the ion concentration in solution, no matter what that particular concentration happens to be. So long as it is an equilibrium concentration, it satisfies K_{sp}. ■

15.2 Calculations from K_{sp}

Once the numerical value of K_{sp} of a given salt has been determined, it can be used from there on to describe any equilibrium system in which the same ions in solution coexist with excess solid phase. Extensive tabulations of K_{sp}'s may be found in the chemistry handbooks as well as in the book *Oxidation Potentials* by W. M. Latimer (Prentice-Hall, 1952). The one special point that deserves emphasizing is that the K_{sp} puts a restriction only on the combined product of ionic concentrations, not on what the individual values must be.

Two principal calculations involving K_{sp} are (1) the prediction of how much salt can dissolve in a given amount of water (or of solution containing a common ion) and (2) the prediction whether precipitation will occur when two solutions are mixed. K_{sp} also enables us to calculate one ionic concentration if the other one is specified.

First we consider the simple case where a solid dissolves in water to give the same number of positive ions as negative ions.

■ EXAMPLE 389.   Given that the K_{sp} of $BaSO_4$ is 1.5×10^{-9}, how many grams of barium sulfate can you dissolve in 1000 liters of water?

SOLUTION: The equilibrium will be

$$BaSO_4(s) \rightleftharpoons Ba^{2+} + SO_4^{2-}$$

for which the equilibrium requirement is

$$K_{ps} = [Ba^{2+}][SO_4^{2-}] = 1.5 \times 10^{-9}$$

This means that the concentration of barium ion, whatever it is, times the concentration of sulfate ion, whatever it is, must come out to equal 1.5×10^{-9}.

There is, however, another requirement in this case. We are dissolving $BaSO_4$ in pure water—that is, in water containing no Ba^{2+} and no SO_4^{2-} to start. The dissolving reaction is $BaSO_4(s) \rightarrow Ba^{2+} + SO_4^{2-}$ and it indicates that we get one Ba^{2+} for each SO_4^{2-}. In other words, we have here also the requirement that the Ba^{2+} concentration be equal to the SO_4^{2-} concentration. This is not part of the K_{sp} requirement; it comes about because of the way we are making our saturated solution.

Let $x =$ moles per liter of solid $BaSO_4$ that dissolves.
This will give x moles/liter of Ba^{2+} and x moles/liter of SO_4^{2-}.

$$[Ba^{2+}] = x; \qquad [SO_4^{2-}] = x$$

$$K_{sp} = [Ba^{2+}][SO_4^{2-}] = 1.5 \times 10^{-9} = (x)(x)$$

Solving for x gives 3.9×10^{-5}.
This gives the moles of $BaSO_4$ that can dissolve per liter.
One mole of $BaSO_4$ weighs 233.4 g, so we can dissolve per liter

$$(3.9 \times 10^{-5} \text{ mole})\left(233.4 \frac{g}{\text{mole}}\right) = 9.1 \times 10^{-3} g$$

In 1000 liters, this amounts to

$$\left(9.1 \times 10^{-3} \frac{g}{\text{liter}}\right)(1000 \text{ liter}) = 9.1 \text{ g } \blacksquare$$

Next we consider a case where a salt dissolves in a solution already containing a common ion.

■ EXAMPLE 390.   Given that the K_{sp} of $BaSO_4$ is 1.5×10^{-9}, how many grams of $BaSO_4$ can you dissolve in 1000 liters of 0.100 M Na_2SO_4 solution? (Compare this problem to Example 389.)

SOLUTION: When you have added enough $BaSO_4$ to saturate the solution, the equilibrium will be

$$BaSO_4(s) \rightleftharpoons Ba^{2+} + SO_4^{2-}$$

for which the equilibrium condition requires that

$$K_{sp} = [Ba^{2+}][SO_4^{2-}] = 1.5 \times 10^{-9}$$

Let $x =$ moles per liter of $BaSO_4$ that dissolve. This will put into the solution x moles/liter of Ba^{2+} and x moles/liter of SO_4^{2-}. But the solution already contains $0.100\ M$ Na_2SO_4 before you dissolve in any $BaSO_4$. Furthermore, since Na_2SO_4 is a strong electrolyte, $0.100\ M$ Na_2SO_4 corresponds to $0.100\ M$ SO_4^{2-}. If we add x moles/liter of SO_4^{2-} to the $0.100\ M$ already there, we will have at equilibrium

$$[SO_4^{2-}] = 0.100 + x; \qquad [Ba^{2+}] = x$$

[*Note:* $[Ba^{2+}]$ is *not* equal to SO_4^{2-}, as was true in Example 399.]
Substitute these values in K_{sp} and solve for x.

$$K_{sp} = [Ba^{2+}][SO_4^{2-}] = 1.5 \times 10^{-9} = (x)(0.100 + x)$$

No point to solving the quadratic for x, since x is obviously small compared to 0.100, so we assume that $(0.100 + x) \cong 0.100$. This leads to $1.5 \times 10^{-9} \cong 0.100x$, which solves to $x = 1.5 \times 10^{-8}$ for the moles per liter of $BaSO_4$ that can be dissolved.

One mole of $BaSO_4$ weighs 233.4 g and we are working with 1000 liters.

$$\text{Total } BaSO_4 \text{ dissolved} = \left(1.5 \times 10^{-8} \frac{\text{mole}}{\text{liter}}\right)\left(233.4 \frac{\text{g}}{\text{mole}}\right)(1000 \text{ liter})$$

$$= 0.0035 \text{ g} \ \blacksquare$$

The cases above correspond to salts dissolving to give an equal number of positive and negative ions. What do we do if the dissolving produces twice as many of one ion as of the other?

■ EXAMPLE 391. Given that the K_{sp} of MgF_2 is 8×10^{-8}, how many grams of MgF_2 can you dissolve in 0.250 liter of water?

SOLUTION: The equilibrium will be

$$MgF_2(s) \ \rightleftharpoons \ Mg^{2+} + 2F^-$$

and the requirement for equilibrium is that

$$K_{sp} = [Mg^{2+}][F^-]^2 = 8 \times 10^{-8}$$

Let $x =$ moles per liter of MgF_2 that dissolve in water.
When x moles/liter of MgF_2 dissolve, then according to the reaction

$MgF_2(s) \rightarrow Mg^{2+} + 2F^-$, we would get x moles/liter of Mg^{2+} and $2x$ moles/liter of F^-. This gives us at equilibrium:

$$[Mg^{2+}] = x; \quad [F^-] = 2x$$

Substitute these in the K_{sp} expression:

$$K_{sp} = [Mg^{2+}][F^-]^2 = (x)(2x)^2 = 8 \times 10^{-8}$$

[*Note:* The square bracket indicates moles per liter of that species. The exponent 2 on $[F^-]^2$ means we have to square whatever goes inside the bracket—that is, whatever the fluoride-ion concentration happens to be. The way we have defined x, the fluoride-ion concentration is $2x$, and we shall have to square that whole business. This appearance of the 2 in two places causes students a great deal of trouble, usually because they forget the meaning of the square brackets and also because they resort to dangerous shortcuts, such as putting $2F^-$ inside the square brackets. You can save yourself much grief if you stick to the notation as used above. If in doubt at this point, review the material in Chapter 11 on writing equilibrium constants.]

We need to solve $4x^3 = 8 \times 10^{-8}$, which can be done by dividing both sides by 4 and taking the cube root.

$$x = (2 \times 10^{-8})^{1/3} = (20 \times 10^{-9})^{1/3} = 3 \times 10^{-3}$$

where we round off to stick with the only-one-allowed significant figure. This gives us how many moles of MgF_2 can dissolve per liter. To get grams per 0.250 liter, we need to multiply by the weight of 1 mole, 62.31 g, and 0.250 liter.

$$\text{Total } MgF_2 \text{ dissolved} = \left(3 \times 10^{-3} \frac{\text{mole}}{\text{liter}}\right)\left(62.31 \frac{\text{g}}{\text{mole}}\right)(0.250 \text{ liter})$$

$$= 0.05\text{g} \quad \blacksquare$$

Now we complicate it a bit. Still working with the case where the dissolving introduces twice as many of one kind of ion as the other, what do we do if the solution already contains a common ion?

■ EXAMPLE 392. (Compare to Example 391.) Given that the K_{sp} of MgF_2 is 8×10^{-8}, how many grams of MgF_2 can you dissolve in 0.250 liter of 0.100 M $Mg(NO_3)_2$?

SOLUTION: Let $x =$ moles per liter of MgF_2 that dissolve to establish equilibrium. This will produce x moles/liter of Mg^{2+} and $2x$ moles/liter of F^- via the reaction $MgF_2 \rightarrow Mg^{2+} + 2F^-$.

But there is already some Mg^{2+} in the solution, corresponding to $0.100M$ $Mg(NO_3)_2$. $Mg(NO_3)_2$ is a strong electrolyte, and 0.100 M $Mg(NO_3)_2$ is the same as 0.100 M Mg^{2+} and 0.200 M NO_3^-.

Therefore, the total Mg^{2+} concentration at equilibrium will be $0.100 + x$; the total F^- concentration will be $2x$.

$$[Mg^{2+}] = 0.100 + x; \qquad [F^-] = 2x$$

The equilibrium

$$MgF_2(s) \rightleftarrows Mg^{2+} + 2F^-$$

requires that

$$K_{sp} = [Mg^{2+}][F^-]^2 = 8 \times 10^{-8}$$

Substituting the foregoing concentrations, we get

$$[Mg^{2+}][F^-]^2 = (0.100 + x)(2x)^2 = 8 \times 10^{-8}$$

Nominally, this is a cubic equation that can be solved exactly. However, since only one significant figure is allowed in the answer, we can make do with an approximate solution. Since x has to be small compared to 0.100, we can assume $0.100 + x \cong 0.100$. This gives us the approximate equation $(0.100)(2x)^2 = 8 \times 10^{-8}$, which quickly solves to $x = 4 \times 10^{-4}$ as the number of moles of MgF_2 that can be dissolved per liter of 0.100 M $Mg(NO_3)_2$. To get grams per 0.250 liter,

$$\left(4 \times 10^{-4} \frac{mole}{liter}\right)(0.250 \text{ liter})\left(62.31 \frac{g\ MgF_2}{mole}\right) = 0.006 \text{ g} \blacksquare$$

■ EXAMPLE 393. (Compare to Example 392.) Given that K_{sp} of MgF_2 is 8×10^{-8}, how many grams of MgF_2 can you dissolve in 0.250 liter of 0.100 M NaF?

SOLUTION: Let $x =$ moles per liter of MgF_2 that can dissolve. This will give x moles/liter of Mg^{2+} and $2x$ moles/liter of F^- via the reaction

$$MgF_2 \longrightarrow Mg^{2+} + 2F^-.$$

But the solution already contains 0.100 M F^-. At equilibrium

$$[Mg^{2+}] = x; \qquad [F^-] = 2x + 0.100$$
$$K_{sp} = [Mg^{2+}][F^-]^2 = (x)(2x + 0.100)^2 = 8 \times 10^{-8}$$

Again, x looks to be small, so we assume $2x + 0.100 \cong 0.100$.

This gives $(x)(0.100)^2 \cong 8 \times 10^{-8}$, which solves to $x = 8 \times 10^{-6}$ mole/liter of MgF_2 that can be dissolved in $0.100\ M$ NaF. For grams in 0.250 liter,

$$\left(8 \times 10^{-6}\ \frac{\text{mole of MgF}_2}{\text{liter}}\right)(0.250\ \text{liter})\left(62.31\ \frac{\text{g of MgF}_2}{\text{mole}}\right) = 1 \times 10^{-4}\text{g} \ \blacksquare$$

It would pay to spend some extra time studying the preceding three problems to see how they compare with each other. In all three cases, the equilibrium condition for saturation is the same: $K_{sp} = [Mg^{2+}][F^-]^2 = 8 \times 10^{-8}$. The only difference has been in what goes inside the brackets. For dissolving MgF_2 in pure water, what goes in the F^- bracket is just twice what goes in for Mg^{2+}. For dissolving MgF_2 in $0.100\ M$ $Mg(NO_3)_2$, we need to allow for an additional $0.100\ M$ Mg^{2+}, so $[Mg^{2+}]$ becomes $(x + 0.100)$. For dissolving MgF_2 in $0.100\ M$ NaF, we need to allow for an additional $0.100\ M$ F^-, so $[F^-]$ becomes $(2x + 0.100)$. The values of x will thus come out to be different in the three cases.

15.3   Precipitation reactions

In the preceding section, we approached equilibrium by adding a slightly soluble salt, such as AB, to water or to a common-ion solution and letting enough ·AB dissolve to give equilibrium concentrations of A^+ and B^-. We could just as well approach equilibrium from the other side by mixing solutions of A^+ and B^- and letting AB precipitate until equilibrium concentrations of A^+ and B^- are left. The two major difficulties that students encounter in the latter type of calculation are (a) proper allowance for the dilution of one solution by the other; (b) proper setting up of the equations so that difficult algebraic computations are avoided.

A practical point to consider is that the calculations will generally tell only whether precipitation is expected. In actual laboratory manipulation, other factors need to be considered, such as how fast does the reaction come to equilibrium, or how much extra solid has to be allowed for in order to make a precipitate "visible."

$\blacksquare$ EXAMPLE 394.   Given that the K_{sp} of $BaSO_4$ is 1.5×10^{-9}, would you expect a precipitate to form if 10.0 ml of $0.0100\ M$ $BaCl_2$ solution were mixed with 30.0 ml of $0.0050\ M$ Na_2SO_4 solution?

SOLUTION: The condition for the equilibrium

$$BaSO_4(s) \ \rightleftharpoons \ Ba^{2+} + SO_4^{2-}$$

is that

$$K_{sp} = [Ba^{2+}][SO_4^{2-}] = 1.5 \times 10^{-9}$$

If the product of the barium-ion concentration times the sulfate concentration is greater than 1.5×10^{-9}, $BaSO_4$ should precipitate so as to reduce the ion concentrations to equilibrium values.

Because $BaCl_2$ and Na_2SO_4 are strong electrolytes, they will be 100% dissociated.

Ten ml of 0.0100 M $BaCl_2$ gives

$$(0.0100 \text{ liter})\left(0.0100 \frac{\text{mole of } Ba^{2+}}{\text{liter}}\right) = 0.000100 \text{ mole of } Ba^{2+}$$

Thirty ml of 0.0050 M Na_2SO_4 gives

$$(0.0300 \text{ liter})\left(0.00500 \frac{\text{mole of } SO_4^{2-}}{\text{liter}}\right) = 1.5 \times 10^{-4} \text{ mole of } SO_4^{2-}$$

Total volume of the solution is 10.0 ml + 30.0 ml = 0.0400 liter.

Therefore, if *no reaction occurred* when solutions are mixed, we could calculate the concentrations as follows:

$$\text{concentration of } Ba^{2+} = \frac{0.000100 \text{ mole of } Ba^{2+}}{0.0400 \text{ liter}}$$

$$= 2.50 \times 10^{-3} \text{ } M \text{ } Ba^{2+}$$

$$\text{concentration of } SO_4^{2-} = \frac{1.5 \times 10^{-4} \text{ mole of } SO_4^{2-}}{0.0400 \text{ liter}}$$

$$= 3.8 \times 10^{-3} \text{ } M \text{ } SO_4^{2-}$$

If no $BaSO_4$ precipitates, $[Ba^{2+}][SO_4^{2-}]$ would equal (2.50×10^{-3}) $(3.8 \times 10^{-3}) = 9.5 \times 10^{-6}$.

This ion product would be considerably higher than the value of K_{sp} (which is 1.5×10^{-9}), so it would not correspond to an equilibrium situation and precipitation *should occur*. ∎

■ EXAMPLE 395. Would you expect BaF_2 to precipitate if 20.0 ml of 0.010 M $BaCl_2$ is mixed with 30.0 ml of 0.010 M NaF? The K_{sp} of BaF_2 is 2.4×10^{-5}.

SOLUTION: $BaCl_2$ and NaF are strong electrolytes, 100% dissociated. Twenty ml of 0.010 M $BaCl_2$ would give

$$(0.0200 \text{ liter})\left(0.010 \, \frac{\text{mole of Ba}^{2+}}{\text{liter}}\right) = 2.0 \times 10^{-4} \text{ mole of Ba}^{2+}$$

Thirty ml of 0.010 M NaF would give

$$(0.0300 \text{ liter})\left(0.010 \, \frac{\text{mole of F}^{-}}{\text{liter}}\right) = 3.0 \times 10^{-4} \text{ mole of F}^{-}$$

If no reaction occurs, the concentrations in the final solution would be as follows, taking into account that the total volume is now 0.0500 liter:

$$\text{concentration of Ba}^{2+} = \frac{2.0 \times 10^{-4} \text{ mole}}{0.0500 \text{ liter}}$$

$$= 4.0 \times 10^{-3} M \text{ Ba}^{2+}$$

$$\text{concentration of F}^{-} = \frac{3.0 \times 10^{-4} \text{ mole}}{0.0500 \text{ liter}} = 6.0 \times 10^{-3} M \text{ F}^{-}$$

To get precipitation would mean setting up the equilibrium

$$BaF_2(s) \rightleftharpoons Ba^{2+} + 2F^{-}$$

for which the equilibrium requirement is

$$K_{sp} = [\text{Ba}^{2+}][\text{F}^{-}]^2 = 2.4 \times 10^{-5}$$

How does our mixture compare with this equilibrium requirement? Calculate $[\text{Ba}^{2+}][\text{F}^{-}]^2$ to find out.

$$[\text{Ba}^{2+}][\text{F}^{-}]^2 = (4.0 \times 10^{-3})(6.0 \times 10^{-3})^2 = 1.4 \times 10^{-7}$$

Since this ion product is appreciably less than K_{sp}, it means we do *not* have enough concentration of Ba^{2+} and F^{-} to form a saturated solution. In other words, we do *not* expect precipitation to occur. ■

■ EXAMPLE 396.   The K_{sp} of $AgOAc$ is 2.3×10^{-3}. Should you get precipitation on mixing 20.0 ml of 1.2 M $AgNO_3$ with 30.0 ml of 1.4 M $HOAc$?

SOLUTION: You have to be careful here since $AgNO_3$ is a strong electrolyte 100% dissociated into Ag^+ and NO_3^-, but HOAc is a weak electrolyte, only slightly dissociated to H_3O^+ and OAc^-.

Calculate first as if no precipitation occurs.

Twenty ml of 1.2 M $AgNO_3$ gives

$$(0.0200 \text{ liter})\left(1.2 \, \frac{\text{mole of Ag}^+}{\text{liter}}\right) = 0.024 \text{ mole of Ag}^+$$

Thirty ml of 1.4 M HOAc gives

$$(0.0300 \text{ liter})\left(1.4 \, \frac{\text{mole of HOAc}}{\text{liter}}\right) = 0.042 \text{ mole of HOAc}$$

Total volume of solution is 0.0500 liter.

Concentration of Ag^+ would be

$$\frac{0.024 \text{ mole}}{0.0500 \text{ liter}} = 0.48 \, M \text{ Ag}^+$$

Concentration of HOAc would be

$$\frac{0.042 \text{ mole}}{0.0500 \text{ liter}} = 0.84 \, M \text{ HOAc}$$

If we get a precipitate of AgOAc, the equilibrium we are interested in would be

$$AgOAc(s) \rightleftharpoons Ag^+ + OAc^-$$

$$K_{sp} = [Ag^+][OAc^-] = 2.3 \times 10^{-3}$$

So we do not really care what the concentration of HOAc is, but what the concentration of free OAc^- would be. We need, therefore, to calculate the concentration of OAc^- in 0.84 M HOAc.

Let x = moles per liter of HOAc that dissociate.

$$HOAc + H_2O \rightleftharpoons H_3O^+ + OAc^-$$

$$K_{diss} = \frac{[H_3O^+][OAc^-]}{[HOAc]} = 1.8 \times 10^{-5} = \frac{(x)(x)}{0.84 - x}$$

$$[OAc^-] = x = 3.9 \times 10^{-3} \, M$$

If now we look at the ion product $[Ag^+][OAc^-]$, we get $(0.48)(3.9 \times 10^{-3}) = 1.9 \times 10^{-3}$, which is smaller than K_{sp}, so no precipitation of AgOAc is expected. ∎

The foregoing problems are not very complicated and in fact are not much more than exercises in calculating concentrations. For this purpose, it is useful to develop shortcuts for speeding up the computations. One of the most obvious is to recognize the dilution factor, whereby the concentration in the final mixture is figured as the initial concentration times a fraction that describes what portion of the final volume is contributed by each solution. For example, when 10.0 ml of 0.10 M $CaCl_2$ is mixed with 40.0 ml of 1.5 M NH_3, the final concentration of $CaCl_2$ is 1/5 of 0.10 M, since 10.0 ml is 10.0/50.0 of the total volume; similarly, the final concentration of NH_3 is 4/5 of 1.5 M, since 40.0 ml is 40.0/50.0 of the total volume. The other point to be especially careful about is to note that the "ion product" that is to be compared with the K_{sp} value must be calculated by taking each concentration to the proper power as shown by the coefficients in the chemical equation for the equilibrium.

A more complex problem appears when we look at a precipitation reaction and try to calculate what concentration of ions should be left in solution after precipitation occurs. There are two different cases: (1) the ions are mixed in the molar ratio tht corresponds to their consumption in the chemical equation for precipitation (e.g., mix Mg^{2+} and F^- in a ratio of 1 to 2); (2) one of the ions is in excess over what the stoichiometry requires.

∎ **EXAMPLE 397.**   If you mix 10.0 ml of 0.100 M $BaCl_2$ with 40.0 ml of 0.0250 M Na_2SO_4, what should be the concentration of Ba^{2+} and of SO_4^{2-} in solution after precipitation has occurred? The K_{sp} of $BaSO_4$ is 1.5×10^{-9}.

SOLUTION: Ten ml of 0.100 M $BaCl_2$ gives

$$(0.0100 \text{ liter})\left(0.100 \; \frac{\text{mole of } Ba^{2+}}{\text{liter}}\right) = 0.00100 \text{ mole of } Ba^{2+}$$

Forty ml of 0.0250 M Na_2SO_4 gives

$$(0.0400 \text{ liter})\left(0.0250 \; \frac{\text{mole of } SO_4^{2-}}{\text{liter}}\right) = 0.00100 \text{ mole of } SO_4^{2-}$$

We thus have equal numbers of moles of Ba^{2+} and of SO_4^{2-}. When they react via the equation $Ba^{2+} + SO_4^{2-} \rightarrow BaSO_4(s)$, neither Ba^{2+} nor SO_4^{2-}

will be left in excess. This means solid $BaSO_4$ is precipitated until equilibrium is established, one in which Ba^{2+} and SO_4^{2-} are left in solution in equal concentrations. But that would be the same problem as dissolving $BaSO_4$ in pure water. The resulting solution there also has Ba^{2+} and SO_4^{2-} in equal concentration. (In the present case, there is also Cl^- and Na^+ in solution but neither of these is involved in the $BaSO_4$ equilibrium.)

So, the simplest way to solve the foregoing problem of mixing Ba^{2+} and SO_4^{2-} in equal quantities is to approach equilibrium from the other side, namely, add $BaSO_4$ to saturate a solution.

$$BaSO_4(s) \rightleftharpoons Ba^{2+} + SO_4^{2-}$$

$$[Ba^{2+}] = x; \qquad [SO_4^{2-}] = x$$

$$K_{sp} = [Ba^{2+}][SO_4^{2-}] = 1.5 \times 10^{-9} = (x)(x)$$

$$x = 3.9 \times 10^{-5} \, M = [Ba^{2+}] = [SO_4^{2-}]$$

The same procedure applies when the stoichiometry of the precipitation reaction requires two of one ion per one of the other. Again, assuming you mix enough to cause precipitation, the easiest thing to do is to approach equilibrium from the other side. ∎

∎ EXAMPLE 398.   Suppose you mix 10.0 ml of 0.25 M $Mg(NO_3)_2$ and 25.0 ml of 0.20 M NaF. What should be the concentration of Mg^{2+} and of F^- in the final solution? The K_{sp} of MgF_2 is 8×10^{-8}.

SOLUTION: First you should see whether precipitation occurs. Just after mixing, before any reaction occurs, the

concentration of Mg^{2+} would be $\left(\dfrac{10.0}{35.0}\right)(0.25 \, M)$ or 0.071 M

concentration of F^- would be $\left(\dfrac{25.0}{35.0}\right)(0.20 \, M)$ or 0.14 M

The ion-product $[Mg^{2+}][F^-]^2$, which equals $(0.071)(0.14)^2 = 1.4 \times 10^{-3}$ exceeds K_{sp}, so precipitation to form solid MgF_2 should occur.

Now we note that Mg^{2+} and F^- are being added together in a 1:2 ratio, the same as required in the chemical equation $Mg^{2+} + 2F^- \rightleftharpoons MgF_2(s)$. The equilibrium system will thus be the same as if approached from the other side—that is, by adding solid MgF_2 to water.

Hence, all we need to do is to solve the regular solubility problem.

$$MgF_2(s) \rightleftharpoons Mg^{2+} + 2F^-$$

$$[Mg^{2+}] = x; \quad [F^-] = 2x$$

$$K_{sp} = [Mg^{2+}][F^-]^2 = 8 \times 10^{-8} = (x)(2x)^2$$

$$x = 3 \times 10^{-3}$$

$$[Mg^{2+}] = x = 3 \times 10^{-3} \, M; \quad [F^-] = 2x = 6 \times 10^{-3} \, M$$

[*Note:* There is another way to solve these problems that may on first sight look more straightforward. Given a solution where you put together 0.071 M Mg^{2+} and 0.14 M F^-, let y = moles per liter of MgF_2 that precipitate. According to the reaction $Mg^{2+} + 2F^- \rightarrow MgF_2(s)$, this would use up y moles/liter of Mg^{2+} and $2y$ moles/liter of F^-. Hence at equilibrium, the concentration of Mg^{2+} would be $(0.071 - y)$ M and the concentration of F^- would be $(0.14 - 2y)$ M. Substitute these as usual:

$$MgF_2(s) \rightleftharpoons Mg^{2+} + 2F^-$$

$$K_{sp} = [Mg^{2+}][F^-]^2 = 8 \times 10^{-8} = (0.071 - y)(0.14 - 2y)^2$$

What is wrong with such a method? Nothing. The only objection is that it gets you into some nasty algebra—a cubic equation and a situation where y cannot be neglected. The reason you get into this mess is that you are approaching equilibrium from the unfavorable side and defining an unknown that is a relatively large number. Always try to rig your calculations so you are calculating a small effect (e.g., a trace of solid dissolving) instead of a big effect (e.g., lots of precipitate forming). If you are clever, the approximations will be clearly evident and the algebra will be simple, but only *if you set up the problem right.*] ■

What happens if one of the ions in a precipitation reaction is in excess over what stoichiometry requires? Then, some of that ion will be left in solution in appreciable quantity and its concentration will effectively control how much of the other ion can be in solution at the same time. The computation is a "common-ion" problem.

■ EXAMPLE 399. Suppose you mix 40.0 ml of 0.10 M $AgNO_3$ and 10.0 ml of 0.15 M NaBr. What should be the final concentrations of Ag^+ and Br^-, given that the K_{sp} of AgBr is 5.0×10^{-13}?

SOLUTION: Forty ml of 0.10 M $AgNO_3$ gives

$$(0.0400 \text{ liter})\left(0.10 \, \frac{\text{mole of } Ag^+}{\text{liter}}\right) = 0.0040 \text{ mole of } Ag^+$$

Ten ml of 0.15 M NaBr gives

$$(0.0100 \text{ liter})\left(0.15 \frac{\text{mole of Br}^-}{\text{liter}}\right) = 0.0015 \text{ mole of Br}^-$$

Total volume of final solution is 50.0 ml.
If no precipitation occurred, there would be

$$\text{concentration of Ag}^+ = \frac{0.0040 \text{ mole}}{0.050 \text{ liter}} = 0.080 \ M \ \text{Ag}^+$$

$$\text{concentration of Br}^- = \frac{0.0015 \text{ mole}}{0.050 \text{ liter}} = 0.030 \ M \ \text{Br}^-$$

The precipitation reaction $\text{Ag}^+ + \text{Br}^- \rightarrow \text{AgBr(s)}$ requires a 1:1 molar ratio, so Ag^+ is evidently in excess. Let us assume that the excess Ag^+ drives *all* the Br^- out of solution. This means that 0.080 M Ag^+ reacts with 0.030 M Br^- to produce AgBr while leaving 0.050 M Ag^+ in excess.

Let us now let as much AgBr dissolve as needed to establish the AgBr saturation equilibrium—let us say, x moles/liter. This will raise the Ag^+ concentration to $0.050 + x$ and will make the Br^- concentration equal to x.

$$\text{AgBr(s)} \ \rightleftharpoons \ \text{Ag}^+ + \text{Br}^-$$

$$K_{sp} = [\text{Ag}^+][\text{Br}^-] = 5.0 \times 10^{-13} = (0.050 + x)(x)$$

Since x is small compared to 0.050, we can estimate $0.050 + x \cong 0.050$. This gives us the approximate equation $5.0 \times 10^{-13} \cong (0.050)(x)$, which solves to $x = 1.0 \times 10^{-11}$. The final concentrations would be

$$[\text{Ag}^+] = 0.050 + x \cong 0.050 \ M; \qquad [\text{Br}^-] = x = 1.0 \times 10^{-11} \ M \ \blacksquare$$

■ EXAMPLE 400. The K_{sp} for $\text{Ag}_2\text{MoO}_4 \rightleftharpoons 2\text{Ag}^+ + \text{MoO}_4^{2-}$ is 2.6×10^{-11}. What should be the final concentration of Ag^+ and of MoO_4^{2-} in a solution made by mixing 25.0 ml of 0.10 M AgNO_3 and 45.0 ml of 0.10 M Na_2MoO_4?

SOLUTION: On mixing, the concentrations, without precipitation, become

$$\text{concentration of Ag}^+ = \left(\frac{25.0 \text{ ml}}{70.0 \text{ ml}}\right)(0.10 \ M) = 0.036 \ M$$

$$\text{concentration of MoO}_4^{2-} = \left(\frac{45.0 \text{ ml}}{70.0 \text{ ml}}\right)(0.10 \ M) = 0.064 \ M$$

The precipitation reaction $2Ag^+ + MoO_4^{2-} \rightarrow Ag_2MoO_4(s)$ requires 2 moles of Ag^+ per 1 mole of MoO_4^{2-}, so the MoO_4^{2-} is present in large excess. If we assume it drives *all* the Ag^+ out of solution, then to remove 0.036 M Ag^+ out of solution we will use up 0.018 M MoO_4^{2-}. This will leave $0.064 - 0.018 = 0.046$ M MoO_4^{2-}.

Let x = moles per liter of the solid Ag_2MoO_4 that will redissolve to set up the Ag_2MoO_4 saturation equilibrium. This will give $2x$ moles/liter of Ag^+ and x moles/liter of MoO_4^{2-} additional to the 0.046 M there already.

$$[Ag^+] = 2x; \qquad [MoO_4^{2-}] = 0.046 + x$$

$$Ag_2MoO_4(s) \; \rightleftharpoons \; 2Ag^+ + MoO_4^{2-}$$

$$K_{sp} = [Ag^+]^2[MoO_4^{2-}] = 2.6 \times 10^{-11}$$

$$(2x)^2(0.046 + x) = 2.6 \times 10^{-11}$$

Neglecting the x where it adds to 0.046, we get the approximate equation $(2x)^2(0.046) \cong 2.6 \times 10^{-11}$, which solves to give $x = 1.2 \times 10^{-5}$. Final concentrations will be

$$[Ag^+] = 2x = 2.4 \times 10^{-5} \, M; \qquad [MoO_4^{2-}] = 0.046 + x = 0.046 \, M \; \blacksquare$$

15.4   Mixing problems

One of the most common problems in freshman chemistry is to mix several reagents in a solution, and to decide what species are present in the final solution, and at what concentrations. To handle these problems successfully it is necessary to be systematic in collecting the information given, taking care of any reactions that occur, and calculating final concentrations with due allowance for dilution effects. If any reactions occur, they can usually be calculated easily if one or more of the reactants is present in excess. If a reactant is left over in excess, then in general its residual concentration can be used as a departure point for calculating the concentrations of other species in equilibrium with it.

In this section we consider primarily "mixing problems" in which precipitation and neutralization reactions are involved. We have already done several of these in preceding sections but not with all the ions in question.

■ EXAMPLE 401. Suppose you mix 25.0 ml of 0.012 M $BaCl_2$ with 50.0 ml of 0.010 M Ag_2SO_4. What will be the final concentrations of the ions in solution? The K_{sp} of AgCl is 1.7×10^{-10}; K_{sp} of $BaSO_4$ is 1.5×10^{-9}.

SOLUTION: One of the most efficient ways to solve a mixing problem such as this is to make a table showing which ions are involved, how many moles of each are available for reaction, how much change occurs due to reaction, what moles are left after reaction, and what the final concentrations will be.

(1) Ion	(2) Moles Available	(3) Change	(4) Moles after	(5) Concentration
Ba^{2+}	0.00030	−0.00030	∼0	$5.6 \times 10^{-7} M$
Cl^-	0.00060	−0.00060	∼0	$3 \times 10^{-8} M$
Ag^+	0.0010	−0.00060	0.0004	$0.005 M$
SO_4^{2-}	0.00050	−0.00030	0.00020	$0.0027 M$

In collecting our information for this table we note that $BaCl_2$ is a strong electrolyte, which is taken to be 100% dissociated into Ba^{2+} and Cl^-. Similarly, Ag_2SO_4 is considered 100% dissociated into Ag^+ and SO_4^{2-}. For calculating column 2, "moles available," we proceed in the usual way of multiplying the volume in liters of each solution by the concentration in moles per liter it provides of each ion.

Twenty-five ml of 0.012 M $BaCl_2$ gives

$$(0.0250 \text{ liter})\left(0.012 \, \frac{\text{mole of } Ba^{2+}}{\text{liter}}\right) = 0.00030 \text{ mole of } Ba^{2+}$$

and

$$(0.0250 \text{ liter})\left(0.024 \, \frac{\text{mole of } Cl^-}{\text{liter}}\right) = 0.00060 \text{ mole of } Cl^-$$

Fifty ml of 0.010 M Ag_2SO_4 gives

$$(0.0500 \text{ liter})\left(0.020 \, \frac{\text{mole of } Ag^+}{\text{liter}}\right) = 0.0010 \text{ mole of } Ag^+$$

and

$$(0.0500 \text{ liter})\left(0.010 \, \frac{\text{mole of } SO_4^{2-}}{\text{liter}}\right) = 0.00050 \text{ mole of } SO_4^{2-}$$

For column 3, labeled "change," we need to know how many moles of each species will be used up in reaction. Two reactions occur here and we consider them independently.

$$Ag^+ + Cl^- \longrightarrow AgCl(s)$$

$$Ba^{2+} + SO_4^{2-} \longrightarrow BaSO_4(s)$$

For the silver chloride precipitation, where we need to use up equimolar amounts of Ag^+ and Cl^-, we note from column 2 that we have available 0.00060 mole of Cl^- and 0.0010 mole of Ag^+—in other words, an excess of Ag^+. We assume as a first approximation that the excess Ag^+ drives *all* the Cl^- out of solution. (We will correct later for the small amount of Cl^- left.) To precipitate 0.00060 mole of Cl^- out of solution requires 0.00060 mole of Ag^+. We have 0.0010 mole of Ag^+ available, and we use up 0.00060; that leaves us 0.0004 mole of Ag^+, which we can enter in column 4, labeled "moles after." Since we are assuming all the Cl^- has been driven out of solution, we put ~ 0 (approximately zero) in column 4 for Cl^-.

Similarly, for the barium sulfate precipitation we assume that the excess SO_4^{2-} drives out all the Ba^{2+}. Starting with 0.00030 mole of Ba^{2+} and 0.00050 mole of SO_4^{2-} (column 2), we see that we need to use up 0.00030 mole of each (column 3), leaving approximately zero moles of Ba^{2+} and 0.00020 mole of SO_4^{2-} (column 4).

To get the values in column 5, labeled "concentration," we simply note that the moles (column 4) are in a total volume of 25.0 ml + 50.0 ml = 0.0750 liter of solution. So each value in column 4 gets divided by 0.0750 liter to go into column 5. This will give us a pretty good idea of the Ag^+ and SO_4^{2-} concentrations—namely, 0.005 M and 0.0027 M, respectively (note the difference in significant figures)—but it does not help for the ~ 0 of Ba^{2+} and Cl^-. There, however, we can quickly calculate $[Cl^-]$ from the K_{sp} of AgCl once we know that $[Ag^+] = 0.005$ M. Similarly, we can use the $[SO_4^{2-}] = 0.0027$ M to fix the concentration of Ba^{2+} via the K_{sp} of $BaSO_4$.

Let x = moles per liter of AgCl that dissolve in the presence of 0.005 M Ag^+.

$$K_{sp} = [Ag^+][Cl^-] = 1.7 \times 10^{-10}$$

$$(0.005 + x)(x) = 1.7 \times 10^{-10}$$

$$x \cong \frac{1.7 \times 10^{-10}}{0.005} = 3 \times 10^{-8} M = [Cl^-]$$

Let z = moles per liter of $BaSO_4$ that can dissolve in the presence of 0.0027 M SO_4^{2-}.

$$K_{sp} = [Ba^{2+}][SO_4^{2-}] = 1.5 \times 10^{-9}$$

$$(z)(z + 0.0027) = 1.5 \times 10^{-9}$$

$$z \cong \frac{1.5 \times 10^{-9}}{0.0027} = 5.6 \times 10^{-7} \, M = [Ba^{2+}] \ \blacksquare$$

■ EXAMPLE 402. What will be the final concentrations of the ions in a solution made by mixing 1.50×10^{-2} mole of $Sr(NO_3)_2$ and 3.0×10^{-3} mole of NaF in enough water to make 0.200 liter of solution? The K_{sp} of SrF_2 is 7.9×10^{-10}.

SOLUTION:

Ion	Moles Available	Change	Moles After	Concentration
Sr^{2+}	0.0150	−0.0015	0.0135	0.0675 M
NO_3^-	0.0300	no	0.0300	0.150 M
Na^+	0.0030	no	0.0030	0.015 M
F^-	0.0030	−0.0030	~ 0	1.1×10^{-4} M

The "change" is the precipitation of $SrF_2(s)$ via the reaction $Sr^{2+} + 2F^- \rightarrow SrF_2(s)$. We have available 0.0150 mole of Sr^{2+} and 0.0030 mole of F^-—in other words, a large excess of Sr^{2+}. Reaction is limited by the F^-. If all the fluoride ion is chased out of solution, 0.0030 mole of F^- requires 0.0015 mole of Sr^{2+}, corresponding to the $2F^-$-to-$1Sr^{2+}$ stoichiometry. This will leave $0.0150 - 0.0015$ mole, or 0.0135 mole, of Sr^{2+}. In a volume of 0.200 liter it corresponds to a concentration of

$$\frac{0.0135 \text{ mole } Sr^{2+}}{0.200 \text{ liter}} = 0.0675 \, M$$

The equilibrium between solid SrF_2 precipitate and the saturated solution is

$$SrF_2(s) \ \rightleftharpoons \ Sr^{2+} + 2F^-$$

$$K_{sp} = [Sr^{2+}][F^-]^2 = 7.9 \times 10^{-10}$$

If the concentration of Sr^{2+} is 0.0675 M (plus a small trace due to some dissolved SrF_2), we can calculate the fluoride-ion concentration by substituting 0.0675 M for $[Sr^{2+}]$.

$$(0.0675)[F^-]^2 = 7.9 \times 10^{-10}$$

$$[F^-] = 1.1 \times 10^{-4} \, M$$

[*Note:* An easy way to get a quick check on these problems is to see if the concentration of positive charge equals the concentration of negative charge. For positive charge, we have a total of $(2) \times (0.0675)$ M from the Sr^{2+} plus 0.015 M from the $Na^+ = 0.150$ M; for negative charge, we have 0.150 M from the NO_3^- and a negligible amount from the F^-.] ∎

■ EXAMPLE 403.   If you mix 10.0 ml of 0.10 M $CaCl_2$ with 20.0 ml of 0.10 M $AgNO_3$, what will be the concentrations of the ions in the final solution?

The K_{sp} of AgCl is 1.7×10^{-10}.

SOLUTION:

Ion	Moles Available	Change	Moles After	Concentration
Ca^{2+}	0.0010	no	0.0010	0.033 M
Cl^-	0.0020	-0.0020	~ 0	1.3×10^{-5} M
Ag^+	0.0020	-0.0020	~ 0	1.3×10^{-5} M
NO_3^-	0.0020	no	0.0020	0.067 M

[*Note:* This problem differs from the preceding ones in that the precipitating ions, Ag^+ and Cl^-, are present in stoichiometric amounts. Neither is in excess, so the final solution is the same as if AgCl(s) were dissolved in pure water.

$[Ag^+] = [Cl^-] =$ square root of $1.7 \times 10^{-10} = 1.3 \times 10^{-5}$ M ∎

■ EXAMPLE 404.   What will be the final concentrations in the solution resulting from the mixing of 0.100 liter of 0.150 M H_2SO_4 with 0.300 liter of 0.200 M $Ba(OH)_2$? The K_{sp} of $BaSO_4$ is 1.5×10^{-9}.

SOLUTION:

Ion	Moles Available	Change	Moles After	Concentration
H_3O^+	0.0300	-0.0300	~ 0	4.3×10^{-14} M
SO_4^{2-}	0.0150	-0.0150	~ 0	1.3×10^{-8} M
Ba^{2+}	0.0600	-0.0150	0.0450	0.113 M
OH^-	0.120	-0.0300	0.090	0.23 M

There are two reactions here:

$$H_3O^+ + OH^- \longrightarrow 2H_2O$$

$$Ba^{2+} + SO_4^{2-} \longrightarrow BaSO_4(s)$$

There is more OH^- available than H_3O^+, so all the acid will be neutralized. This means we do not have to worry about HSO_4^- in the solution. Once the acid–base neutralization has occurred, we can calculate the residual OH^- concentration and get from that the concentration of H_3O^+.

$$[H_3O^+] = \frac{K_w}{[OH^-]} = \frac{1.00 \times 10^{-14}}{0.23} = 4.3 \times 10^{-14} \ M$$

The excess Ba^{2+} drives most of the SO^{2-} out of the solution; the trace that is left can be computed from the residual Ba^{2+} concentration and the K_{sp} of $BaSO_4$. ∎

■ EXAMPLE 405. A solution is made by mixing 0.10 liter of 0.12 M NaCl, 0.20 liter of 0.14 M NaBr, and 0.30 liter of 0.10 M AgNO$_3$. What will be the concentration of each ion in the final solution? The K_{sp} of AgCl is 1.7×10^{-10}; the K_{sp} of AgBr is 5.0×10^{-13}.

SOLUTION:

(1)	(2)	(3)	(4)	(5)
	Moles			
Ion	Available	Change	Moles After	Concentration
Na$^+$	0.012 + 0.028	no	0.040	0.067 M
Cl$^-$	0.012	−0.002	0.010	0.017 M
Br$^-$	0.028	−0.028	∼0	5.0×10^{-5} M
Ag$^+$	0.030	−0.030	∼0	1.0×10^{-8} M
NO$_3^-$	0.030	no	0.030	0.050 M

In working up the data, we note for column 2 that Na$^+$ is contributed from two sources: 0.012 mole from 0.10 liter of 0.12 M NaCl plus 0.028 mole from 0.20 liter of 0.14 M NaBr.

In deciding on the change for column 3, we note that two possible reactions can occur:

$$Ag^+ + Cl^- \longrightarrow AgCl(s)$$

$$Ag^+ + Br^- \longrightarrow AgBr(s)$$

We do not have enough Ag^+ to complete both of these reactions; only 0.030 mole of Ag^+ is available, but 0.12 mole of Cl^- plus 0.28 mole of Br^- would need 0.040 mole of Ag^+ for complete precipitation. The question is, which reaction is preferred. As a rule, the least soluble substance is the one most likely to be produced. (You can justify this by noting that if the more soluble salt deposited first, then it would have in equilibrium with itself a *higher* concentration of Ag^+ than the less soluble salt. With a *higher* Ag^+ concentration in solution we would be more likely to exceed the K_{sp} of the less soluble salt and therefore preferentially drive the less soluble salt out of solution.)

Since the K_{sp} of AgBr is smaller than the K_{sp} of AgCl, we expect AgBr to precipitate first. Starting with the initially available moles, we would have 0.030 mole of Ag^+ + 0.028 mole of $Br^- \rightarrow 0.028$ mole of AgBr precipitated and 0.002 mole of Ag^+ left over.

Then, the remainder of the Ag^+ can operate on the Cl^-: 0.002 mole of Ag^+ + 0.012 mole of $Cl^- \rightarrow 0.002$ mole of AgCl precipitated and 0.010 mole of Cl^- left over.

The net result is to use up all the available Ag^+ and drive out of solution all the available Br^- and part of the available Cl^-.

Knowing how many moles of Cl^- are left in the final solution (column 4), we can calculate the final concentration of Cl^- as

$$\frac{0.010 \text{ mole}}{0.60 \text{ liter}} = 0.017 \ M \text{ (column 5)}$$

Once we have $[Cl^-] = 0.017 \ M$, we can calculate $[Ag^+]$.

$$[Ag^+] = \frac{K_{sp} \text{ of AgCl}}{[Cl^-]} = \frac{1.7 \times 10^{-10}}{0.017} = 1.0 \times 10^{-8} \ M$$

Once we have the $[Ag^+]$, we can calculate the trace of Br^- that is left in solution.

$$[Br^-] = \frac{K_{sp} \text{ of AgBr}}{[Ag^+]} = \frac{5.0 \times 10^{-13}}{1.0 \times 10^{-8}} = 5.0 \times 10^{-5} \ M \ \blacksquare$$

Exercises

15.1. Compute the K_{sp} of AgBr given the fact that the solubility of AgBr in water is 1.3×10^{-4} g/liter.

15.2. Calculate the K_{sp} of $Ba(OH)_2$ given the fact that the solubility of $Ba(OH)_2$ in water is 4.6 g per 0.250 liter.

*15.3. Given that the K_{sp} of AgI is 8.5×10^{-17}, calculate how many grams of AgI can dissolve per liter of water. (*Answer*: 2.2×10^{-6} g.)

**15.4. Given that the K_{sp} of $Sr(OH)_2$ is 3.2×10^{-4}, calculate how many grams of $Sr(OH)_2$ can dissolve per liter of water.

**15.5. Given that K_{sp} for the equilibrium $Hg_2Cl_2(s) \rightleftarrows Hg_2^{2+} + 2Cl^-$ is equal to 1.1×10^{-18}, calculate what would be the concentration of chloride ion in water that is saturated with Hg_2Cl_2.

**15.6. Given that the K_{sp} of $Ca(OH)_2$ is 1.3×10^{-6}, what would be the concentration of calcium ion in $0.10\ M$ NaOH solution that has been saturated with calcium hydroxide? (*Answer*: $1.3 \times 10^{-4}\ M$.)

**15.7. If 2.6 g of Ag_2SO_4 can dissolve in 100.0 ml of pure water, how many grams of Ag_2SO_4 should be able to dissolve in 100.0 ml of $1.0\ M$ $AgNO_3$ solution?

***15.8. If 2.6 g of Ag_2SO_4 can dissolve in 100.0 ml of pure water, how many grams of Ag_2SO_4 should be able to dissolve in 100.0 ml of $0.10\ M$ $AgNO_3$? (*Answer*: 1.7 g.)

**15.9. If 2.4×10^{-3} g of $Cu(IO_3)_2$ can dissolve in a liter of $0.150\ M$ $NaIO_3$ solution, then how many grams should be able to dissolve in a liter of water?

*15.10. Given that the K_{sp} of $AgNO_2$ is 1.2×10^{-4}, what will be the final concentration of Ag^+ and of NO_2^- on mixing 100.0 ml of $0.010\ M$ $AgNO_3$ and 100.0 ml of $0.010\ M$ $NaNO_2$?

**15.11. Given that the K_{sp} of $AgBrO_3$ is 5.4×10^{-5}, what will be the final concentration of Ag^+ and of BrO_3^- on mixing 100.0 ml of $0.10\ M$ $AgNO_3$ and 100.0 ml of $0.10\ M$ $NaBrO_3$?

**15.12. Given that the K_{sp} of AgCNS is 1.0×10^{-12}, how many milliliters of $0.10\ M$ $AgNO_3$ do you have to add to 100.0 ml of $0.10\ M$ NaCNS to make the final concentration of CNS^- equal to $2.0 \times 10^{-10}\ M$? (*Answer*: 111 ml.)

**15.13. What will be the final concentration of Ag^+ and of CrO_4^{2-} in a solution made by mixing 10.0 ml of $0.050\ M$ $AgNO_3$ with 5.0 ml of $0.050\ M$ Na_2CrO_4? The K_{sp} of Ag_2CrO_4 is 1.9×10^{-12}.

**15.14. What will be the final concentration of Ag^+ and of CrO_4^{2-} in a solution made by mixing 5.0 ml of $0.050\ M$ $AgNO_3$ with 10.0 ml of $0.050\ M$ Na_2CrO_4? The K_{sp} of Ag_2CrO_4 is 1.9×10^{-12}.

**15.15. Given that the K_{sp} of $Mg(OH)_2$ is 8.9×10^{-12}, what will be the final concentration of Mg^{2+} when 0.300 liter of $0.10\ M$ $Mg(NO_3)_2$ is mixed with 0.100 liter of $0.50\ M$ NaOH plus 0.100 liter of $0.30\ M$ KOH? (*Answer*: $5.6 \times 10^{-9}\ M$.)

**15.16. What must be the K_{sp} of an MX salt such that mixing 0.100 liter of $0.050\ M$ M^+ with 0.200 liter of $0.075\ M$ X^- produces a solution containing M^+ at a concentration of $9.0 \times 10^{-9}\ M$?

**15.17. What must be the K_{sp} of an MX_2 salt such that mixing 0.100 liter of 0.050 M M^{2+} with 0.200 liter of 0.075 M X^- produces a solution containing M^{2+} at a concentration of 9.0×10^{-9} M?

15.18. Given that $K_{sp} = 1.5 \times 10^{-9}$ for $BaSO_4$ and $K_{sp} = 1.7 \times 10^{-10}$ for AgCl, what is the maximum volume of 0.0010 M $BaCl_2$ that can be added to 0.500 liter of 0.0010 M Ag_2SO_4 before something starts to precipitate? (*Answer:* 0.021 ml.)

15.19. What will be the final concentration of each ion in a solution made by mixing 0.10 liter of 0.35 M $AgNO_3$ with 0.20 liter of 0.25 M HCl? The K_{sp} of AgCl is 1.7×10^{-10}.

15.20. Suppose you add dropwise 0.10 M NaF solution to 0.100 liter of a solution that is 0.050 M $Ca(NO_3)_2$ and 0.050 M $Mg(NO_3)_2$. What will be the concentration of Ca^{2+} when Mg^{2+} first starts to precipitate? The K_{sp} of CaF_2 is 1.7×10^{-10}; the K_{sp} of MgF_2 is 8×10^{-8}.

15.21. The respective K_{sp}'s of AgCl, AgBr, and AgI are 1.7×10^{-10}, 5.0×10^{-13}, and 8.5×10^{-17}. A mixture is made of 0.0010 mole of NaI, 0.0020 mole of NaBr, 0.0030 mole of NaCl, 0.0040 mole of $AgNO_3$, and 100.0 ml of water. What is the concentration of I^- in the final mixture? (*Answer:* 1.0×10^{-8} M.)

15.22. How many milliliters of 1.0 M $AgNO_3$ would you need to add to start precipitating $AgNO_2$ from 0.100 liter of a solution that is 0.10 M HNO_3 and 0.10 M HNO_2? The K_{sp} of $AgNO_2$ is 1.2×10^{-4}; the K_{diss} of HNO_2 is 4.5×10^{-4}.

15.23. What would be the final concentration of Ag^+ and of NO_2^- in a solution made by mixing 0.010 mole of $AgNO_2$ with 0.100 liter of 0.50 M HCl? The K_{sp} of AgCl is 1.7×10^{-10}, the K_{sp} of $AgNO_2$ is 1.2×10^{-4}, and the K_{diss} of HNO_2 is 4.5×10^{-4}.

***15.24.** Radium nitrate is unusual in that it is relatively insoluble compared to most nitrates, which are generally quite soluble. The K_{sp} of $Ra(NO_3)_2$ is 6.3×10^{-3}. What should be the final concentration of Ra^{2+} and of NO_3^- in a solution made by mixing 30 ml of 1 M $NaNO_3$ with 30 ml of 0.20 M $Ra(OH)_2$? Assume $Ra(OH)_2$ to be a soluble, strong electrolyte. (*Answer:* $[Ra^{2+}] = 0.043$ M; $[NO_3^-] = 0.38$ M.)

16

HYDROLYSIS AND OTHER CASES OF SIMULTANEOUS EQUILIBRIA

IN THE LAST several chapters we have been flirting with the rather complex problem of handling several equilibria at the same time. So far, we have been able to avoid serious difficulties by introducing appropriate approximations when necessary. Now we need to look more closely at what approximations can be made and why they can be made. There is nothing wrong with a procedure based on valid approximations and, in fact, the chemist normally approaches his systems of study with a highly developed sense of what is the important process in a given system, and what can be ignored. It takes considerable training to develop such a feeling for what can be neglected, and frequently it comes only after extensive experience in carrying through hundreds of laborious calculations. The purpose of working out the preceding problems of this book was largely to acquire such an experience.

16.1 Hydrolysis of NaOAc

Sodium acetate is typical of a salt that is derived from a strong base (NaOH) and a weak acid (HOAc). When NaOAc is placed in water, we get a basic solution and the problem is to account for this basicity in a quantitative fashion. The qualitative reason why NaOAc gives a basic solution is this: The added OAc^- disturbs the water equilibrium $H_2O + H_2O \rightleftarrows H_3O^+ + OH^-$ by reacting with some H_3O^+ and thereby reducing the concentration of hydronium ion below the concentration of hydroxide ion.

■ EXAMPLE 406. The dissociation constant of HOAc is 1.8×10^{-5}. Calculate the species concentrations and the pH of 1.00 M NaOAc solution.

SOLUTION: There are three ways we can solve this problem.

(a) *Method I.* Let us try to solve the problem exactly. When we put 1.00 M NaOAc in water, we get 1.00 M Na$^+$ and 1.00 M OAc$^-$. The Na$^+$ remains free, but some of the OAc$^-$ will combine with H_3O^+ to form HOAc. Thus we will have in the solution two equilibria to worry about:

$$H_2O + H_2O \;\rightleftharpoons\; H_3O^+ + OH^- \qquad \text{for which } K_w$$
$$= [H_3O^+][OH^-]$$

$$H_2O + HOAc \;\rightleftharpoons\; H_3O^+ + OAc^- \qquad \text{for which } K_{diss}$$
$$= \frac{[H_3O^+][OAc^-]}{[HOAc]}$$

Let the concentrations at equilibrium be represented as follows:

$[H_3O^+] = w$

$[OH^-] = x$

$[OAc^-] = y$

$[HOAc] = z$

What relations must exist between these four unknowns?

First, we need to satisfy $K_w = [H_3O^+][OH^-] = 1.00 \times 10^{-14}$. This gives us the equation $(w)(x) = 1.00 \times 10^{-14}$.

Second, we need to satisfy the dissociation constant.

$$K_{diss} = \frac{[H_3O^+][OAc^-]}{[HOAc]} = 1.8 \times 10^{-5}$$

this gives us the equation

$$\frac{(w)(y)}{(z)} = 1.8 \times 10^{-5}$$

Third, the solution must remain electrically neutral. This requires that the sum of the positive-charge concentrations equals the sum of the negative-charge concentrations. The positive species in the solution are Na$^+$ and H_3O^+; the negative, OAc$^-$ and OH$^-$. This means that the sum of the Na$^+$ and H_3O^+ concentrations equals the sum of the OAc$^-$ and OH$^-$ concentrations.

$$[Na^+] + [H_3O^+] = [OAc^-] + [OH^-]$$

or

$$1.00 + w = y + x$$

Fourth, we need to maintain conservation of mass. There can be no loss or gain of atoms as equilibrium gets established. So far as we are concerned, this means that the total acetate we put into the solution must show up as HOAc or OAc⁻. We put in 1.00 mole/liter, so we have to account for that much.

$$[HOAc] + [OAc^-] = 1.00$$

or

$$z + y = 1.00$$

That is all we need. We have four unknowns w, x, y, and z and we have four independent relations between them:

$$(w)(x) = 1.00 \times 10^{-14}$$

$$\frac{(w)(y)}{(z)} = 1.8 \times 10^{-5}$$

$$1.00 + w = y + x$$

$$z + y = 1.00$$

All we need to do is to solve these four simultaneous equations. It is messy and tedious but it can be done to give the following results:

$$w = 4.2 \times 10^{-10} M = [H_3O^+]$$

$$x = 2.4 \times 10^{-5} M = [OH^-]$$

$$y = 1.00\ M = [OAc^-]$$

$$z = 2.4 \times 10^{-5}\ M = [HOAc]$$

Try the method from scratch and you will see that this is obviously not a method to use when you are in a hurry.

(b) *Method II.* This is a method where we introduce the approximation that for every OAc⁻ that combines with an H_3O^+ from the H_2O dissociation, an OH⁻ is released into the solution. This means that if x moles/liter of OAc⁻ pick up H_3O^+ to form HOAc, we will get x moles/liter of HOAc and x

moles/liter of OH^-, as well as reduce the OAc^- concentration from 1.00 to $1.00 - x$.

At equilibrium, we would have

$$[OAc^-] = 1.00 - x; [HOAc] = x; [OH^-] = x$$

We know that we must satisfy K_{diss}:

$$K_{diss} = \frac{[H_3O^+][OAc^-]}{[HOAc]} = 1.8 \times 10^{-5}$$

If we express $[H_3O^+] = \dfrac{K_w}{[OH^-]} = \dfrac{1.00 \times 10^{-14}}{x}$, which simply amounts to relating x moles/liter of OH^- via the water constant to

$$\left(\frac{1.00 \times 10^{-14}}{x}\right) \text{ moles/liter of } H_3O^+$$

then on substituting in K_{diss}, we get

$$K_{diss} = \frac{[H_3O^+][OAc^-]}{[HOAc]} = \frac{\left(\dfrac{1.00 \times 10^{-14}}{x}\right)(1.00 - x)}{x} = 1.8 \times 10^{-5}$$

Rewriting, we get

$$\frac{(1.00 \times 10^{-14})(1.00 - x)}{x^2} = 1.8 \times 10^{-5}$$

This can easily be solved for x, since x can be neglected when subtracted from 1.00 and we can assume $1.00 - x \cong 1.00$.

The approximate equation

$$\frac{(1.00 \times 10^{-14})(1.00)}{x^2} \cong 1.8 \times 10^{-5}$$

solves to give $x = 2.4 \times 10^{-5}$. This leads to the following equilibrium concentrations:

$$[HOAc] = x = 2.4 \times 10^{-5} \, M$$

$$[OAc^-] = 1.00 - x = 1.00 \, M$$

$$[OH^-] = x = 2.4 + 10^{-5} \, M$$

$$[H_3O^+] = \frac{K_w}{[OH^-]} = \frac{1.00 \times 10^{-14}}{2.4 \times 10^{-5}} = 4.2 \times 10^{-10} \, M$$

which answers are the same as those obtained, more tediously, in method I.

(c) *Method III (recommended)*. When OAc^- is put in water, it reacts with H_3O^+ to form HOAc. We can write this reaction

$$OAc^- + H_3O^+ \longrightarrow HOAc + H_2O$$

But as soon as some H_3O^+ is withdrawn from the solution this way, the H_2O dissociates some more to try to replace that H_3O^+. The reaction would be

$$H_2O + H_2O \longrightarrow H_3O^+ + OH^-$$

If we assume that the H_2O thus furnishes all the H_3O^+ used up in converting OAc^- to HOAc, then we can add the two reactions above to get the *net reaction*

$$OAc^- + H_2O \rightleftharpoons HOAc + OH^-$$

We have canceled the H_3O^+ from both sides of the equation and have put in the double arrows to emphasize that this net reaction as well as the individual steps, must come to equilibrium. This is the so-called *hydrolysis reaction*, and its equilibrium constant, sometimes designated K_h (called the *hydrolysis constant*), can be written as follows:

$$K_h = \frac{[HOAc][OH^-]}{[OAc^-]}$$

We leave out the water from the denominator because as usual its activity stays constant.

The numerical value of K_h can be shown to be equal to K_w/K_{diss} by the simple expedient of multiplying the numerator and the denominator by $[H_3O^+]$.

$$K_h = \frac{[HOAc][OH^-]}{[OAc^-]} \frac{[H_3O^+]}{[H_3O^+]} = \frac{[H_3O^+][OH^-]}{[H_3O^+][OAc^-]/[HOAc]} = \frac{K_w}{K_{diss}}$$

To solve our problem, we now compute as with any equilibrium situation.

Let $x =$ moles per liter of OAc^- that hydrolyze by the net reaction

$$OAc^- + H_2O \rightleftharpoons HOAc + OH^-$$

This will give us at equilibrium x moles/liter of HOAc and x moles/liter of OH^- while reducing the concentration of OAc^- from $1.00\ M$ to $(1.00 - x)\ M$.

$$[HOAc] = x; \qquad [OH^-] = x; \qquad [OAc^-] = 1.00 - x$$

$$K_h = \frac{[HOAc][OH^-]}{[OAc^-]} = \frac{(x)(x)}{1.00 - x} = \frac{K_w}{K_{diss}} = \frac{1.00 \times 10^{-14}}{1.8 \times 10^{-5}} = 5.6 \times 10^{-10}$$

If we solve the equation

$$\frac{(x)(x)}{1.00 - x} = 5.6 \times 10^{-10}$$

we get $x = 2.4 \times 10^{-5}$, which leads to the following:

$$[HOAc] = x = 2.4 \times 10^{-5} \ M$$

$$[OH^-] = x = 2.4 \times 10^{-5} \ M$$

$$[OAc^-] = 1.00 - x = 1.00 \ M$$

$$[H_3O^+] = \frac{K_w}{[OH^-]} = \frac{1.00 \times 10^{-14}}{2.4 \times 10^{-5}} = 4.2 \times 10^{-10} \ M$$

These values are the same as those obtained by method I or II.

$$pH = -\log [H_3O^+] = -\log (4.2 \times 10^{-10}) = -(0.62 - 10) \ \blacksquare$$
$$= 9.38$$

■ EXAMPLE 407.   How concentrated a solution of NaOAc must you take in order to have 0.015% of the acetate ion in the "hydrolyzed" state?

SOLUTION: Let c be the overall concentration of the desired solution. This will give us c moles/liter of Na^+ and c moles/liter of OAc^-. But 0.015% of c, or $0.00015c$, will be converted to HOAc and OH^- by the reaction

$$OAc^- + H_2O \longrightarrow HOAc + OH^-$$

This will leave 99.985% of c, or $0.99985c$, in the unhydrolyzed state. So, at equilibrium we can write the concentrations as follows:

$$[OAc^-] = 0.99985c; \qquad [HOAc] = 0.00015c; \qquad [OH^-] = 0.00015c$$

Substitute these values in the equilibrium condition for the hydrolysis equilibrium:

$$OAc^- + H_2O \rightleftharpoons HOAc + OH^-$$

$$K_h = \frac{[HOAc][OH^-]}{[OAc^-]} = \frac{K_w}{K_{diss}} = \frac{1.00 \times 10^{-14}}{1.8 \times 10^{-5}} = 5.6 \times 10^{-10}$$

$$\frac{(0.00015c)(0.00015c)}{0.99985c} = 5.6 \times 10^{-10}$$

$$c = 2.5 \times 10^{-2} \text{ mole/liter} \blacksquare$$

When can we expect the above approximation, as outlined in methods II and III, to break down? The assumption we are making is that the H_3O^+ picked up by the OAc^- is just matched by dissociation of $H_2O + H_2O \rightarrow H_3O^+ + OH^-$. This leads to the net equation $OAc^- + H_2O \rightarrow HOAc + OH^-$, which shows that HOAc and OH^- are produced in equal amounts. Under what circumstances will we *not* be able to say that the concentration of HOAc equals the concentration of OH^-? The answer: When the solution is very dilute. Given a very dilute solution of NaOAc—say, 10^{-7} M—then the amount of OH^- produced by the hydrolysis is actually less than the amount of OH^- produced by the normal water dissociation, and we cannot assume that $[HOAc] = [OH^-]$. In fact, the normal $H_2O + H_2O \rightleftharpoons H_3O^+ + OH^-$ equilibrium would be the dominant equilibrium and would set $[H_3O^+] \cong 10^{-7}$ M, to which the $[OAc^-]/[HOAc]$ would have to conform. To solve the problem exactly, we would need to follow the procedure as outlined in method I with all its tedious algebra. Fortunately, freshman chemistry problems routinely avoid asking for precise hydrolysis calculations in very dilute solutions.

16.2   Hydrolysis of NH₄Cl

Ammonium chloride is typical of the salts derived from a weak base (NH_3) and a strong acid (HCl). There is the minor additional complication that the weak base is not of the general type BOH, although all the following computations can be done just as well assuming that NH_3 in H_2O is NH_4OH. However, we shall try to stick to modern thinking on ammonia solutions as much as possible and avoid postulating presence of NH_4OH.

When ammonium chloride is added to water, we get 100% dissociation into NH_4^+ and Cl^-, as is generally assumed for any strong electrolyte. The Cl^- does not affect the water equilibrium $H_2O + H_2O \rightleftharpoons H_3O^+ + OH^-$ but the NH_4^+ does, because some of it can combine with OH^- to produce NH_3 and H_2O by the reaction

$$NH_4^+ + OH^- \longrightarrow NH_3 + H_2O$$

If we again make the assumption that H_2O dissociation furnishes all the OH^- used in this reaction, we can add the equation

$$H_2O + H_2O \longrightarrow H_3O^+ + OH^-$$

and get a total equation

$$NH_4^+ + H_2O \longrightarrow NH_3 + H_3O$$

This looks like a simple dissociation of a protonic acid and in fact you have a choice of considering NH_4Cl solutions as representing hydrolysis of NH_4^+ or dissociation of NH_4^+, provided you use the right equilibrium constants and interpret them correctly. This illustrates, incidentally, the very fundamental point that *hydrolysis is not really a phenomenon separate from dissociation but that it represents a special way of looking at the competition between H_2O and other proton-donating or proton-accepting species.*

What is the equilibrium constant for $NH_4^+ + H_2O \rightleftarrows NH_3 + H_3O^+$? It has the form

$$K = \frac{[NH_3][H_3O^+]}{[NH_4^+]}$$

If we multiply numerator and denominator by $[OH^-]$, which does not change any numerical values, because we are multiplying the whole business by one, we get

$$K = \frac{[NH_3][H_3O^+]}{[NH_4^+]} \frac{[OH^-]}{[OH^-]} = \frac{[H_3O^+][OH^-]}{[NH_4^+][OH^-]/[NH_3]}$$

The terms have been rearranged to emphasize certain groupings and we see immediately that $[H_3O^+][OH^-]$ is just K_w and the denominator $[NH_4^+]$ $[OH^-]/[NH_3]$ is just what we have been calling the "dissociation constant of aqueous ammonia" (see Section 12.3). Putting in the numerical values $K_w = 1.00 \times 10^{-14}$ and $K_{diss} = 1.81 \times 10^{-5}$, we get $K = 5.52 \times 10^{-10}$. This is frequently called the hydrolysis constant of ammonium ion but it could just as well be called the "dissociation constant of ammonium ion."

■ EXAMPLE 408.   Compute the pH of 1.00 M NH₄Cl.

SOLUTION: NH₄Cl is a strong electrolyte; we take it to be 100% dissociated into NH_4^+ and Cl^-. Of the 1.00 M NH_4^+ introduced into the solution, let x moles/liter convert by the reaction

$$NH_4^+ + H_2O \longrightarrow NH_3 + H_3O^+$$

to give x moles/liter of NH_3, x moles/liter of H_3O^+, and leave $(1.00 - x)$ moles/liter of NH_4^+.

$$[NH_4^+] = 1.00 - x; \quad [NH_3] = x; \quad [H_3O^+] = x$$

$$NH_4^+ + H_2O \rightleftharpoons NH_3 + H_3O^+$$

$$K = \frac{[NH_3][H_3O^+]}{[NH_4^+]} = 5.52 \times 10^{-10} = \frac{(x)(x)}{1.00 - x}$$

$$x = 2.35 \times 10^{-5}$$

$$[NH_4^+] = 1.00\,M; \quad [NH_3] = 2.35 \times 10^{-5}M; \quad [H_3O^+] = 2.35 \times 10^{-5}M$$

$$pH = -\log [H_3O^+] = -\log (2.35 \times 10^{-5}) = -(0.371 - 5)$$

$$= 4.629 \ ■$$

■ EXAMPLE 409.   How many grams of NH₄Cl would you need to dissolve in 0.200 liter of water to provide a solution having a pH of 4.75?

SOLUTION:

$$pH = 4.75 = -\log [H_3O^+]$$

$$\log [H_3O^+] = -4.75 = -5.00 + 0.25$$

$$[H_3O^+] = (10^{-5})(10^{0.25}) = 1.8 \times 10^{-5}$$

If this H_3O^+ is to come from NH_4^+ dissociation, there must be an equal concentration of NH_3 in solution.

$$[H_3O^+] = 1.8 \times 10^{-5}\ M; \quad [NH_3] = 1.8 \times 10^{-5}\ M$$

What must be the concentration of ammonium ion to be consistent with these? We substitute in K to find out.

$$K = \frac{[NH_3][H_3O^+]}{[NH_4^+]} = 5.52 \times 10^{-10} = \frac{(1.8 \times 10^{-5})(1.8 \times 10^{-5})}{[NH_4^+]}$$

$$[NH_4^+] = \frac{(1.8 \times 10^{-5})(1.8 \times 10^{-5})}{5.52 \times 10^{-10}} = 0.59\ M$$

Total concentration needed is 0.59 M (to take care of NH_4^+) plus 1.8×10^{-5} M (to take care of NH_3), or 0.59 M.

To answer the question of how many grams per 0.200 liter, we need to multiply molarity by the number of liters and by the weight of 1 mole of NH_4Cl (53.49 g).

$$\text{Grams needed} = \left(0.59 \frac{\text{mole}}{\text{liter}}\right)(0.200 \text{ liter})\left(53.49 \frac{\text{g}}{\text{mole}}\right) = 6.3 \text{ g} \blacksquare$$

As in any salt solution, the degree of hydrolysis can be repressed by having some of the weak electrolyte added to the solution. Thus we can diminish the extent of NH_4^+ hydrolysis by having some additional NH_3 in the solution. Then of course, the problem becomes similar to the buffer problems we considered in Chapter 14.

$\blacksquare$ EXAMPLE 410.   Suppose you have 1.00 liter of 1.50 M NH_4Cl solution and 1.00 liter of 1.50 M NH_3 solution. You wish to make 0.200 liter of a solution that has pH = 7.00. What is the recipe?

SOLUTION:

$$pH = 7.00 = -\log [H_3O^+]$$
$$\log [H_3O^+] = -7.00 = 0.00 - 7$$
$$[H_3O^+] = (10^{0.000})(10^{-7}) = 1.0 \times 10^{-7} \ M$$

Use this $[H_3O^+]$ to fix the ratio of NH_3 to NH_4^+ needed. The equilibrium is

$$NH_4^+ + H_2O \ \rightleftharpoons \ NH_3 + H_3O^+$$

The requirement for equilibrium is

$$K = \frac{[NH_3][H_3O^+]}{[NH_4^+]} = 5.52 \times 10^{-10} = \frac{[NH_3][1.0 \times 10^{-7}]}{[NH_4^+]}$$

which leads to

$$\frac{[NH_3]}{[NH_4^+]} = \frac{5.52 \times 10^{-10}}{1.0 \times 10^{-7}} = 5.5 \times 10^{-3}$$

This fraction is so much less than one that we can be quite sure the solution is essentially all NH_4^+ with just a trace of NH_3.

Let V_1 = volume of 1.50 M NH₄Cl solution to be taken; V_2 = volume of 1.50 M NH₃ solution to be taken.

$V_1 + V_2$ = total volume = 0.200 liter

V_1 liters of 1.50 M NH₄Cl will give 1.50V_1 moles of NH_4^+.

V_2 liters of 1.50 M NH₃ will give 1.50V_2 moles of NH₃.

Let x = moles of NH_4^+ that will dissociate to NH₃ and H_3O^+ to fix the final equilibrium. This will leave us with $(1.50V_1 - x)$ moles of NH_4^+ and $(1.50V_2 + x)$ moles of NH₃ in the final solution. Dividing by 0.200 liter will give us concentrations. The ratio of the final concentrations must come out to be the number 5.5×10^{-3} as figured above.

$$\frac{[NH_3]}{[NH_4^+]} = 5.5 \times 10^{-3} = \frac{(1.50V_2 + x) \text{ moles of NH}_3 \text{ per 0.200 liter}}{(1.50V_1 - x) \text{ moles of NH}_4^+ \text{ per 0.200 liter}}$$

Since the 0.200 liter divides into the numerator and into the denominator, it cancels out. We are left with

$$5.5 \times 10^{-3} = \frac{1.50V_2 + x}{1.50V_1 - x}$$

where $V_1 + V_2 = 0.200$ liter; x can be neglected, since it is a very small number, being equal to the number of moles of H_3O^+ produced in the solution— namely, $(1.0 \times 10^{-7} M)(0.200 \text{ liter}) = 2 \times 10^{-8}$.

Substituting $V_1 = 0.200 - V_2$ in the equation above gives us

$$5.5 \times 10^{-3} = \frac{1.50V_2}{(1.50)(0.200 - V_2)}$$

which solves to $V_2 = 1.1 \times 10^{-3}$ liter and $V_1 = 0.199$ liter.

So, to make our solution we need to add 1.1 ml of 1.50 M NH₃ solution to 0.199 liter of 1.50 M NH₄Cl solution. ■

When will these equations for calculating the hydrolysis of NH_4^+ break down? Again the answer is: in dilute solution. At first sight, this may be surprising, since the NH_4^+ hydrolysis looks like a dissociation ($NH_4^+ + H_2O$ $\rightleftarrows NH_3 + H_3O^+$) and in Chapter 12 we did not encounter any troubles with dissociation of weak acids. True, but it is also true that we did not consider in Chapter 12 any very dilute solutions. If we had, we would have been in trouble, because it stands to reason that in a very dilute solution of HX the amount of H_3O^+ coming from $HX + H_2O \rightarrow H_3O^+ + X^-$ may be comparable or even less than that which comes from $H_2O + H_2O \rightleftarrows H_3O^+ + OH^-$. In such a case, we cannot just work with $K_{diss} = [H_3O^+][X^-]/[HX]$ but must also consider $K_w = [H_3O^+][OH^-]$ as a simultaneous equilibrium. The quantitative calculations again get to be very tedious, requiring an exact method such as method I (Section 16.1) for NaOAc solutions. Fortunately, we run into this kind of trouble only with very dilute solutions (e.g., more dilute than 10^{-5} M NH_4^+).

■ EXAMPLE 411.   Calculate the concentration of H_3O^+, OH^-, NH_4^+, and NH_3 in 1.00×10^{-6} M NH_4Cl solution. $K_w = 1.00 \times 10^{-14}$ and $K_{diss} = 5.52 \times 10^{-10}$ for $NH_4^+ + H_2O \rightleftarrows NH_3 + H_3O^+$.

SOLUTION: Set up four unknowns for the four species mentioned.

$$[H_3O^+] = w; \quad [OH^-] = x; \quad [NH_4^+] = y; \quad [NH_3] = z$$

Then set up the four equations that relate these concentrations.
(a) Water equilibrium: $H_2O + H_2O \rightleftarrows H_3O^+ + OH^-$

$$K_w = [H_3O^+][OH^-] = 1.00 \times 10^{-14} = (w)(x)$$

(b) Ammonium dissociation: $NH_4^+ + H_2O \rightleftarrows NH_3 + H_3O^+$

$$K = \frac{[NH_3][H_3O^+]}{[NH_4^+]} = 5.52 \times 10^{-10} = \frac{(z)(w)}{y}$$

(c) Neutrality condition: $[NH_4^+] + [H_3O^+] = [Cl^-] + [OH^-]$

$$y + w = 1.00 \times 10^{-6} + x$$

(d) Mass conservation: $[NH_4^+] + [NH_3] = 1.00 \times 10^{-6}$

$$y + z = 1.00 \times 10^{-6}$$

Solving these four simultaneous equations gives the following results:

$$[H_3O^+] = 1.03 \times 10^{-7} \ M; \quad [OH^-] = 9.75 \times 10^{-8} \ M$$

$$[NH_4^+] = 9.95 \times 10^{-7} \ M; \quad [NH_3] = 5.35 \times 10^{-9} \ M \ ■$$

16.3   *Hydrolysis of NH₄OAc*

Ammonium acetate represents a salt derived from a weak base (NH_3) and a weak acid (HOAc), with the added twist that the constant for the hydrolysis of NH_4^+ just matches the constant for the hydrolysis of OAc^-. Therefore, it should be no surprise that the solutions of NH_4OAc come out to be neutral, because the pickup of H_3O^+ by OAc^- just matches the pickup of OH^- by NH_4^+. What is surprising is the large extent to which the ions get hydrolyzed in this solution—much more so than if the other ion were not present. The problem then is to calculate the extent of hydrolysis of NH_4^+ (or of OAc^-) in a solution of NH_4OAc.

The hydrolysis of NH_4^+ can be written as we saw in Section 16.2 in the following way:

$$NH_4^+ + H_2O \longrightarrow NH_3 + H_3O^+$$

The hydrolysis of OAc^- can be written

$$OAc^- + H_2O \longrightarrow HOAc + OH^-$$

The point that is of interest in a solution of NH_4OAc is that each of these reactions helps the other to go further to the right. This comes about because the H_3O^+ from the first reaction gets neutralized by the OH^- from the second reaction. The LeChatelier principle says that if an ion is removed from an equilibrium, the system tends to readjust to re-form that ion. So, we get more NH_4^+ hydrolysis in the presence of OAc^- than we would in its absence.

To handle this problem mathematically we make the same assumption as before as to the mechanism of hydrolysis and assume in addition that the H_3O^+ from the first reaction combines with the OH^- from the second reaction by the reaction

$$H_3O^+ + OH^- \longrightarrow H_2O + H_2O$$

This is a very good assumption here, because the constants for the two hydrolysis reactions are equal ($K = 5.5 \times 10^{-10}$ for each); but it turns out to be a rather good assumption more generally, even when the K of hydrolysis of the cation does not equal the K of hydrolysis of the anion. What saves us is that neither of the hydrolysis reactions can predominate very much. If it did, it would generate an excess of H_3O^+ (or OH^-), which would have the immediate effect of pulling the other reaction further to the right to compensate.

If we add up the three reactions

$$NH_4^+ + H_2O \longrightarrow NH_3 + H_3O^+$$

$$OAc^- + H_2O \longrightarrow HOAc + OH^-$$

$$H_3O^+ + OH^- \longrightarrow H_2O + H_2O$$

and cancel duplications, we get a net reaction as follows:

$$NH_4^+ + OAc^- \longrightarrow NH_3 + HOAc$$

The K for this reaction is given by

$$K = \frac{[NH_3][HOAc]}{[NH_4^+][OAc^-]} = \frac{K_w}{K_{NH_3} K_{HOAc}}$$

and is called the hydrolysis constant of ammonium acetate. Its numerical value can be calculated from the dissociation constants of water, aqueous ammonia, and acetic acid. That this is so can be proved by multiplying numerator and denominator by $[H_3O^+]$ times $[OH^-]$ and regrouping the terms.

$$K = \frac{[NH_3][HOAc]}{[NH_4^+][OAc^-]} \frac{[H_3O^+][OH^-]}{[H_3O^+][OH^-]} = \frac{[H_3O^+][OH^-]}{\dfrac{[NH_4^+][OH^-]}{[NH_3]} \dfrac{[H_3O^+][OAc^-]}{[HOAc]}}$$

Since $K_{NH_3} = 1.81 \times 10^{-5}$ and $K_{HOAc} = 1.8 \times 10^{-5}$, we can calculate

$$K = \frac{K_w}{K_{NH_3} K_{HOAc}} = \frac{1.00 \times 10^{-14}}{(1.81 \times 10^{-5})(1.8 \times 10^{-5})} = 3.1 \times 10^{-5}$$

■ EXAMPLE 412.   Calculate the per cent hydrolysis of OAc^- in 1.00 M NH_4OAc and compare it to the per cent hydrolysis of OAc^- in 1.00 M NaOAc.

SOLUTION: NH_4OAc is a strong electrolyte and is taken to be 100% dissociated into 1.00 M NH_4^+ and 1.00 M OAc^-. Let x = moles per liter of OAc^- that hydrolyzes by the reaction

$$NH_4^+ + OAc^- \rightleftharpoons NH_3 + HOAc$$

Then according to the stoichiometry, x also represents the moles per liter of NH_4^+ that are used up in the hydrolysis to form x moles/liter of NH_3 and of HOAc. At equilibrium, we have

$$[NH_4^+] = 1.00 - x; \quad [OAc^-] = 1.00 - x; \quad [NH_3] = x; \quad [HOAc] = x$$

$$K = \frac{[NH_3][HOAc]}{[NH_4^+][OAc^-]} = \frac{K_w}{K_{NH_3} K_{HOAc}} = 3.1 \times 10^{-5}$$

$$= \frac{(x)(x)}{(1.00 - x)(1.00 - x)}$$

The equation

$$3.1 \times 10^{-5} = \frac{x^2}{(1.00 - x)^2}$$

is simply solved by taking the square root of both sides of the equation. The result is $x = 5.5 \times 10^{-3}$.

To calculate the per cent hydrolysis, we need to divide the amount actually hydrolyzed by the amount available.

$$\text{Per cent hydrolysis} = \frac{5.5 \times 10^{-3}}{1.00} \times 100 = 0.55\%$$

In 1.00 M NaOAc, if we let z moles/liter hydrolyze by the reaction

$$OAc^- + H_2O \longrightarrow HOAc + OH^-$$

we have at equilibrium:

$$[OAc^-] = 1.00 - z; \quad [HOAc] = z; \quad [OH^-] = z$$

$$K_h = \frac{[HOAc][OH^-]}{[OAc^-]} = \frac{(z)(z)}{1.00 - z}$$

$$K_h = \frac{K_w}{K_{HOAc}} = \frac{1.00 \times 10^{-14}}{1.8 \times 10^{-5}} = 5.6 \times 10^{-10} = \frac{z^2}{1.00 - z}$$

$$z = 2.4 \times 10^{-5}$$

$$\text{Per cent hydrolysis} = \frac{\text{amount of } OAc^- \text{ hydrolyzed}}{\text{total } OAc^- \text{ available}}$$

$$= \frac{2.4 \times 10^{-5}}{1.00} \times 100$$

$$= 2.4 \times 10^{-3}\%$$

Thus, the per cent hydrolysis of OAc^- in $1.00\ M$ NH_4OAc is about 230 times as great as it is in $1.00\ M$ NaOAc. ∎

■ EXAMPLE 413. What will be the concentration of hydronium ion in $0.15\ M$ NH_4OAc solution?

SOLUTION: Let $x =$ moles per liter of NH_4^+ that hydrolyze by the reaction

$$NH_4^+ + OAc^- \longrightarrow NH_3 + HOAc$$

Then, according to the equation, x is also the moles per liter of OAc^- that hydrolyze, giving x M NH_3 and x M HOAc.

$$[NH_4^+] = 0.15 - x; \quad [OAc^-] = 0.15 - x; \quad [NH_3] = x; \quad [HOAc] = x$$

For the equilibrium:

$$NH_4^+ + OAc^- \rightleftharpoons NH_3 + HOAc$$

$$K = \frac{[NH_3][HOAc]}{[NH_4^+][OAc^-]} = \frac{(x)(x)}{(0.15 - x)(0.15 - x)}$$

$$K = \frac{K_w}{K_{NH_3}K_{HOAc}} = \frac{1.00 \times 10^{-14}}{(1.81 \times 10^{-5})(1.8 \times 10^{-5})} = 3.1 \times 10^{-5}$$

$$= \frac{x^2}{(0.15 - x)^2}$$

$$x = 8.3 \times 10^{-4}\ M$$

Substituting this in the foregoing concentration expressions, we get

$$[NH_4^+] = 0.15\ M; \quad [OAc^-] = 0.15\ M; \quad [NH_3] = 8.3 \times 10^{-4}\ M$$

$$[HOAc] = 8.3 \times 10^{-4}\ M$$

We can calculate the $[H_3O^+]$ from the dissociation constant of acetic acid, since we know all but one of the concentration terms.

$$K_{HOAc} = \frac{[H_3O^+][OAc^-]}{[HOAc]} = 1.8 \times 10^{-5}$$

$$\frac{[H_3O^+](0.15)}{8.3 \times 10^{-4}} = 1.8 \times 10^{-5}$$

$$[H_3O^+] = 1.0 \times 10^{-7}\ M\ ∎$$

16.4   Hydrolysis of Na$_2$CO$_3$ and Na$_2$S

Sodium carbonate and sodium sulfide represent salts in which dinegative anions, derived from diprotic acids, can hydrolyze to a considerable extent. The way of handling the problem mathematically is similar to that used for acetate ion in Section 16.1 except that the dissociation constant in question is K_{II}.

■ EXAMPLE 414.   Calculate the pH of 0.150 M Na$_2$CO$_3$ solution and the per cent hydrolysis. The $K_{II} = 4.84 \times 10^{-11}$.

SOLUTION: The hydrolysis reaction can be written

$$CO_3^{2-} + H_2O \longrightarrow HCO_3^- + OH^-$$

corresponding to the pickup of a proton from H$_2$O by CO$_3^{2-}$ to form HCO$_3^-$ and set free OH$^-$. The equilibrium constant for this reaction has the form

$$K_h = \frac{[HCO_3^-][OH^-]}{[CO_3^{2-}]}$$

If we multiply numerator and denominator by [H$_3$O$^+$] we get

$$K_h = \frac{[HCO_3^-][OH^-]}{[CO_3^{2-}]} \frac{[H_3O^+]}{[H_3O^+]} = \frac{[H_3O^+][OH^-]}{[H_3O^+][CO_3^{2-}]/[HCO_3^-]} = \frac{K_w}{K_{II}}$$

where the numerator is K_w and the denominator is the second dissociation constant of "carbonic acid." Thus, we can evaluate $K_h = 1.00 \times 10^{-14}/ 4.84 \times 10^{-11} = 2.07 \times 10^{-4}$.

To solve the problem of 0.150 M Na$_2$CO$_3$, let us put x as the moles per liter of CO$_3^{2-}$ that get hydrolyzed over to HCO$_3^-$ and OH$^-$. This will decrease the concentration of CO$_3^{2-}$ from 0.150 to $(0.150 - x)$ and will raise the concentration of HCO$_3^-$ and OH$^-$ from zero to x. At equilibrium we will then have

$$[CO_3^{2-}] = 0.150 - x; \qquad [HCO_3^-] = x; \qquad [OH^-] = x$$

$$K_h = \frac{[HCO_3^-][OH^-]}{[CO_3^{2-}]} = \frac{K_w}{K_{II}} = \frac{1.00 \times 10^{-14}}{4.84 \times 10^{-11}} = 2.07 \times 10^{-4}$$

$$\frac{(x)(x)}{0.150 - x} = 2.07 \times 10^{-4}$$

In solving this equation, x is not quite negligible compared to 0.150; but neither is it important enough to justify using the quadratic formula. Use successive approximations: first, estimate $(0.150 - x) \cong 0.150$ and solve the rest of the equation to get $x \cong 0.558 \times 10^{-2}$; then, estimate $(0.150 - x) \cong (0.150 - 0.558 \times 10^{-2}) = 0.144$ and solve the rest to get $x = 0.546 \times 10^{-2}$; finally, try $(0.150 - x) \cong (0.150 - 0.546 \times 10^{-2}) = 0.145$. We get $x = 5.48 \times 10^{-3}$. This gives us the concentration of hydroxide ion. We want the concentration of hydronium ion.

$$[H_3O^+] = \frac{K_w}{[OH^-]} = \frac{1.00 \times 10^{-14}}{5.48 \times 10^{-3}} = 1.82 \times 10^{-12}$$

$$pH = -\log [H_3O^+] = -\log (1.82 \times 10^{-12}) = -(0.260 - 12)$$

$$= 11.740$$

The per cent hydrolysis is the amount hydrolyzed divided by what was available originally. The amount hydrolyzed is $5.48 \times 10^{-3} \ M$; the amount available was $0.150 \ M$.

$$\text{Per cent hydrolysis} = \frac{5.48 \times 10^{-3}}{0.150} \times 100 = 3.65\% \ \blacksquare$$

For the preceding problems, we might raise the question: How come we worry only about picking up one proton? Is it not possible that a second proton will be picked up as well? The anwer is: Although in principle a second proton can be picked up by a reaction such as

$$HCO_3^- + H_2O \longrightarrow H_2O + CO_2 + OH^-$$

not much of this reaction will occur. There are two reasons for this: (1) there is not much HCO_3^- to start with, only the small amount formed by the first stage of hydrolysis $CO_3^{2-} + H_2O \rightleftharpoons HCO_3^- + OH^-$; (2) the dissociation K of $H_2O + CO_2$ (which can also be written H_2CO_3) is relatively large, which means HCO_3^- does not have nearly the affinity for protons that CO_3^{2-} does. Therefore, the second stage of hydrolysis is less likely to occur. We will generally ignore it in computing the OH^-. That is not to say, however, that we cannot compute the concentrations of the additional species involved. Once we have calculated the principal equilibrium, we can use the results to find out about the other equilibria in the same solution.

■ EXAMPLE 415.   Given that the "first dissociation constant of carbonic acid" is 4.16×10^{-7} for $2H_2O + CO_2 \rightleftharpoons H_3O^+ + HCO_3^-$, calculate the concentration of CO_2 in the $0.150 \ M \ Na_2CO_3$ solution we examined in Example 414.

SOLUTION: In Example 414 we had

$$[CO_3^{2-}] = 0.150 - x; \qquad [HCO_3^-] = x; \qquad [OH^-] = x$$

where x turned out to be 5.48×10^{-3} M. From this we calculated the hydronium-ion concentration $[H_3O^+] = K_w/[OH^-] = 1.82 \times 10^{-12}$ M. So, we know that in a solution of 0.150 M Na$_2$CO$_3$, we must have $[H_3O^+] = 1.82 \times 10^{-12}$ M and $[HCO_3^-] = x = 5.48 \times 10^{-3}$ M. But that is all we need to fix $[CO_2]$ since the equilibrium

$$CO_2 + 2H_2O \;\rightleftharpoons\; H_3O^+ + HCO_3^-$$

requires

$$K_I = \frac{[H_3O^+][HCO_3^-]}{[CO_2]} = 4.16 \times 10^{-7}$$

If we substitute into this expression $[H_3O^+] = 1.82 \times 10^{-12}$ M and $[HCO_3^-]$ $= 5.48 \times 10^{-3}$ M, we get

$$\frac{(1.82 \times 10^{-12})(5.48 \times 10^{-3})}{[CO_2]} = 4.16 \times 10^{-7}$$

which leads to $[CO_2] = 2.40 \times 10^{-8}$ M. ∎

∎ EXAMPLE 416.   For the diprotic acid H$_2$S, the successive dissociation constants are $K_I = 1.1 \times 10^{-7}$ and $K_{II} = 1.0 \times 10^{-14}$. Calculate the concentrations of all species in 0.15 M Na$_2$S solution.

SOLUTION: Take 0.15 M Na$_2$S to be 100% dissociated into 0.30 M Na$^+$ and 0.15 M S^{2-}.

Let x = moles per liter of S^{2-} that hydrolyze

$$S^{2-} + H_2O \;\rightleftharpoons\; HS^- + OH^-$$

This will give at equilibrium the following concentrations:

$$[S^{2-}] = 0.15 - x; \qquad [HS^-] = x; \qquad [OH^-] = x$$

$$K_h = \frac{[HS^-][OH^-]}{[S^{2-}]} = \frac{K_w}{K_{II}} = \frac{1.00 \times 10^{-14}}{1.0 \times 10^{-14}} = 1.0$$

$$\frac{(x)(x)}{0.15 - x} = 1.0$$

Solve this by the quadratic formula to give $x = 0.13$.
The corresponding concentrations will be

$$[S^{2-}] = 0.15 - x = 0.15 - 0.13 = 0.02 \ M$$

$$[HS^-] = x = 0.13 \ M$$

$$[OH^-] = x = 0.13 \ M$$

We can also calculate the hydrogen-ion concentration:

$$[H_3O^+] = \frac{K_w}{[OH^-]} = \frac{1.00 \times 10^{-14}}{0.13} = 7.7 \times 10^{-14} \ M$$

Furthermore, knowing both $[H_3O^+]$ and $[HS^-]$, we can get $[H_2S]$ from the first dissociation constant corresponding to

$$H_2S + H_2O \ \rightleftharpoons \ H_3O^+ + HS^-$$

$$K_I = \frac{[H_3O^+][HS^-]}{[H_2S]} = 1.1 \times 10^{-7}$$

$$[H_2S] = \frac{[H_3O^+][HS^-]}{1.1 \times 10^{-7}} = \frac{(7.7 \times 10^{-14})(0.13)}{1.1 \times 10^{-7}} = 9.1 \times 10^{-8} \ M$$

Thus, we have, in summary:

$$[Na^+] = 0.30 \ M; \quad [HS^-] = 0.13 \ M; \quad [OH^-] = 0.13 \ M$$
$$[H_3O^+] = 7.7 \times 10^{-14} \ M; \quad [S^{2-}] = 0.02 \ M$$
$$[H_2S] = 9.1 \times 10^{-8} \ M$$

[Note the interesting point that according to the calculation above, the major sulfur-containing species in 0.15 M Na_2S is *not* sulfide ion but is actually HS^-.] ∎

16.5 Hydrolysis of $(NH_4)_2CO_3$

Aqueous solutions of ammonium carbonate, $(NH_4)_2CO_3$, are important for two reasons: (1) they represent a typical case of complex simultaneous equilibria where both cation and anion undergo extensive hydrolysis; (2) when mixed with aqueous NH_3, they provide an important reagent for precipitating alkaline earth carbonates under buffered conditions.

■ EXAMPLE 417. Given that $K_{NH_3} = 1.81 \times 10^{-5}$ for $NH_3 + H_2O \rightleftarrows$ $NH_4^+ + OH^-$ and $K_{II} = 4.84 \times 10^{-11}$ for $HCO_3^- + H_2O \rightleftarrows H_3O^+ + CO_3^{2-}$, calculate the concentrations of the species involved in 0.500 M $(NH_4)_2CO_3$ solution.

SOLUTION: We assume that the 0.500 M $(NH_4)_2CO_3$ is 100% dissociated into 1.000 M NH_4^+ and 0.500 CO_3^{2-}.

The NH_4^+ ion will hydrolyze by the reaction

$$NH_4^+ + H_2O \longrightarrow NH_3 + H_3O^+$$

The CO_3^{2-} ion will hydrolyze according to the reaction

$$CO_3^{2-} + H_2O \longrightarrow HCO_3^- + OH^-$$

We assume that the H_3O^+ created by the first reaction combines with OH^- from the second reaction

$$H_3O^+ + OH^- \longrightarrow 2H_2O$$

The *net reaction*, obtained as the sum of these three equations, is

$$NH_4^+ + CO_3^{2-} \longrightarrow NH_3 + HCO_3^-$$

It assumes that NH_3 and HCO_3^- are produced in about equal amounts, based on the reasoning that neither H_3O^+ nor OH^- can pile up in the solution in large excess.

Let x = moles per liter of NH_4^+ that hydrolyze by this reaction. That will make the final equilibrium concentration of NH_4^+ equal to $(1.000 - x)$ M and will make the CO_3^{2-} go to $(0.500 - x)$ M. At the same time, NH_3 and HCO_3^- will be created in the solution to the extent of x moles/liter.

$$[NH_4^+] = 1.000 - x; \qquad [CO_3^{2-}] = 0.500 - x; \qquad [NH_3] = x$$

$$[HCO_3^-] = x$$

These concentrations must satisfy the equilibrium condition for

$$NH_4^+ + CO_3^{2-} \rightleftharpoons NH_3 + HCO_3^-$$

which is given by

$$K = \frac{[NH_3][HCO_3^-]}{[NH_4^+][CO_3^{2-}]} = \frac{(x)(x)}{(1.000 - x)(0.500 - x)}$$

What is the numerical value of K? We can show that $K = K_w/K_{NH_3}K_{II}$ by multiplying numerator and denominator by $[H_3O^+] \times [OH^-]$ and regrouping the terms:

$$K = \frac{[NH_3][HCO_3^-]}{[NH_4^+][CO_3^{2-}]} \frac{[H_3O^+][OH^-]}{[H_3O^+][OH^-]} = \frac{[H_3O^+][OH^-]}{\dfrac{[NH_4^+][OH^-]}{[NH_3]} \dfrac{[H_3O^+][CO_3^{2-}]}{[HCO_3^-]}}$$

$$= \frac{K_w}{K_{NH_3}K_{II}}$$

Putting in $K_{NH_3} = 1.81 \times 10^{-5}$ and $K_{II} = 4.84 \times 10^{-11}$, we get $K = 11.4$. This gives us the equation

$$11.4 = \frac{(x)(x)}{(1.000 - x)(0.500 - x)}$$

which is best solved by the quadratic formula. It gives two roots, $x = 1.18$ or 0.465. The former is physically impossible since it would lead to negative values of the NH_4^+ concentration (i.e., $1.000 - x$) and of the CO_3^{2-} concentration (i.e., $0.500 - x$). So, the root $x = 0.465$ must be the correct one. Substituting it in our expressions for concentrations above, we get

$$[NH_4^+] = 0.535\ M; \qquad [CO_3^{2-}] = 0.035\ M; \qquad [NH_3] = 0.465\ M$$

$$[HCO_3^-] = 0.465\ M$$

We can calculate the hydronium-ion concentration from K_{II} by substituting in the values for CO_3^{2-} and HCO_3^-.

$$K_{II} = \frac{[H_3O^+][CO_3^{2-}]}{[HCO_3^-]} = 4.84 \times 10^{-11} = \frac{[H_3O^+](0.035)}{(0.465)}$$

It gives us $[H_3O^+] = 6.4 \times 10^{-10}\ M$, which leads to $[OH^-] = K_w/[H_3O^+]$ $= (1.00 \times 10^{-14})/(6.4 \times 10^{-10}) = 1.6 \times 10^{-5}\ M$.

We could, of course, also calculate the hydroxide-ion concentration from K_{NH_3} by substituting in the values for the concentrations of NH_3 and NH_4^+.

$$K_{NH_3} = \frac{[NH_4^+][OH^-]}{[NH_3]} = 1.81 \times 10^{-5} = \frac{(0.535)[OH^-]}{0.465}$$

It gives us $[OH^-] = 1.57 \times 10^{-5}\ M$, agreeing with the above.

In summary, we have $[H_3O^+] = 6.4 \times 10^{-10}\ M$; $[OH^-] = 1.57 \times 10^{-5}$

M; $[NH_4^+] = 0.535$ M; $[CO_3^{2-}] = 0.035$; M; $[NH_3] = 0.465$ M; $[HCO_3^-] = 0.465$ M. ∎

■ **EXAMPLE 418.** Suppose you make up a solution that is simultaneously 0.500 M $(NH_4)_2CO_3$ and 1.00 M NH_3. Calculate the equilibrium concentrations of H_3O^+, OH^-, NH_4^+, CO_3^{2-}, NH_3, and HCO_3^-, given $K_{NH_3} = 1.81 \times 10^{-5}$ for $NH_3 + H_2O \rightleftarrows NH_4^+ + OH^-$ and $K_{II} = 4.84 \times 10^{-11}$ for $HCO_3^- + H_2O \rightleftarrows H_3O^+ + CO_3^{2-}$.

SOLUTION: Consider 0.500 M $(NH_4)_2CO_3$ to be 100% dissociated into 1.00 M NH_4^+ and 0.500 M CO_3^{2-}.

As in Example 417, the net reaction will be taken to be

$$NH_4^+ + CO_3^{2-} \longrightarrow NH_3 + HCO_3^-$$

Let $x =$ moles per liter of NH_4^+ that hydrolyze this way. This will decrease the concentration of NH_4^+ from 1.000 to $(1.000 - x)$ M. It will also decrease the concentration of CO_3^{2-} from 0.500 to $(0.500 - x)$ M. At the same time, it will form x moles/liter of NH_3 and x moles/liter of HCO_3^-.

However, we start out with 1.00 M NH_3, so the concentration of NH_3 goes from 1.00 to $(1.00 + x)$ M while the concentration of HCO_3^- goes from zero to x M. At equilibrium, the concentrations will be $[NH_4^+] = 1.000 - x$; $[CO_3^{2-}] = 0.500 - x$; $[NH_3] = 1.00 + x$; $[HCO_3^-] = x$.

For the equilibrium:

$$NH_4^+ + CO_3^{2-} \rightleftarrows NH_3 + HCO_3^-$$

$$K = \frac{[NH_3][HCO_3^-]}{[NH_4^+][CO_3^{2-}]} = \frac{K_w}{K_{NH_3} K_{II}} = \frac{1.00 \times 10^{-14}}{(1.81 \times 10^{-5})(4.84 \times 10^{-11})} = 11.4$$

Substituting the concentrations above in the equilibrium condition gives

$$\frac{(1.00 + x)(x)}{(1.000 - x)(0.500 - x)} = 11.4$$

Solving this by the quadratic formula, we get $x = 0.413$, which leads to

$$[NH_4^+] = 0.587 \ M; \qquad [CO_3^{2-}] = 0.087 \ M; \qquad [NH_3] = 1.41 \ M$$
$$[HCO_3^-] = 0.413 \ M$$

To get the hydronium-ion concentration, we substitute these equilibrium concentrations of CO_3^{2-} and HCO_3^- into K_{II} for $HCO_3^- + H_2O \rightleftarrows H_3O^+ + CO_3^{2-}$:

$$K_{II} = \frac{[H_3O^+][CO_3^{2-}]}{[HCO_3^-]} = 4.84 \times 10^{-11} = \frac{[H_3O^+](0.087)}{(0.413)}$$

$$[H_3O^+] = 2.3 \times 10^{-10} \ M$$

from which

$$[OH^-] = \frac{K_w}{[H_3O^+]} = \frac{1.00 \times 10^{-14}}{2.3 \times 10^{-10}} = 4.3 \times 10^{-5} \ M$$

The summary is $[H_3O^+] = 2.3 \times 10^{-10} \ M$; $[OH^-] = 4.3 \times 10^{-5} \ M$; $[NH_4^+] = 0.587 \ M$; $[CO_3^{2-}] = 0.087 \ M$; $[NH_3] = 1.41 \ M$; $[HCO_3^-] = 0.413 \ M$.

Comparing these numbers to those obtained in Example 417, we see that the main effect of added NH_3 is to boost the carbonate concentration considerably and make the solution a bit more basic. ∎

16.6   Solutions of NaHCO₃

Solutions of sodium bicarbonate, $NaHCO_3$, are typical of those cases where the anion of a salt can act as an acid in dissociating off a proton ($HCO_3^- + H_2O \rightarrow H_3O^+ + CO_3^{2-}$) and simultaneously as a base in picking up a proton ($HCO_3^- + H_3O^+ \rightarrow 2H_2O + CO_2$). The same problem was raised in Section 12.5 in connection with $NaHSO_4$, but there it was pointed out that HSO_4^- has no appreciable tendency to react with H_3O^+, since H_2SO_4 is a strong electrolyte (100% dissociated) for the first step of dissociation. The exact mathematical analysis of $NaHCO_3$ solutions can be done by defining five unknowns for H_3O^+, OH^-, CO_2, HCO_3^-, and CO_3^{2-} and writing five simultaneous equations (three from the three equilibrium conditions set by K_I, K_{II}, and K_w and two from the electrical neutrality requirement and mass conservation condition). Solving these five simultaneous equations is extraordinarily difficult, so we cast about for some simplifying assumptions. We get a major assist from noting that

$$HCO_3^- + H_2O \longrightarrow H_3O^+ + CO_3^{2-}$$

cannot "go" much farther to the right than does the reaction

$$HCO_3^- + H_3O^+ \longrightarrow 2\,H_2O + CO_2$$

the reason being that we cannot greatly pile up excess H_3O^+ in the solution (otherwise, the second reaction would sponge it up) or seriously deplete the solution of H_3O^+ (otherwise, the first reaction would tend to replenish it). The upshot is that these two reactions must go along to about the same extent—that is, the H_3O^+ produced by the first is used up by the second reaction. The situation is not unlike that encountered previously with the dual hydrolysis of NH_4^+ and OAc^- (Section 16.3), where the two reactions substantially cancel each other. If we can assume that the overall reaction is the sum of the two equations written above, then we can add the two equations to get the net reaction

$$2HCO_3^- \longrightarrow CO_3^{2-} + H_2O + CO_2$$

to describe the *principal net equilibrium* in a $NaHCO_3$ solution. The reaction can be visualized as a transfer of a proton from one HCO_3^- to another HCO_3^-, followed by immediate breakup of the H_2CO_3 formed into H_2O and CO_2.

■ EXAMPLE 419. Given that $K_I = 4.16 \times 10^{-7}$ for $CO_2 + 2H_2O \rightleftarrows H_3O^+ + HCO_3^-$ and $K_{II} = 4.84 \times 10^{-11}$ for $HCO_3^- + H_2O \rightleftarrows H_3O^+ + CO_3^{2-}$ calculate the K for $2HCO_3^- \rightleftarrows H_2O + CO_2 + CO_3^{2-}$.

SOLUTION: For the equilibrium

$$2HCO_3^- \rightleftarrows H_2O + CO_2 + CO_3^{2-}$$

the equilibrium constant has the form

$$K = \frac{[CO_2][CO_3^{2-}]}{[HCO_3^-)]^2}$$

leaving out the water because its activity stays constant.
   If we multiply the numerator and denominator by $[H_3O^+]$ we get

$$K = \frac{[CO_2][CO_3^{2-}]}{[HCO_3^-]^2} \frac{[H_3O^+]}{[H_3O^+]} = \frac{[H_3O^+][CO_3^{2-}]}{[HCO_3^-]} \frac{[CO_2]}{[H_3O^+][HCO_3^-]}$$

The grouping $[H_3O^+][CO_3^{2-}]/[HCO_3^-]$ is just equal to K_{II} for the equilibrium $HCO_3^- + H_2O \rightleftarrows H_3O^+ + CO_3^{2-}$.
   The grouping $[CO_2]/[H_3O^+][HCO_3^-]$ is equal to the inverse of K_I for the equilibrium $CO_2 + 2H_2O \rightleftarrows H_3O^+ + HCO_3^-$.
   We have, thus,

$$K = K_{II} \frac{1}{K_I} = \frac{K_{II}}{K_I} = \frac{4.84 \times 10^{-11}}{4.16 \times 10^{-7}} = 1.16 \times 10^{-4} \; ■$$

■ EXAMPLE 420.   Calculate the concentrations of the various species present in a solution that is labeled 0.500 M $NaHCO_3$, using the constants given in Example 419.

SOLUTION: The net reaction is $2HCO_3^- \rightarrow H_2O + CO_2 + CO_3^{2-}$. Let x = moles per liter of CO_2 and CO_3^{2-} formed this way. Then $2x$ represents the moles per liter of HCO_3^- that have disappeared in the process. This means the HCO_3^- concentration has gone from 0.500 M (which was put into the solution) to $(0.500 - 2x)$ M at equilibrium. We can summarize the equilibrium concentrations as follows:

$$[HCO_3^-] = 0.500 - 2x; \qquad [CO_2] = x; \qquad [CO_3^{2-}] = x$$

The condition for equilibrium for the system

$$2HCO_3^- \; \rightleftharpoons \; H_2O + CO_2 + CO_3^{2-}$$

is that

$$K = \frac{[CO_2][CO_3^{2-}]}{[HCO_3^-]^2} = \frac{K_{II}}{K_I} = \frac{4.84 \times 10^{-11}}{4.16 \times 10^{-7}} = 1.16 \times 10^{-4}$$

$$\frac{(x)(x)}{(0.500 - 2x)^2} = 1.16 \times 10^{-4}$$

Solving this equation by taking the square root of each side gives

$$\frac{x}{0.500 - 2x} = 1.08 \times 10^{-2}$$

or $x = 5.28 \times 10^{-3}$. Substituting in the equilibrium concentrations as summarized above, we get

$$[HCO_3^-] = 0.489 \; M; \qquad [CO_2] = 5.28 \times 10^{-3} \; M;$$
$$[CO_3^{2-}] = 5.28 \times 10^{-3} \; M$$

The $[H_3O^+]$ we can calculate either from K_I or from K_{II}.

$$K_I = \frac{[H_3O^+][HCO_3^-]}{[CO_2]} = 4.16 \times 10^{-7}$$

$$[H_3O^+] = \frac{[CO_2]}{[HCO_3^-]}(4.16 \times 10^{-7}) = \frac{5.28 \times 10^{-3}}{0.489}(4.16 \times 10^{-7})$$
$$= 4.49 \times 10^{-9}$$

$$K_{II} = \frac{[H_3O^+][CO_3^{2-}]}{[HCO_3^-]} = 4.84 \times 10^{-11}$$

$$[H_3O^+] = \frac{[HCO_3^-]}{[CO_3^{2-}]}(4.84 \times 10^{-11}) = \frac{0.489}{5.28 \times 10^{-3}}(4.84 \times 10^{-11})$$

$$= 4.48 \times 10^{-9}$$

To get $[OH^-]$, we use the water constant K_w:

$$[OH^+] = \frac{K_w}{[H_3O^+]} = \frac{1.00 \times 10^{-14}}{4.48 \times 10^{-9}} = 2.23 \times 10^{-6} \ M$$

So, in summary, we have for 0.500 M NaHCO₃ solution:

$[Na^+] = 0.500 \ M;$ $\quad$ $[HCO_3^-] = 0.489 \ M;$ $\quad$ $[CO_2] = 5.28 \times 10^{-3} \ M$

$[CO_3^{2-}] = 5.28 \times 10^{-3} \ M;$ $\quad$ $[H_3O^+] = 4.48 \times 10^{-9} \ M$

$[OH^-] = 2.23 \times 10^{-6} \ M$ ∎

■ EXAMPLE 421. (Compare to Example 420.) Suppose you have 0.280 liter of 0.500 M NaHCO₃ and you add to it 24.0 ml of 0.500 M NaOH. What happens to all the concentrations in the solution? The $K_I = 4.16 \times 10^{-7}$; the $K_{II} = 4.84 \times 10^{-11}$.

SOLUTION: Twenty-four ml of 0.500 M NaOH supplies 0.0120 mole of Na^+ and 0.0120 mole of OH^-.

0.280 liter of 0.500 M NaHCO₃ supplies 0.140 mole of Na^+ and 0.140 mole of HCO_3^-.

Let us assume as a first approximation that the OH^- is completely used up by HCO_3^- in the reaction

$$OH^- + HCO_3^- \longrightarrow H_2O + CO_3^{2-}$$

This will use up 0.0120 mole of HCO_3^-, leaving 0.140 − 0.012, or 0.128 mole of HCO_3^-; it will form 0.0120 mole of CO_3^{2-}.

The total volume of the solution is 0.304 liter.

The following concentrations are available:

$$Na^+: \frac{0.0120 \ \text{mole} + 0.140 \ \text{mole}}{0.304 \ \text{liter}} = 0.500 \ M$$

$$HCO_3^-: \frac{0.128 \ \text{mole}}{0.304 \ \text{liter}} = 0.421 \ M$$

$$CO_3^{2-}: \frac{0.0120 \ \text{mole}}{0.304 \ \text{liter}} = 0.0395 \ M$$

This, however, does not allow for the reaction

$$2HCO_3^- \rightleftharpoons H_2O + CO_2 + CO_3^{2-}$$

Let x = moles per liter of CO_2 formed this way. It will be accompanied by formation of x moles/liter more of CO_3^{2-} and a decrease of $2x$ moles/liter of HCO_3^-. So, at equilibrium the concentrations will be

$$[HCO_3^-] = 0.421 - 2x; \quad [CO_3^{2-}] = 0.0395 + x; \quad [CO_2] = x$$

Substitute these in the equilibrium condition:

$$K = \frac{[CO_2][CO_3^{2-}]}{[HCO_3^-]^2} = \frac{K_{II}}{K_I} = \frac{4.84 \times 10^{-11}}{4.16 \times 10^{-7}} = 1.16 \times 10^{-4}$$

$$\frac{(x)(0.0395 + x)}{(0.421 - 2x)^2} = 1.16 \times 10^{-4}$$

This looks like a hard one to solve, except that we might guess x to be small since the effect of added CO_3^{2-} will be to repress the tendency to form more. If we neglect the x terms where added or subtracted, we get $x \cong 5.20 \times 10^{-4}$ as a first approximation. If we put this x in place of the dropped terms, we get $x \cong 5.12 \times 10^{-4}$ as a second approximation. A third approximation gives also $x = 5.12 \times 10^{-4}$. Substituting this value of x in the equilibrium concentrations above, we get

$$[HCO_3^-] = 0.420\ M; \quad [CO_3^{2-}] = 0.0400\ M; \quad [CO_2] = 5.12 \times 10^{-4}\ M$$

To get $[H_3O^+]$, we use K_{II}:

$$[H_3O^+] = \frac{[HCO_3^-]}{[CO_3^{2-}]}(4.84 \times 10^{-11}) = 5.08 \times 10^{-10}\ M$$

To get $[OH^-]$, we use K_w:

$$[OH^-] = \frac{K_w}{[H_3O^+]} = \frac{1.00 \times 10^{-14}}{5.08 \times 10^{-10}} = 1.97 \times 10^{-5}\ M$$

So, for the final concentrations, we would have the following:

$$[Na^+] = 0.500\ M; \quad [HCO_3^-] = 0.420\ M; \quad [CO_2] = 5.12 \times 10^{-14}\ M$$
$$[CO_3^{2-}] = 0.0400\ M; \quad [H_3O^+] = 5.08 \times 10^{-10}\ M$$
$$[OH^-] = 1.97 \times 10^{-5}\ M \blacksquare$$

16.7   *Hydrolysis of aluminum salts*

Many of the cations of metals, especially the small-size and large-charge ones, tend to make solutions acidic because they again disturb the water equilibrium $H_2O + H_2O \rightleftarrows H_3O^+ + OH^-$. The aluminum ion Al^{3+} is a good example of this. In aqueous solution it can be considered to react with the hydroxide of water to form $AlOH^{2+}$ and set free H_3O^+ by the net reaction

$$Al^{3+} + 2H_2O \rightleftharpoons AlOH^{2+} + H_3O^+$$

■ EXAMPLE 422.   The dissociation constant of $AlOH^{2+} \rightleftarrows Al^{3+} + OH^-$ has a value of 7.1×10^{-10}. Calculate the hydrolysis constant corresponding to $Al^{3+} + 2H_2O \rightleftarrows AlOH^{2+} + H_3O^+$.

SOLUTION: The hydrolysis constant in this case has the usual form of species on the right in the numerator and species on the left in the denominator.

$$K_h = \frac{[AlOH^{2+}][H_3O^+]}{[Al^{3+}]}$$

If we multiply numerator and denominator by $[OH^-]$ we get

$$K_h = \frac{[AlOH^{2+}][H_3O^+]}{[Al^{3+}]} \frac{[OH^-]}{[OH^-]} = \frac{[H_3O^+][OH^-]}{[Al^{3+}][OH^-]/[AlOH^{2+}]}$$

The numerator is just K_w and has the value 1.00×10^{-14}. The denominator is the equilibrium constant for the reaction $AlOH^{2+} \rightleftarrows Al^{3+} + OH^-$, which is the dissociation of hydroxyaluminum ion. Thus, the denominator has the value 7.1×10^{-10}, as given in the problem.

$$K_h = \frac{K_w}{K_{AlOH^{2+}}} = \frac{1.00 \times 10^{-14}}{7.1 \times 10^{-10}} = 1.4 \times 10^{-5} \ ■$$

■ EXAMPLE 423.   Assuming the main equilibrium is $Al^{3+} + 2H_2O \rightleftarrows AlOH^{2+} + H_3O^+$, calculate the concentrations of the ions in a solution made by dissolving 6.37 g of $Al(NO_3)_3 \cdot 9H_2O$ in enough water to make 0.250 liter of solution.

SOLUTION: One mole of $Al(NO_3)_3 \cdot 9H_2O$ weighs 375.13 g.

$$\frac{6.37 \text{ g of } Al(NO_3)_3 \cdot 9H_2O}{375.13 \text{ g/mole}} = 0.0170 \text{ mole}$$

Concentration would be

$$\frac{0.0170 \text{ mole}}{0.250 \text{ liter}} = 0.0680 \ M.$$

Aluminum nitrate is a strong electrolyte, so we assume that the 0.0680 M $Al(NO_3)_3$ is 100% dissociated to give 0.0680 M Al^{3+} and 0.204 M NO_3^-. Let x = moles per liter of Al^{3+} that hydrolyze via the reaction

$$Al^{3+} + 2H_2O \ \rightleftharpoons \ AlOH^{2+} + H_3O^+$$

At equilibrium:

$$[Al^{3+}] = 0.0680 - x; \quad [AlOH^{2+}] = x; \quad [H_3O^+] = x$$

$$K_h = \frac{[AlOH^{2+}][H_3O^+]}{[Al^{3+}]} = \frac{K_w}{K_{AlOH^{2+}}} = \frac{1.00 \times 10^{-14}}{7.1 \times 10^{-10}}$$

$$= 1.4 \times 10^{-5}$$

$$\frac{(x)(x)}{0.0680 - x} = 1.4 \times 10^{-5}$$

$$x = 9.7 \times 10^{-4}$$

Substituting this value of x in the equilibrium concentrations we get

$$[Al^{3+}] = 0.0670 \ M; \quad [AlOH^{2+}] = 9.7 \times 10^{-4} \ M$$
$$[H_3O^+] = 9.7 \times 10^{-4} \ M$$

Furthermore, we can solve for the hydroxide-ion concentration by using the water constant:

$$[OH^-] = \frac{K_w}{[H_3O^+]} = \frac{1.00 \times 10^{-14}}{9.7 \times 10^{-4}} = 1.0 \times 10^{-11} \ M$$

Summarizing our results for all the species present, we have

$$[Al^{3+}] = 0.0670 \ M; \quad [NO_3^-] = 0.204 \ M$$
$$[AlOH^{2+}] = 9.7 \times 10^{-4} \ M; \quad [H_3O^+] = 9.7 \times 10^{-4} \ M$$
$$[OH^-] = 1.0 \times 10^{-11} \ M \ \blacksquare$$

■ EXAMPLE 424. Potassium alum is $KAl(SO_4)_2 \cdot 12H_2O$. As a strong electrolyte, it is considered to be 100% dissociated into K^+, Al^{3+}, and SO_4^{2-}. The solutions are acidic because of the hydrolysis of Al^{3+}, but not so acidic as might be expected, because the sulfate ion can sponge up some of the hydronium ion by forming HSO_4^-. Given a solution made by dissolving 11.4 g of $KAl(SO_4)_2 \cdot 12H_2O$ in enough water to make 0.100 liter of solution, calculate its hydronium-ion concentration: (a) just considering the hydrolysis $Al^{3+} + 2H_2O \rightleftarrows AlOH^{2+} + H_3O^+$ with $K_h = 1.4 \times 10^{-5}$; and (b) allowing also for the equilibrium $HSO_4^- + H_2O \rightleftarrows H_3O^+ + SO_4^{2-}$ with $K_{II} = 1.26 \times 10^{-2}$.

SOLUTION:
(a) One mole of $KAl(SO_4)_2 \cdot 12\ H_2O$ weighs 474.38 g.

$$\frac{11.4 \text{ g of } KAl(SO_4)_2 \cdot 12H_2O}{474.38 \text{ g/mole}} = 0.0240 \text{ mole}$$

Dissolved in 0.100 liter, this would give a concentration of

$$\frac{0.0240 \text{ mole}}{0.100 \text{ liter}} = 0.240\ M$$

Assuming 100% dissociation, 0.240 M $KAl(SO_4)_2$ would correspond to

0.240 M K^+, 0.240 M Al^{3+}, 0.480 M SO_4^{2-}.

Considering only the hydrolysis of Al^{3+} we can write

$$Al^{3+} + 2H_2O \rightleftarrows AlOH^{2+} + H_3O^+$$

If x = moles per liter of Al^{3+} so converted to $AlOH^{2+}$, we would have at equilibrium:

$$[Al^{3+}] = 0.240 - x; \qquad [AlOH^{2+}] = x; \qquad [H_3O^+] = x$$

Substituting into the hydrolysis constant, we get

$$K_h = \frac{[AlOH^{2+}][H_3O^+]}{[Al^{3+}]} = \frac{(x)(x)}{0.240 - x} = 1.4 \times 10^{-5}$$

This equation solves to give $x = 1.82 \times 10^{-3}\ M = [H_3O^+]$.

(b) Now take into account that some of this H_3O^+ just calculated will combine with SO_4^{2-} to form HSO_4^-. The simplest thing to do would be to assume that, because of the relatively large SO_4^{2-} concentration in the solution, *all* the H_3O^+ set free by the Al^{3+} hydrolysis gets picked up by SO_4^{2-}. This means adding the two reactions:

$$Al^{3+} + 2H_2O \rightleftharpoons AlOH^{2+} + H_3O^+$$

$$\underline{H_3O^+ + SO_4^{2-} \rightleftharpoons HSO_4^- + H_2O}$$
$$H_2O + Al^{3+} + SO_4^{2-} \rightleftharpoons AlOH^{2+} + HSO_4^-$$

The K for this net reaction has the form

$$K = \frac{[AlOH^{2+}][HSO_4^-]}{[Al^{3+}][SO_4^{2-}]} = \frac{K_h}{K_{II}} = \frac{1.4 \times 10^{-5}}{1.26 \times 10^{-2}} = 1.1 \times 10^{-3}$$

Let y = moles per liter of Al^{3+} that convert to $AlOH^{2+}$ in this new set-up. This will produce y moles/liter of $AlOH^{2+}$ and y moles/liter of HSO_4^-.

$$[Al^{3+}] = 0.240 - y; \qquad [AlOH^{2+}] = y; \qquad [SO_4^{2-}] = 0.480 - y$$
$$[HSO_4^-] = y$$

$$K = \frac{[AlOH^{2+}][HSO_4^-]}{[Al^{3+}][SO_4^{2-}]} = \frac{(y)(y)}{(0.240 - y)(0.480 - y)} = 1.1 \times 10^{-3}$$

Solving this equation by successive approximation leads to $y = 1.1 \times 10^{-2}$, and substituting this y in the concentration expressions leads to the following:

$$[SO_4^{2-}] = 0.480 - y = 0.480 - 1.1 \times 10^{-2} = 0.469 \ M$$
$$[HSO_4^-] = y = 1.1 \times 10^{-2} \ M$$

We can find the hydronium-ion concentration by putting these values into

$$K_{II} \text{ for } HSO_4^- + H_2O \rightleftharpoons H_3O^+ + SO_4^{2-}:$$

$$K_{II} = \frac{[H_3O^+][SO_4^{2-}]}{[HSO_4^-]} = 1.26 \times 10^{-2}$$

$$[H_3O^+] = \frac{[HSO_4^-]}{[SO_4^{2-}]}(1.26 \times 10^{-2}) = \frac{1.1 \times 10^{-2}}{0.469}(1.26 \times 10^{-2})$$

$$[H_3O^+] = 3.0 \times 10^{-4} \ M$$

Comparing this result with that obtained in part (a) we note that the presence of the SO_4^{2-} reduces the hydronium-ion concentration to about one sixth of the previous value. (Incidentally, an exact solution gives very nearly the same result.) ∎

16.8   Mixtures of weak acids

How do we handle the problem of two or more weak acids in the same solution? We have already worked on the problem in two of its applications: (1) any solution of a weak acid in water is really a two-electrolyte problem, because we have the dissociation $H_2O + H_2O \rightleftharpoons H_3O^+ + OH^-$ to consider as well as the dissociation $HX + H_2O \rightleftharpoons H_3O^+ + X^-$. In general, we have been able to assume that the water equilibrium is a subsidiary equilibrium that is fixed after the $HX + H_2O \rightleftharpoons H_3O^+ + X^-$ equilibrium has been computed. This procedure will break down for acids with K_{diss} of the order of 10^{-12} or smaller and also in very dilute solutions of HX; (2) in the case of polyprotic acids, we have had several proton donors able to dissociate to give H_3O^+ (e.g., H_3PO_4, $H_2PO_4^-$, HPO_4^{2-}). We have avoided difficulties (see Sections 12.5 and 12.6) because in general the K's for the successive dissociation steps usually go down by several orders of magnitude (e.g., H_3PO_4 has $K_I = 7.5 \times 10^{-3}$, $K_{II} = 6.2 \times 10^{-8}$, and $K_{III} = 1 \times 10^{-12}$), so that one of the proton donors is clearly dominant; furthermore, as a result of stepwise reaction, the concentration of each successive proton donor is much less than that of a preceding one (e.g., in a solution of H_3PO_4, $[H_3PO_4] > [H_2PO_4^-] > [HPO_4^{2-}]$). Thus, we have taken the first dissociation step as the principal equilibrium and treated the second and third as subsidiary. This procedure will break down if the successive K's are large and not very different from each other—for example, in the case of pyrophosphoric acid, $H_4P_2O_7$, where K_I is 1.4×10^{-1} and K_{II} is 1.1×10^{-2}.

Now we have to face up to the general problem of several weak acids in equilibrium simultaneously in the same solution.

∎ **EXAMPLE 425.**   Suppose you have a solution that is simultaneously 0.150 M HNO_2 ($K_{diss} = 4.5 \times 10^{-4}$) and 0.200 M HOAc ($K_{diss} = 1.8 \times 10^{-5}$) Calculate the concentrations of H_3O^+, NO_2^-, and OAc^- in the solution.

SOLUTION: We assume that the dissociation of water contributes a negligible concentration of hydronium ion and worry only about that which comes from HNO_2 and from HOAc.

Let x = moles per liter of HNO_2 that dissociate. This will give x moles/liter of H_3O^+ and x moles/liter of NO_2^- and leave $(0.150 - x)$ moles/liter of undissociated HNO_2.

Let y = moles per liter of HOAc that dissociate. This will give y moles/liter of H_3O^+ and y moles/liter of OAc^-, leaving $(0.200 - y)$ moles/liter of undissociated HOAc.

The equilibrium concentration of hydronium ion will be $x + y$—that is, the sum of what is contributed by each acid.

The equilibrium concentrations of all the species are as follows:

$$[H_3O^+] = x + y; \quad [NO_2^-] = x; \quad [OAc^-] = y; \quad [HNO_2] = 0.150 - x$$
$$[HOAc] = 0.200 - y$$

Both equilibria need to be satisfied:

$$HNO_2 + H_2O \rightleftharpoons H_3O^+ + NO_2^-$$
$$HOAc + H_2O \rightleftharpoons H_3O^+ + OAc^-$$

The conditions required for equilibrium are

$$K_{HNO_2} = \frac{[H_3O^+][NO_2^-]}{[HNO_2]} = 4.5 \times 10^{-4} = \frac{(x + y)(x)}{0.150 - x}$$

$$K_{HOAc} = \frac{[H_3O^+][OAc^-]}{[HOAc]} = 1.8 \times 10^{-5} = \frac{(x + y)(y)}{0.200 - y}$$

Obviously there is only one hydronium-ion concentration characteristic of the whole solution, and it must satisfy each equilibrium condition. The problem now boils down to solving the two simultaneous equations

$$4.5 \times 10^{-4} = \frac{(x + y)(x)}{0.150 - x} \quad \text{and} \quad 1.8 \times 10^{-5} = \frac{(x + y)(y)}{0.200 - y}$$

(Generally, one of x or y would be significantly smaller than the other and could be neglected in the sum $x + y$. However, we have deliberately picked a case where this is not true.)

To solve, we first note that x must be small compared to 0.150, so the denominator $(0.150 - x)$ can be approximated as 0.150. Similarly, y must be small compared to 0.200, so the denominator $(0.200 - y)$ can be approximated as 0.200. These substitutions give us

$$4.5 \times 10^{-4} \cong \frac{(x + y)(x)}{0.150} \quad \text{and} \quad 1.8 \times 10^{-5} \cong \frac{(x + y)(y)}{0.200}$$

Next, we get y in terms of x from one equation and substitute it in the other. From the equation on the left, we get

$$y = \frac{6.7_5 \times 10^{-5}}{x} - x$$

and substituting this into the equation on the right, we get $x = 8.0_2 \times 10^{-3}$. Putting this value of x back into the equation on the left, we get $y = 0.40 \times 10^{-3}$.

We can improve the answers somewhat by using these values of x and y in a second approximation, where in the original equation the denominator $(0.150 - x)$ is approximated by $(0.150 - 8 \times 10^{-3})$, or 0.142. The other denominator, $0.200 - y$, was well estimated as 0.200. The second cycle of approximation leads to $x = 7.8 \times 10^{-3}$ and $y = 4.3 \times 10^{-4}$.

Substituting these numbers in our concentration expressions gives us the final equilibrium concentrations:

$$[H_3O^+] = x + y = 7.8 \times 10^{-3} + 4.3 \times 10^{-4} = 8.2 \times 10^{-3} \ M$$
$$[NO_2^-] = x = 7.8 \times 10^{-3} \ M$$
$$[OAc^-] = y = 4.3 \times 10^{-4} \ M \ \blacksquare$$

■ EXAMPLE 426.   What will be the concentrations of all the species present in a solution made by mixing 0.200 liter of 0.400 M HOAc and 0.600 liter of 0.800 M HCN? The dissociation constant of HOAc is 1.8×10^{-5}; the dissociation constant of HCN is 4×10^{-10}.

SOLUTION: Although HOAc and HCN are both weak acids, HOAc is so much stronger than HCN that its dissociation dominates the situation. In other words, the concentration of hydronium ion coming from the dissociation of HCN is negligible compared to what comes from the dissociation of HOAc.

First calculate the concentration of HOAc and HCN in the mixture. The HOAc gets diluted from 0.200 liter to 0.800 liter, so its concentration falls to (0.200/0.800) of 0.400 M, or 0.100 M. The HCN gets diluted from 0.600 liter to 0.800 liter, so its concentration goes to (0.600/0.800) of 0.800 M, or 0.600 M.

Let x = moles per liter of HOAc that dissociate. This will give x moles/liter of H_3O^+ and x moles/liter of OAc^-, leaving $(0.100 - x)$ moles/liter of undissociated HOAc.

$$[H_3O^+] = x; \quad [OAc^-] = x; \quad [HOAc] = 0.100 - x$$
$$HOAc + H_2O \ \rightleftharpoons \ H_3O^+ + OAc^-$$
$$K = \frac{[H_3O^+][OAc^-]}{[HOAc]} = \frac{(x)(x)}{0.100 - x} = 1.8 \times 10^{-5}$$
$$x = 1.3 \times 10^{-3} \ M$$

Let y = moles per liter of HCN that dissociate. This will produce y moles/liter of H_3O^+ and y moles/liter of CN^-, leaving $(0.600 - y)$ moles/liter of undissociated HCN. The H_3O^+ adds on to the $1.3 \times 10^{-3}\,M$ that comes from the HOAc to make $[H_3O^+] = 1.3 \times 10^{-3} + y$, but y is such a small number that it is negligible compared to 1.3×10^{-3}. Therefore, we can write for the equilibrium concentrations:

$$[H_3O^+] = 13. \times 10^{-3}; \qquad [CN^-] = y; \qquad [HCN] = 0.600 - y$$

$$HCN + H_2O \;\rightleftharpoons\; H_3O^+ + CN^-$$

$$K = \frac{[H_3O^+][CN^-]}{[HCN]} = \frac{(1.3 \times 10^{-3})(y)}{0.600 - y} = 4 \times 10^{-10}$$

$$y = 2 \times 10^{-7}$$

Note that this small value of y confirms our assumption it was small compared to 1.3×10^{-3}.

Collecting our results, we have

$$[H_3O^+] = x = 1.3 \times 10^{-3}\,M$$
$$[OAc^-] = x = 1.3 \times 10^{-3}\,M$$
$$[HOAc] = 0.100 - x = 0.099\,M$$
$$[CN^-] = y = 2 \times 10^{-7}\,M$$
$$[HCN] = 0.600 - y = 0.600\,M$$

Finally, we can also calculate the hydroxide-ion concentration, once we know $[H_3O^+]$:

$$[OH^-] = \frac{K_w}{[H_3O^+]} = \frac{1.00 \times 10^{-14}}{1.3 \times 10^{-3}} = 7.7 \times 10^{-12}\,M$$

In the calculations above, we have ignored the water equilibrium $H_2O + H_2O \rightleftharpoons H_3O^+ + OH^-$ except as an afterthought. However, as noted at the beginning of this section, it becomes important in very dilute solutions. The following problem illustrates how a simple calculation such as the dissociation of acetic acid can become enormously complicated when the solute is present in very low concentration ■

■ EXAMPLE 427.   Calculate the concentration of H_3O^+ and the per cent dissociation of acetic acid in $1.00 \times 10^{-7}\,M$ HOAc.

SOLUTION: The first impulse is to set up the problem in the usual way: let x = moles per liter of HOAc that dissociate to give x moles/liter of H_3O^+

and x moles/liter of OAc$^-$, leaving $(1.00 \times 10^{-7} - x)$ moles/liter of undissociated HOAc. Substituting these concentrations in K_{diss} would give

$$K_{diss} = \frac{[H_3O^+][OAc^-]}{[HOAc]} = \frac{(x)(x)}{1.00 \times 10^{-7} - x} = 1.8 \times 10^{-5}$$

and the quadratic formula would lead to $x = 0.995 \times 10^{-7}$. We *cannot* accept this result because it would give us a hydronium-ion concentration *less than that of pure water*. What happened to the added acetic acid, which appears to be 99.5% dissociated?

So now we start to worry about the water dissociation. We really ought to allow for some H_3O^+ to come from $H_2O + H_2O \rightleftarrows H_3O^+ + OH^-$ as well as from $HOAc + H_2O \rightleftarrows H_3O^+ + OAc^-$. Perhaps we can set up the problem in the same way we handled two weak acids in Example 425. Let $x =$ moles per liter of H_3O^+ that come from $H_2O + H_2O \rightleftarrows H_3O^+ + OH^-$ dissociation and let $y =$ moles per liter of H_3O^+ that come from the $HOAc + H_2O \rightleftarrows H_3O^+ + OAc^-$ dissociation. The total concentration of hydronium ion would then be $(x + y)$ moles/liter and the other concentrations would be as follows:

$$[OH^-] = x; \qquad [OAc^-] = y; \qquad [HOAc] = 1.00 \times 10^{-7} - y$$

Both equilibrium conditions must be satisfied simultaneously:

$$K_w = [H_3O^+][OH^-] = 1.00 \times 10^{-14} = (x + y)(x)$$

$$K_{HOAc} = \frac{[H_3O^+][OAc^-]}{[HOAc]} = 1.8 \times 10^{-5} = \frac{(x + y)(y)}{1.00 \times 10^{-7} - y}$$

With two independent equations for two unknowns, we should be able to solve for x and y. Unfortunately, y is almost certainly not negligible with respect to 1.00×10^{-7}, so we appear to have some difficult algebra in prospect. We can solve the two equations by getting x in terms of y from the second equation:

$$x = \left(\frac{1.8 \times 10^{-5}}{y}\right)(1.00 \times 10^{-7} - y) - y$$

and substituting the result for x in the first equation. We get

$$0 = y^3 + 1.79 \times 10^{-5}y^2 - 3.60 \times 10^{-12}y + 1.8 \times 10^{-19}$$

Undaunted by the third-order equation, we solve it using any method given in the handbooks for cubic equations, only to discover to our horror that its only acceptable root $y = 1.00 \times 10^{-7}$ is paired with $x = -1.00 \times 10^{-7}$. Since x represents the hydroxide-ion concentration, we are faced with the awkward question, what does a negative hydroxide-ion concentration mean? Actually, the situation is not all that bad, because we can repeat the calculation carrying through more digits—more than apparently allowed by significant figure rules. But it is still a mess, and we ask whether there might be another way to set up the problem.

There is one last hope. Let us set up the problem in the most general way possible. (Compare Section 16.1, where the same method was used for hydrolysis.) We have four species in solution: H_3O^+, OH^-, OAc^-, and HOAc. Let us define these as our four unknowns:

$$[H_3O^+] = w; \qquad [OH^-] = x; \qquad [OAc^-] = y; \qquad [HOAc] = z$$

We need four independent equations in order to be able to evaluate all four of these unknowns:

(a) The water equilibrium must be satisfied.

$$H_2O + H_2O \; \rightleftharpoons \; H_3O^+ + OH^-$$
$$K_w = [H_3O^+][OH^-] = 1.00 \times 10^{-14} = (w)(x)$$

(b) The acetic acid dissociation equilibrium must also be satisfied.

$$HOAc + H_2O \; \rightleftharpoons \; H_3O^+ + OAc^-$$
$$K_{diss} = \frac{[H_3O^+][OAc^-]}{[HOAc]} = 1.8 \times 10^{-5} = \frac{(w)(y)}{z}$$

(c) The solution must remain electrically neutral. This means the positive charge concentration must equal the negative charge concentration.

$$[H_3O^+] = [OH^-] + [OAc^-]$$

$$w = x + y$$

(d) All the acetic acid put into the solution (1.00×10^{-7} mole/liter) has to be accounted for. It must show up either as acetate ion or as undissociated acetic acid.

$$1.00 \times 10^{-7} = [OAc^-] + [HOAc]$$
$$1.00 \times 10^{-7} = y + z$$

So here we have our four equations:

$$(w)(x) = 1.00 \times 10^{-14} \tag{1}$$

$$\frac{(w)(y)}{z} = 1.8 \times 10^{-5} \tag{2}$$

$$w = x + y \tag{3}$$

$$1.00 \times 10^{-7} = y + z \tag{4}$$

All we need to do is to solve them as a simultaneous set. We will minimize our troubles if we try to solve for the most abundant species first. This will probably be the hydronium ion, so we ought to combine our equations in such a way that we end up with only the one variable w. (In this way we will avoid the hassle we got into in the preceding method we tried, where we were trying to solve for acetate concentration, which is probably small). The best way to tackle the problem is to solve equation (1) for x in terms of w and substitute the result in equation (3). Then solve equation (4) for z in terms of y and substitute the result in equation (2). Finally, combine the two resulting equations to eliminate y. We end up with a cubic equation for w:

$$0 = w^3 + 1.8 \times 10^{-5}w^2 - 1.81 \times 10^{-12}w - 1.8 \times 10^{-19}$$

Using the method for third-order equations, we get $w = 1.61 \times 10^{-7}$. (In using the formulas given in handbooks for cubic or quadratic equations, it is wise to carry along several extra digits, disregarding significant-figure problems until the very end. Both the quadratic formula and the cubic equation formulas demand this concession; otherwise, you may lose your answer on the way.) Feeding w back into equations (1), (2), (3), and (4), we can get values for x, y, and z. The final results look like this:

$$[H_3O^+] = w = 1.6 \times 10^{-7} \ M$$
$$[OH^-] = x = 6.2 \times 10^{-8} \ M$$
$$[OAc^-] = y = 9.9 \times 10^{-8} \ M$$
$$[HOAc] = z = 0.090 \times 10^{-8} \ M$$

Thus, the acetic acid in $1.00 \times 10^{-7} \ M$ HOAc solution turns out to be 99% dissociated. ∎

16.9 *Mixtures of insoluble salts*

Suppose we try to dissolve two slightly soluble salts in the same solution. If there is no common ion (e.g., the case of AgCl and $BaSO_4$), we assume each salt establishes its own equilibrium independent of the other salt present. Thus, in the case of AgCl and $BaSO_4$ mixtures we have

$$AgCl(s) \rightleftharpoons Ag^+ + Cl^- \qquad K_{sp} = [Ag^+][Cl^-]$$
$$BaSO_4(s) \rightleftharpoons Ba^{2+} + SO_4^{2-} \qquad K_{sp} = [Ba^{2+}][SO_4^{2-}]$$

In certain cases, not this one, we may also have to worry about the "cross-products" (e.g. Ag_2SO_4 and $BaCl_2$). If their K_{sp}'s are exceeded, the solid phase will be the less soluble salt, which may not be the one dumped into the solution.

■ EXAMPLE 428.   A mixture of solid AgCl and solid $BaSO_4$ is shaken up with water until saturation equilibrium is established. Given that $K_{sp} = 1.7 \times 10^{-10}$ for AgCl and $K_{sp} = 1.5 \times 10^{-9}$ for $BaSO_4$, calculate the concentrations of Ag^+, Cl^-, Ba^{2+}, and SO_4^{2-} in the final solution.

SOLUTION: Let x = moles per liter of AgCl that dissolve.
Let y = moles per liter of $BaSO_4$ that dissolve.
This will give x moles/liter of Ag^+ and x moles/liter of Cl^- by the reaction $AgCl(s) \rightarrow Ag^+ + Cl^-$ and y moles/liter of Ba^{2+} and y moles/liter of SO_4^{2-} by the reaction $BaSO_4(s) \rightarrow Ba^{2+} + SO_4^{2-}$.
At equilibrium, then, the concentrations will be as follows:

$$[Ag^+] = x; \qquad [Cl^-] = x; \qquad [Ba^{2+}] = y; \qquad [SO_4^{2-}] = y$$

The two equilibria that must be satisfied are

$$AgCl(s) \rightleftharpoons Ag^+ + Cl^- \quad \text{with} \quad K_{sp} = [Ag^+][Cl^-] = 1.7 \times 10^{-10}$$
$$BaSO_4(s) \rightleftharpoons Ba^{2+} + SO_4^{2-} \quad \text{with} \quad K_{sp} = [Ba^{2+}][SO_4^{2-}]$$
$$= 1.5 \times 10^{-9}$$

Substituting the concentrations above in these equations gives

$$(x)(x) = 1.7 \times 10^{-10}$$
$$(y)(y) = 1.5 \times 10^{-9}$$

Solving these gives $x = 1.3 \times 10^{-5}$ and $y = 3.9 \times 10^{-5}$, or

$$[Ag^+] = 1.3 \times 10^{-5} \, M; \qquad [Cl^-] = 1.3 \times 10^{-5} \, M$$
$$[Ba^{2+}] = 3.9 \times 10^{-5} \, M; \qquad [SO_4^{2-}] = 3.9 \times 10^{-5} \, M$$

These results are, of course, the same as if each salt were by itself in the solution. ■

What do we do if we have a common ion? Then we have to allow for the fact that the concentration of the common ion is made up of the two contributions. For example, when $SrSO_4$ and $BaSO_4$ are dissolved in the same solution, the sulfate concentration is the sum of what comes from $SrSO_4$ plus what comes from $BaSO_4$.

■ EXAMPLE 429. A mixture of solid $SrSO_4$ and solid $BaSO_4$ is shaken up with water until saturation equilibrium is established. Given that $K_{sp} = 7.6 \times 10^{-7}$ for $SrSO_4$ and $K_{sp} = 1.5 \times 10^{-9}$ for $BaSO_4$, calculate the concentrations of Sr^{2+}, Ba^{2+}, and SO_4^{2-} in the final solution.

SOLUTION: Let x = moles per liter of $SrSO_4$ that dissolve.
This will give x moles/liter Sr^{2+}, y moles/liter of Ba^{2+}, and $(x + y)$ moles/liter of SO_4^{2-}.

$$[Sr^{2+}] = x; \qquad [Ba^{2+}] = y; \qquad [SO_4^{2-}] = x + y$$

The two equilibrium conditions are

$$SrSO_4(s) \rightleftharpoons Sr^{2+} + SO_4^{2-} \quad \text{with} \quad K_{sp} = [Sr^{2+}][SO_4^{2-}]$$
$$= 7.6 \times 10^{-7}$$

$$BaSO_4(s) \rightleftharpoons Ba^{2+} + SO_4^{2-} \quad \text{with} \quad K_{sp} = [Ba^{2+}][SO_4^{2-}]$$
$$= 1.5 \times 10^{-9}$$

Substituting the concentrations above, we get

$$[Sr^{2+}][SO_4^{2-}] = (x)(x + y) = 7.6 \times 10^{-7}$$
$$[Ba^{2+}][SO_4^{2-}] = (y)(x + y) = 1.5 \times 10^{-9}$$

We can solve these two simultaneous equations exactly as follows: First solve the first equation for y in terms of x:

$$y = \frac{7.6 \times 10^{-7}}{x} - x$$

Then substitute for y in the second equation to get

$$\left(\frac{7.6 \times 10^{-7}}{x} - x\right)\left(\frac{7.6 \times 10^{-7}}{x}\right) = 1.5 \times 10^{-9}$$

This last equation solves to give $x = 8.71 \times 10^{-4}$, and putting that into the preceding equation gives $y = 1.67 \times 10^{-6}$. To two significant figures, we have

$$[Sr^{2+}] = x = 8.7 \times 10^{-4} \ M$$

$$[Ba^{2+}] = y = 1.7 \times 10^{-6} \ M$$

$$[SO_4^{2-}] = x + y = 8.7 \times 10^{-4} + 1.7 \times 10^{-6} = 8.7 \times 10^{-4} \ M \ \blacksquare$$

These results deserve some scrutiny. Note that the sulfate concentration, 8.7×10^{-4}, is essentially decided by x alone—that is, by the dissolving of the $SrSO_4$, which is the more soluble salt. Note further that the strontium-ion concentration is also 8.7×10^{-4}, which turns out to be the square root of K_{sp} of $SrSO_4$. In other words, *the more soluble salt behaves as if it were alone in the system.* Finally, note that the barium-ion concentration, 1.7×10^{-6}, is just equal to 1.5×10^{-9} (the K_{sp} of $BaSO_4$) divided by 8.7×10^{-4} (which is the sulfate concentration formed by dissolving $SrSO_4$). Thus, *the less soluble salt has its equilibrium decided by the more soluble salt.*

In retrospect, we could have solved the above problem in two simple steps. First calculate $[Sr^{2+}]$ and $[SO_4^{2-}]$ as if only $SrSO_4$ were dissolving in the system. Then, use the $[SO_4^{2-}]$ so obtained to find out what $[Ba^{2+}]$ is consistent with it. Such an approximate procedure will work, except when the amount of common ion coming from the two salts is roughly comparable.

You may be tempted to conclude that if the K_{sp} is about equal for two insoluble salts, then we have to work out the problem by the exact procedure, but if the K_{sp}'s are grossly different from each other, we can use the salt with the larger K_{sp} to establish the principal equilibrium. Such a rule is acceptable if *each salt has the same number of ions*, but breaks down if the number of ions is different. As an illustration, the K_{sp} of SrF_2 is not directly comparable to the K_{sp} of $SrSO_4$, because $SrSO_4$ gives two ions (Sr^{2+} and SO_4^{2-}), whereas SrF_2 gives three (Sr^{2+} and $2F^-$).

■ EXAMPLE 430.   The K_{sp} of SrF_2 is 7.9×10^{-10}; the K_{sp} of $SrSO_4$ is 7.6×10^{-7}. What will be the concentrations of Sr^{2+}, F^-, and SO_4^{2-} in a solution that is simultaneously in equilibrium with solid SrF_2 and $SrSO_4$?

SOLUTION: The temptation is very great to say that $SrSO_4$ is by far the more soluble (because it has by far the bigger K_{sp}) and establishes the principal equilibrium. We might then approximately solve for the Sr^{2+} concentration as the square root of 7.6×10^{-7} ($= 8.7 \times 10^{-4}$) and use this in the K_{sp} of SrF_2 ($= [Sr^{2+}][F^-]^2$) to fix $[F^-]$ at 9.5×10^{-4}. Such a procedure is *wrong*, as we can show by doing the exact problem.

Let x = moles per liter of $SrSO_4$ that dissolve via the reaction $SrSO_4(s) \rightarrow Sr^{2+} + SO_4^{2-}$.

Let y = moles per liter of SrF_2 that dissolve via the reaction $SrF_2(s) \rightarrow Sr^{2+} + 2F^-$.

These two reactions will give us $(x + y)$ moles/liter of Sr^{2+}, x moles/liter of SO_4^{2-}, and $2y$ moles/liter of F^-. The equilibrium concentrations will be

$$[Sr^{2+}] = x + y; \qquad [SO_4^{2-}] = x; \qquad [F^-] = 2y$$

The equilibrium conditions are

$$SrSO_4(s) \;\rightleftharpoons\; Sr^{2+} + SO_4^{2-} \quad \text{with} \quad K_{sp} = [Sr^{2+}][SO_4^{2-}]$$
$$= 7.6 \times 10^{-7}$$

$$SrF_2(s) \;\rightleftharpoons\; Sr^{2+} + 2F^- \quad \text{with} \quad K_{sp} = [Sr^{2+}][F^-]^2$$
$$= 7.9 \times 10^{-10}$$

Putting in the concentrations above, we get

$$(x + y)(x) = 7.6 \times 10^{-7}$$
$$(x + y)(2y)^2 = 7.9 \times 10^{-10}$$

Solve these two equations simultaneously by getting y in terms of x from the first and putting it into the second. We get a fourth-order equation:

$$x^4 - 2.60 \times 10^{-4}x^3 - 1.52 \times 10^{-6}x^2 + 5.78 \times 10^{-13} = 0$$

that can be solved by successive approximations. (First neglect the x^4 and x^3 terms; solve the rest for x. Feed this value in place of the x^4 and x^3 terms; solve the rest again for x. Keep doing this until the value of x settles down.) The final result gives $x = 6.8_5 \times 10^{-4}$, which corresponds to $y = 4.2_4 \times 10^{-4}$. Putting these values into the expressions above, we get finally

$$[Sr^{2+}] = x + y = 1.1 \times 10^{-3} \ M$$
$$[SO_4^{2-}] = x = 6.9 \times 10^{-4} \ M$$
$$[F^-] = 2y = 8.5 \times 10^{-4} \ M$$

Checking our results, we get

$$[Sr^{2+}][SO_4^{2-}] = (1.1 \times 10^{-3})(6.9 \times 10^{-4}) = 7.6 \times 10^{-7}$$
$$[Sr^{2+}][F^-]^2 = (1.1 \times 10^{-3})(8.5 \times 10^{-4})^2 = 7.9 \times 10^{-10}$$

which is in good agreement with the respective K_{sp} values.

Note again that, since x and y come out to be roughly equal, each salt is contributing about the same amount to the Sr^{2+} concentration, even though the K_{sp}'s are quite different. It is easy enough to foresee this kind of situation:

Calculate the solubilities for separate solutions and see if they lead to roughly equal results. In the case of $SrSO_4$, we would get $x^2 = 7.6 \times 10^{-7}$, or $x = 8.7 \times 10^{-4}$; in the case of SrF_2, we would get $4x^3 = 7.9 \times 10^{-10}$, which leads to $x = 5.8 \times 10^{-4}$. These separate solubilities are close enough to each other to hint at trouble. ∎

What do we do for the case where two slightly soluble salts do not have an ion in common but where a K_{sp} for a cross product is exceeded? For example, if AB(s) and CD(s) are insoluble salts added to the same solution, what happens if the concentrations calculated for "equilibrium" are such that the K_{sp} of AD would be exceeded? The answer is simple: AD should precipitate and both AB and CD should dissolve to a larger extent than if AD(s) did not form. The following problem illustrates such a case.

∎ **EXAMPLE 431.** Suppose you shake up 0.100 mole of AgCl and 0.200 mole of TlI with a liter of water until equilibrium is established. What will be the final equilibrium concentration of each ion? The K_{sp} values are 1.7×10^{-10} for AgCl, 8.9×10^{-8} for TlI, 1.9×10^{-4} for TlCl, and 8.5×10^{-17} for Ag I.

SOLUTION: By itself, AgCl would dissolve as follows:

$$AgCl(s) \rightleftharpoons Ag^+ + Cl^-$$
$$K_{sp} = [Ag^+][Cl^-] = 1.7 \times 10^{-10} = (x)(x)$$
$$x = [Ag^+] = [Cl^-] = 1.3 \times 10^{-5} \ M$$

Similarly, TlI by itself would dissolve as follows:

$$TlI(s) \rightleftharpoons Tl^+ + I^-$$
$$K_{sp} = [Tl^+][I^-] = 8.9 \times 10^{-8} = (y)(y)$$
$$y = [Tl^+] = [I^-] = 3.0 \times 10^{-4} \ M$$

The trouble is that these concentrations $[Ag^+] = 1.3 \times 10^{-5} \ M$ and $[I^-] = 3.0 \times 10^{-4} \ M$, are incompatible. They multiply together to give 3.9×10^{-9}, which is way over the K_{sp} of AgI (which is 8.5×10^{-17}). Therefore, AgI should precipitate. But as it does, the AgCl and the TlI dissolve some more to replace the Ag^+ and I^- that are being removed from solution. Consequently, the Cl^- and the Tl^+ concentrations increase, until TlCl starts to precipitate when its K_{sp} is exceeded.

The easiest way to handle this problem is as a mixing problem, where we bring the various ions together and allow precipitation to occur of the least soluble substance at that stage. First the I^- drives out essentially all the Ag^+, because AgI is the least soluble of the various combinations. Then at stage I, with all the Ag^+ out of solution, the I^- proceeds to chase out TlI, which is the next-least soluble substance. However, since there is 0.200 mole of Tl^+ to 0.100 mole of I^- at this stage, it is the I^- that is depleted to zero. This leaves stage II, with 0.100 mole of Tl^+ in solution, which then combines with 0.100 mole of Cl^- to precipitate TlCl.

Species	Moles Available	Stage I Moles After $Ag^+ + I^- \rightarrow AgI$	Stage II Moles After $Tl^+ + I^- \rightarrow TlI$	Stage III Moles After $Tl^+ + Cl^- \rightarrow TlCl$
Ag^+	0.100	~0	~0	~0
Cl^-	0.100	0.100	0.100	~0
Tl^+	0.200	0.200	0.100	~0
I^-	0.200	0.100	~0	~0

To calculate the equilibrium concentrations, we work backward. The TlCl, being the most soluble, fixes the Tl^+ and Cl^- concentration.

$$TlCl(s) \rightleftharpoons Tl^+ + Cl^-$$
$$K_{sp} = [Tl^+][Cl^-] = 1.9 \times 10^{-4} = (x)(x)$$
$$x = [Tl^+] = [Cl^-] = 1.3_8 \times 10^{-2}$$

Then we use this thallium-ion concentration to fix the iodide concentration, the TlI being the next more soluble of the solids actually there. [*Note:* There is *no* AgCl(s) there in the equilibrium state!)

$$TlI(s) \rightleftharpoons Tl^+ + I^-$$
$$K_{sp} = [Tl^+][I^-] = 8.9 \times 10^{-8}$$
$$[I^-] = \frac{8.9 \times 10^{-8}}{[Tl^+]} = \frac{8.9 \times 10^{-8}}{1.3_8 \times 10^{-2}} = 6.4_5 \times 10^{-6} \ M$$

Finally, we use this iodide-ion concentration to find out what the silver-ion concentration is.

$$AgI(s) \rightleftharpoons Ag^+ + I^-$$
$$K_{sp} = [Ag^+][I^-] = 8.5 \times 10^{-17}$$
$$[Ag^+] = \frac{8.5 \times 10^{-17}}{[I^-]} = \frac{8.5 \times 10^{-17}}{6.4_5 \times 10^{-6}} = 1.3_2 \times 10^{-11}$$

So, summarizing, we have as solid phases AgI, TlI, and TlCl, and the concentrations in solution are

$$[Ag^+] = 1.3 \times 10^{-11} \, M; \qquad [Tl^+] = 1.4 \times 10^{-2} \, M$$
$$[Cl^-] = 1.4 \times 10^{-2} \, M; \qquad [I^-] = 6.5 \times 10^{-6} \, M$$

Note especially that the K_{sp} of AgCl is *not* exceeded. ■

16.10   Hydrolysis of slightly soluble salts

In our solubility calculations we have assumed that hydrolysis can be neglected when computing how much salt can go into solution. This can produce a rather serious error and in fact is one of the main reasons why there is frequently a discrepancy between the observed solubility and that calculated from K_{sp}. We have avoided the issue by largely confining our attention so far to solids whose ions do not hydrolyze very much. But what happens, for example, with carbonates where CO_3^{2-} would be expected to interact appreciably with the water equilibrium? The following problem shows how to handle such a situation.

■ EXAMPLE 432.   The K_{sp} of $CaCO_3$ is 4.7×10^{-9}. Calculate its solubility (a) ignoring hydrolysis and (b) taking hydrolysis into account. K_{II} for $HCO_3^- + H_2O \rightleftarrows H_3O^+ + CO_3^{2-}$ is 4.84×10^{-11}.

SOLUTION:

(a) $CaCO_3(s) \rightleftharpoons Ca^{2+} + CO_3^{2-}$.
$K_{sp} = [Ca^{2+}][CO_3^{2-}] = 4.7 \times 10^{-9}$
$(x)(x) = 4.7 \times 10^{-9}$
$x = [Ca^{2+}] = [CO_3^{2-}] = 6.9 \times 10^{-5} \, M$

(b) Let $x =$ moles of $CaCO_3$ that dissolve per liter. This would give x moles/liter of Ca^{2+} and x moles/liter of CO_3^{2-}. However, we shall need to let y moles/liter of this CO_3^{2-} get converted to HCO_3^- by the hydrolysis reaction

$$CO_3^{2-} + H_2O \longrightarrow HCO_3^- + OH^-$$

This will give us y moles/liter of HCO_3^-, y moles/liter of OH^-, and leave $(x - y)$ moles/liter of CO_3^{2-} unhydrolyzed. The equilibrium concentrations will therefore be as follows:

$$[Ca^{2+}] = x; \qquad [CO_3^{2-}] = x - y; \qquad [HCO_3^-] = y; \qquad [OH^-] = y$$

We need to satisfy both equilibria:

$$CaCO_3(s) \rightleftharpoons Ca^{2+} + CO_3^{2-}$$

$$K_{sp} = [Ca^{2+}][CO_3^{2-}] = 4.7 \times 10^{-9}$$

$$(x)(x-y) = 4.7 \times 10^{-9}$$

$$CO_3^{2-} + H_2O \rightleftharpoons HCO_3^- + OH^-$$

$$K_h = \frac{[HCO_3^-][OH^-]}{[CO_3^{2-}]} = \frac{K_w}{K_{II}} = \frac{1.00 \times 10^{-14}}{4.84 \times 10^{-11}} = 2.07 \times 10^{-4}$$

$$\frac{(y)(y)}{x-y} = 2.07 \times 10^{-4}$$

The problem then is to solve the two simultaneous equations

$$(x)(x-y) = 4.7 \times 10^{-9}$$

$$\frac{(y)(y)}{x-y} = 2.07 \times 10^{-4}$$

This is no easy job, since we end up with a fourth-order equation no matter which variable we choose to eliminate. If we solve the second equation for x in terms of y we get

$$x = \frac{y^2}{2.07 \times 10^{-4}} + y$$

Putting this into the first equation gives us

$$\left(\frac{y^2}{2.07 \times 10^{-4}} + y\right)\left(\frac{y^2}{2.07 \times 10^{-4}}\right) = 4.7 \times 10^{-9}$$

Combining the parentheses gives the following:

$$\frac{y^4}{4.28 \times 10^{-8}} + \frac{y^3}{2.07 \times 10^{-4}} = 4.7 \times 10^{-9}$$

We can solve this last equation by successive approximation: First let $y = 0$ in the first term and solve for y from the second term. This gives us $y = 9.9 \times 10^{-5}$. We then feed this value into the first term to evaluate it and again solve for y from the second term. The process is repeated (about 10 times) until the y value we get from using the second term matches what was fed into the first term. The final value turns out to be $y = 8.8_1 \times 10^{-5}$. When this is put into the equation $x = [y^2/(2.07 \times 10^{-4})] + y$, it leads to $x = 1.2_6 \times 10^{-4}$. Thus, the equilibrium concentrations turn out to be

$$[Ca^{2+}] = x = 1.3 \times 10^{-4} \ M$$
$$[CO_3^{2-}] = x - y = 1.2_6 \times 10^{-4} - 8.8_1 \times 10^{-5} = 3.8 \times 10^{-5} \ M$$
$$[HCO_3^-] = y = 8.8 \times 10^{-5} \ M$$
$$[OH^-] = y = 8.8 \times 10^{-5} \ M$$

So far as the solubility of $CaCO_3$ is concerned, taking account of hydrolysis, it is given by $x = 1.3 \times 10^{-4} \ M$. This value is about twice as great as the one calculated in part (a), where hydrolysis was completely neglected. ∎

Exercises

16.1. Given that K_{diss} of HClO is 3.2×10^{-8}, calculate the pH of 1.0 M NaClO solution.

16.2. Given that K_{diss} of HClO is 3.2×10^{-8}, calculate the per cent hydrolysis of ClO^- in 0.50 M NaClO and in 0.050 M NaClO.

16.3. What must be the K_{diss} of a weak acid HX such that the X^- in 0.50 M NaX solution is 0.1 % hydrolyzed? (*Answer*: 2×10^{-8}).

16.4. How many grams of NaOAc must you dissolve in 100.0 ml of water to produce a solution with pH = 9.00?

16.5. How much water do you need to add to 0.100 ml of 0.750 M NH_4Cl to change the pH from 4.692 to 5.000?

16.6. How many grams of NH_3 do you need to add to 0.250 liter of 0.750 M NH_4Cl to adjust the pH to 5.000? (*Answer*: 0.134 mg.)

***16.7.** How many grams of HCl do you need to add to 0.250 liter of 0.750 M NH_4Cl to adjust the pH to 4.000?

16.8. What K_{diss} would you need for the weak base $BOH \rightleftarrows B^+ + OH^-$ so that 1.0% of the B^+ in a 1.0 M B^+ solution would be hydrolyzed?

16.9. What is the pH of a solution made by mixing 10.0 ml of 0.750 M NH_3 with 30.0 ml of 0.250 M HOAc?

16.10. What is the pH of a solution made by mixing 10.0 ml of 0.750 M NH_3 with 30.0 ml of 0.750 M HOAc? (*Answer*: 4.44.)

16.11. Calculate the per cent hydrolysis of NH_4^+ and of OAc^- in 0.50 M NH_4OAc.

16.12. Suppose you want to make a Na_2CO_3 solution that has a pH of 12.00. How many grams of Na_2CO_3 do you need per liter?

16.13. Given that $K = 5.6 \times 10^{-8}$ for $HSO_3^- + H_2O \rightleftarrows H_3O^+ + SO_3^{2-}$, calculate the pH of 0.250 M Na_2SO_3 solution. (*Answer*: 10.32).

***16.14.** For H_2SO_3, $K_I = 1.25 \times 10^{-2}$ and $K_{II} = 5.6 \times 10^{-8}$. What would be the concentrations of H_2SO_3, HSO_3^-, SO_3^{2-}, and H_3O^+ in 1.50 M $NaHSO_3$?

16.15. Calculate the change in pH that occurs when 10.0 ml of 1.00 M HCl is added to 0.250 liter of 0.500 M NaHCO$_3$. The $K_I = 4.16 \times 10^{-7}$; the $K_{II} = 4.8 \times 10^{-11}$. (*Answer*: From 8.35 to 7.43.)

16.16. Given that $K = 1 \times 10^{-7}$ for the hydrolysis $Fe^{3+} + 2H_2O \rightleftarrows FeOH^{2+} + H_3O^+$, how many ml of 1.0 M HNO$_3$ would you have to add to a liter of 0.10 M Fe(NO$_3$)$_3$ solution to make it neutral?

16.17. Given that $K = 1.5 \times 10^{-4}$ for the hydrolysis $Cr^{3+} + 2H_2O \rightleftarrows CrOH^{2+} + H_3O^+$, what would be the pH of a chrome alum solution containing 50.0 g of KCr(SO$_4$)$_2 \cdot 12H_2O$ per liter. Take into account the SO_4^{2-} interference. The $K_{II} = 1.26 \times 10^{-2}$ for HSO_4^-.

16.18. The dissociation constants of HF and HNO$_2$ are 6.71×10^{-4} and 4.5×10^{-4}, respectively. Calculate the pH and the ratio of $[F^-]$ to $[NO_2^-]$ in a solution that is simultaneously 0.50 M HF and 0.50 M HNO$_2$. (*Answer*: pH = 1.607, $[F^-]/[NO_2^-] = 1.5$.)

***16.19.** Calculate the per cent dissociation of acetic acid in 5.0×10^{-7} M HOAc.

16.20. The K_{sp} of AgNO$_2$(s) is 1.2×10^{-4}; that of AgOAc(s) is 2.3×10^{-3}. Calculate the concentration of Ag^+, NO_2^-, and OAc^- in a solution that is simultaneously saturated with both solids.

***16.21.** The K_{sp} of Ag$_2$CrO$_4$(s) is 1.9×10^{-12}; that of AgIO$_3$(s) is 3.1×10^{-8}. What change in Ag^+ concentration occurs when excess solid AgIO$_3$ is stirred up with an equilibrium system consisting of solid Ag$_2$CrO$_4$ in contact with its saturated solution? (*Answer*: From 1.6×10^{-4} M to 2.2×10^{-4} M.)

***16.22.** Suppose you shake up 0.100 mole of BaSO$_4$ ($K_{sp} = 1.5 \times 10^{-9}$) and 0.200 mole of SrCrO$_4$ ($K_{sp} = 3.6 \times 10^{-5}$) with a liter of water until equilibrium is established. What will be the concentration of each ion in the final solution? Ignore hydrolysis. The K_{sp} of BaCrO$_4$ is 8.5×10^{-11}; that of SrSO$_4$ is 7.6×10^{-7}.

***16.23.** What will be the concentration of each ion in the final equilibrium solution resulting from the mixing of 10.0 ml of 0.10 M Sr(NO$_3$)$_2$, 20.0 ml of 0.10 M Ba(NO$_3$)$_2$, 30.0 ml of 0.20 M NaF, and 10.0 ml of 0.15 M Na$_2$SO$_4$. The K_{sp} values are as follows: 2.4×10^{-5} for BaF$_2$; 7.9×10^{-10} for SrF$_2$; 7.6×10^{-7} for SrSO$_4$; and 1.5×10^{-9} for BaSO$_4$. Ignore hydrolysis.

***16.24.** The observed solubility of BaCO$_3$ is 0.0022 g per 100 ml. Calculate the K_{sp} of BaCO$_3$, allowing for hydrolysis of the carbonate ion. The K_{diss} of HCO$_3^-$ is 4.84×10^{-11}.

****16.25.** The K_{sp} of Ag$_2$CO$_3$ is 8.2×10^{-12}. Calculate its solubility allowing for hydrolysis of the carbonate ion. The K_{diss} of HCO$_3^-$ is 4.84×10^{-11}.

APPENDIXES

A

LOGARITHMS

	0	1	2	3	4	5	6	7	8	9
10	0000	0043	0086	0128	0170	0212	0253	0294	0334	0374
11	0414	0453	0492	0531	0569	0607	0645	0682	0719	0755
12	0792	0828	0864	0899	0934	0969	1004	1038	1072	1106
13	1139	1173	1206	1239	1271	1303	1335	1367	1399	1430
14	1461	1492	1523	1553	1584	1614	1644	1673	1703	1732
15	1761	1790	1818	1847	1875	1903	1931	1959	1987	2014
16	2041	2068	2095	2122	2148	2175	2201	2227	2253	2279
17	2304	2330	2355	2380	2405	2430	2455	2480	2504	2529
18	2553	2577	2601	2625	2648	2672	2695	2718	2742	2765
19	2788	2810	2833	2856	2878	2900	2923	2945	2967	2989
20	3010	3032	3054	3075	3096	3118	3139	3160	3181	3201
21	3222	3243	3263	3284	3304	3324	3345	3365	3385	3404
22	3424	3444	3464	3483	3502	3522	3541	3560	3579	3598
23	3617	3636	3655	3674	3692	3711	3729	3747	3766	3784
24	3802	3820	3838	3856	3874	3892	3909	3927	3945	3962
25	3979	3997	4014	4031	4048	4065	4082	4099	4116	4133
26	4150	4166	4183	4200	4216	4232	4249	4265	4281	4298
27	4314	4330	4346	4362	4378	4393	4409	4425	4440	4456
28	4472	4487	4502	4518	4533	4548	4564	4579	4594	4609
29	4624	4639	4654	4669	4683	4698	4713	4728	4742	4757
30	4771	4786	4800	4814	4829	4843	4857	4871	4886	4900
31	4914	4928	4942	4955	4969	4983	4997	5011	5024	5038
32	5051	5065	5079	5092	5105	5119	5132	5145	5159	5172
33	5185	5198	5211	5224	5237	5250	5263	5276	5289	5302
34	5315	5328	5340	5353	5366	5378	5391	5403	5416	5428
35	5441	5453	5465	5478	5490	5502	5514	5527	5539	5551
36	5563	5575	5587	5599	5611	5623	5635	5647	5658	5670
37	5682	5694	5705	5717	5729	5740	5752	5763	5775	5786
38	5798	5809	5821	5832	5843	5855	5866	5877	5888	5899
39	5911	5922	5933	5944	5955	5966	5977	5988	5999	6010
40	6021	6031	6042	6053	6064	6075	6085	6096	6107	6117
41	6128	6138	6149	6160	6170	6180	6191	6201	6212	6222
42	6232	6243	6253	6263	6274	6284	6294	6304	6314	6325
43	6335	6345	6355	6365	6375	6385	6395	6405	6415	6425
44	6435	6444	6454	6464	6474	6484	6493	6503	6513	6522
45	6532	6542	6551	6561	6571	6580	6590	6599	6609	6618
46	6628	6637	6646	6656	6665	6675	6684	6693	6702	6712
47	6721	6730	6739	6749	6758	6767	6776	6785	6794	6803
48	6812	6821	6830	6839	6848	6857	6866	6875	6884	6893
49	6902	6911	6920	6928	6937	6946	6955	6964	6972	6981
50	6990	6998	7007	7016	7024	7033	7042	7050	7059	7067
51	7076	7084	7093	7101	7110	7118	7126	7135	7143	7152
52	7160	7168	7177	7185	7193	7202	7210	7218	7226	7235
53	7243	7251	7259	7267	7275	7284	7292	7300	7308	7316
54	7324	7332	7340	7348	7356	7364	7372	7380	7388	7396

	0	1	2	3	4	5	6	7	8	9
55	7404	7412	7419	7427	7435	7443	7451	7459	7466	7474
56	7482	7490	7497	7505	7513	7520	7528	7536	7543	7551
57	7559	7566	7574	7582	7589	7597	7604	7612	7619	7627
58	7634	7642	7649	7657	7664	7672	7679	7686	7694	7701
59	7709	7716	7723	7731	7738	7745	7752	7760	7767	7774
60	7782	7789	7796	7803	7810	7818	7825	7832	7839	7846
61	7853	7860	7868	7875	7882	7889	7896	7903	7910	7917
62	7924	7931	7938	7945	7952	7959	7966	7973	7980	7987
63	7993	8000	8007	8014	8021	8028	8035	8041	8048	8055
64	8062	8069	8075	8082	8089	8096	8102	8109	8116	8122
65	8129	8136	8142	8149	8156	8162	8169	8176	8182	8189
66	8195	8202	8209	8215	8222	8228	8235	8241	8248	8254
67	8261	8267	8274	8280	8287	8293	8299	8306	8312	8319
68	8325	8331	8338	8344	8351	8357	8363	8370	8376	8382
69	8388	8395	8401	8407	8414	8420	8426	8432	8439	8445
70	8451	8457	8463	8470	8476	8482	8488	8494	8500	8506
71	8513	8519	8525	8531	8537	8543	8549	8555	8561	8567
72	8573	8579	8585	8591	8597	8603	8609	8615	8621	8627
73	8633	8639	8645	8651	8657	8663	8669	8675	8681	8686
74	8692	8698	8704	8710	8716	8722	8727	8733	8739	8745
75	8751	8756	8762	8768	8774	8779	8785	8791	8797	8802
76	8808	8814	8820	8825	8831	8837	8842	8848	8854	8859
77	8865	8871	8876	8882	8887	8893	8899	8904	8910	8915
78	8921	8927	8932	8938	8943	8949	8954	8960	8965	8971
79	8976	8982	8987	8993	8998	9004	9009	9015	9020	9025
80	9031	9036	9042	9047	9053	9058	9063	9069	9074	9079
81	9085	9090	9096	9101	9106	9112	9117	9122	9128	9133
82	9138	9143	9149	9154	9159	9165	9170	9175	9180	9186
83	9191	9196	9201	9206	9212	9217	9222	9227	9232	9238
84	9243	9248	9253	9258	9263	9269	9274	9279	9284	9289
85	9294	9299	9304	9309	9315	9320	9325	9330	9335	9340
86	9345	9350	9355	9360	9365	9370	9375	9380	9385	9390
87	9395	9400	9405	9410	9415	9420	9425	9430	9435	9440
88	9445	9450	9455	9460	9465	9469	9474	9479	9484	9489
89	9494	9499	9504	9509	9513	9518	9523	9528	9533	9538
90	9542	9547	9552	9557	9562	9566	9571	9576	9581	9586
91	9590	9595	9600	9605	9609	9614	9619	9624	9628	9633
92	9638	9643	9647	9652	9657	9661	9666	9671	9675	9680
93	9685	9689	9694	9699	9703	9708	9713	9717	9722	9727
94	9731	9736	9741	9745	9750	9754	9759	9763	9768	9773
95	9777	9782	9786	9791	9795	9800	9805	9809	9814	9818
96	9823	9827	9832	9836	9841	9845	9850	9854	9859	9863
97	9868	9872	9877	9881	9886	9890	9894	9899	9903	9908
98	9912	9917	9921	9926	9930	9934	9939	9943	9948	9952
99	9956	9961	9965	9969	9974	9978	9983	9987	9991	9996

B

EQUILIBRIUM VAPOR PRESSURE OF H₂0

Temp., °C	Press., mm Hg	Temp., °C	Press., mm Hg	Temp., °C	Press., mm Hg	Temp., °C	Press., mm Hg
0	4.579						
1	4.926	26	25.209	51	97.20	76	301.4
2	5.294	27	26.739	52	102.09	77	314.1
3	5.685	28	28.349	53	107.20	78	327.3
4	6.101	29	30.043	54	112.51	79	341.0
5	6.543	30	31.824	55	118.04	80	355.1
6	7.013	31	33.695	56	123.80	81	369.7
7	7.513	32	35.663	57	129.82	82	384.9
8	8.045	33	37.729	58	136.08	83	400.6
9	8.609	34	39.898	59	142.60	84	416.8
10	9.209	35	42.175	60	149.38	85	433.6
11	9.844	36	44.563	61	156.43	86	450.9
12	10.518	37	47.067	62	163.77	87	468.7
13	11.231	38	49.692	63	171.38	88	487.1
14	11.987	39	52.442	64	179.31	89	506.1
15	12.788	40	55.324	65	187.54	90	525.8
16	13.634	41	58.34	66	196.09	91	546.0
17	14.530	42	61.50	67	204.96	92	567.0
18	15.477	43	64.80	68	214.17	93	588.6
19	16.477	44	68.26	69	223.73	94	610.9
20	17.535	45	71.88	70	233.7	95	633.9
21	18.650	46	75.65	71	243.9	96	657.6
22	19.827	47	79.60	72	254.6	97	682.1
23	21.068	48	83.71	73	265.7	98	707.3
24	22.377	49	88.02	74	277.2	99	733.2
25	23.756	50	92.51	75	289.1	100	760.0

C

OXIDATION POTENTIALS AT 25°C

$Na(s) \rightarrow Na^+ + e^-$	$E° = +2.71$ V
$Mg(s) \rightarrow Mg^{2+} + 2e^-$	$E° = +2.37$ V
$Al(s) \rightarrow Al^{3+} + 3e^-$	$E° = +1.66$ V
$Zn(s) \rightarrow Zn^{2+} + 2e^-$	$E° = +0.76$ V
$Fe(s) \rightarrow Fe^{2+} + 2e^-$	$E° = +0.44$ V
$Co(s) \rightarrow Co^{2+} + 2e^-$	$E° = +0.28$ V
$Ni(s) \rightarrow Ni^{2+} + 2e^-$	$E° = +0.25$ V
$Sn(s) \rightarrow Sn^{2+} + 2e^-$	$E° = +0.14$ V
$Pb(s) \rightarrow Pb^{2+} + 2e^-$	$E° = +0.13$ V
$H_2(g) + 2H_2O \rightarrow 2H_3O^+ + 2e^-$	$E° = $ zero V
$Sn^{2+} \rightarrow Sn^{4+} + 2e^-$	$E° = -0.15$ V
$2I^- \rightarrow I_2(s) + 2e^-$	$E° = -0.54$ V
$4H_2O + HAsO_2 \rightarrow H_3AsO_4 + 2H_3O^+ + 2e^-$	$E° = -0.56$ V
$H_2O_2 + 2H_2O \rightarrow O_2(g) + 2H_3O^+ + 2e^-$	$E° = -0.68$ V
$Fe^{2+} \rightarrow Fe^{3+} + e^-$	$E° = -0.77$ V
$Ag(s) \rightarrow Ag^+ + e^-$	$E° = -0.80$ V
$6H_2O + NO \rightarrow NO_3^- + 4H_3O^+ + 3e^-$	$E° = -0.96$ V
$4Cl^- + Au(s) \rightarrow AuCl_4^- + 3e^-$	$E° = -1.00$ V
$2Br^- \rightarrow Br_2 + 2e^-$	$E° = -1.07$ V
$18H_2O + I_2 \rightarrow 2IO_3^- + 12H_3O^+ + 10e^-$	$E° = -1.20$ V
$6H_2O \rightarrow O_2(g) + 4H_3O^+ + 4e^-$	$E° = -1.23$ V
$21H_2O + 2Cr^{3+} \rightarrow Cr_2O_7^{2-} + 14H_3O^+ + 6e^-$	$E° = -1.33$ V
$2Cl^- \rightarrow Cl_2(g) + 2e^-$	$E° = -1.36$ V
$12H_2O + Mn^{2+} \rightarrow MnO_4^- + 8H_3O^+ + 5e^-$	$E° = -1.51$ V

SOME AQUEOUS DISSOCIATION CONSTANTS AT 25°C

$HF + H_2O \rightleftarrows H_3O^+ + F^-$	$K = 6.71 \times 10^{-4}$
$HClO + H_2O \rightleftarrows H_3O^+ + ClO^-$	$K = 3.2 \times 10^{-8}$
$HClO_2 + H_2O \rightleftarrows H_3O^+ + ClO_2^-$	$K = 1.1 \times 10^{-2}$
$HBrO + H_2O \rightleftarrows H_3O^+ + BrO^-$	$K = 2.06 \times 10^{-9}$
$H_2S + H_2O \rightleftarrows H_3O^+ + HS^-$	$K = 1.1 \times 10^{-7}$
$HS^- + H_2O \rightleftarrows H_3O^+ + S^{2-}$	$K = 1 \times 10^{-14}$
$H_2SO_3 + H_2O \rightleftarrows H_3O^+ + HSO_3^-$	$K = 1.25 \times 10^{-2}$
$HSO_3^- + H_2O \rightleftarrows H_3O^+ + SO_3^{2-}$	$K = 5.6 \times 10^{-8}$
$HSO_4^- + H_2O \rightleftarrows H_3O^+ + SO_4^{3-}$	$K = 1.26 \times 10^{-2}$
$H_2Se + H_2O \rightleftarrows H_3O^+ + HSe^-$	$K = 1.88 \times 10^{-4}$
$HSe^- + H_2O \rightleftarrows H_3O^+ + Se^{2-}$	$K = 1 \times 10^{-14}$
$H_2SeO_3 + H_2O \rightleftarrows H_3O^+ + HSeO_3^-$	$K = 2.7 \times 10^{-3}$
$HSeO_3^- + H_2O \rightleftarrows H_3O^+ + SeO_3^{2-}$	$K = 2.5 \times 10^{-7}$
$HSeO_4^- + H_2O \rightleftarrows H_3O^+ + SeO_4^{2-}$	$K = 8.9 \times 10^{-3}$
$NH_3 + H_2O \rightleftarrows NH_4^+ + OH^-$	$K = 1.81 \times 10^{-5}$
$HNO_2 + H_2O \rightleftarrows H_3O^+ + NO_2^-$	$K = 4.5 \times 10^{-4}$
$H_3PO_4 + H_2O \rightleftarrows H_3O^+ + H_2PO_4^-$	$K = 7.5 \times 10^{-3}$
$H_2PO_4^- + H_2O \rightleftarrows H_3O^+ + HPO_4^{2-}$	$K = 6.2 \times 10^{-8}$
$HPO_4^{2-} + H_2O \rightleftarrows H_3O^+ + PO_4^{3-}$	$K = 1 \times 10^{-12}$
$CO_2 + 2H_2O \rightleftarrows H_3O^+ + HCO_3^-$	$K = 4.16 \times 10^{-7}$
$HCO_3^- + H_2O \rightleftarrows H_3O^+ + CO_3^{2-}$	$K = 4.84 \times 10^{-11}$
$HCN + H_2O \rightleftarrows H_3O^+ + CN^-$	$K = 4 \times 10^{-10}$
$HCNO + H_2O \rightleftarrows H_3O^+ + CNO^-$	$K = 1.2 \times 10^{-4}$
$H_3AsO_4 + H_2O \rightleftarrows H_3O^+ + H_2AsO_4^-$	$K = 2.5 \times 10^{-4}$
$H_2AsO_4^- + H_2O \rightleftarrows H_3O^+ + HAsO_4^{2-}$	$K = 5.6 \times 10^{-8}$
$HAsO_4^- + H_2O \rightleftarrows H_3O^+ + AsO_4^{3-}$	$K = 3 \times 10^{-13}$